Lecture Notes in Networks and Systems 847

The series "Lecture Notes in Networks and Systems" publishes the latest developments in Networks and Systems—quickly, informally and with high quality. Original research reported in proceedings and post-proceedings represents the core of LNNS.

Volumes published in LNNS embrace all aspects and subfields of, as well as new challenges in, Networks and Systems.

The series contains proceedings and edited volumes in systems and networks, spanning the areas of Cyber-Physical Systems, Autonomous Systems, Sensor Networks, Control Systems, Energy Systems, Automotive Systems, Biological Systems, Vehicular Networking and Connected Vehicles, Aerospace Systems, Automation, Manufacturing, Smart Grids, Nonlinear Systems, Power Systems, Robotics, Social Systems, Economic Systems and other. Of particular value to both the contributors and the readership are the short publication timeframe and the world-wide distribution and exposure which enable both a wide and rapid dissemination of research output.

The series covers the theory, applications, and perspectives on the state of the art and future developments relevant to systems and networks, decision making, control, complex processes and related areas, as embedded in the fields of interdisciplinary and applied sciences, engineering, computer science, physics, economics, social, and life sciences, as well as the paradigms and methodologies behind them.

Indexed by SCOPUS, INSPEC, WTI Frankfurt eG, zbMATH, SCImago.

All books published in the series are submitted for consideration in Web of Science.

For proposals from Asia please contact Aninda Bose (aninda.bose@springer.com).

Phung Trung Nghia · Vu Duc Thai ·
Nguyen Thanh Thuy · Le Hoang Son ·
Van-Nam Huynh
Editors

Advances in Information and Communication Technology

Proceedings of the 2nd International Conference ICTA 2023
Volume 1

Springer

Editors
Phung Trung Nghia
Thai Nguyen University of Information and Communication Technology
Thai Nguyen, Vietnam

Vu Duc Thai
Thai Nguyen University of Information and Communication Technology
Thai Nguyen, Vietnam

Nguyen Thanh Thuy
VNU University of Engineering and Technology
Vietnam National University
Ha Noi, Vietnam

Le Hoang Son
Information Technology Institute
Vietnam National University
Ha Noi, Vietnam

Van-Nam Huynh
Japan Advanced Institute of Science and Technology
Nomi, Ishikawa, Japan

ISSN 2367-3370 ISSN 2367-3389 (electronic)
Lecture Notes in Networks and Systems
ISBN 978-3-031-49528-1 ISBN 978-3-031-49529-8 (eBook)
https://doi.org/10.1007/978-3-031-49529-8

This Springer imprint is published by the registered company Springer Nature Switzerland AG
The registered company address is: Gewerbestrasse 11, 6330 Cham, Switzerland

Paper in this product is recyclable.

Organization

Honorary Co-chairs

Pham Hong Quang, TNU, Vietnam
Ho Tu Bao, JAIST, Japan
Nguyen Truong Thang, IoIT, Vietnam
Nguyen Van Tao, ICTU, Vietnam

General Co-chairs

Hoang Van Hung, TNU, Vietnam
Nguyen Thanh Thuy, FISU, Vietnam
Tien-Yin Chou, FCU, Taiwan
Phung Trung Nghia, ICTU, Vietnam

Program Co-chairs

Nguyen Huu Cong, TNU, Vietnam
Dongkyun Kim, KNU, Korea
Van-Nam Huynh, JAIST, Japan
Le Hoang Son, VNU, Vietnam
Nguyen Long Giang, IoIT, Vietnam
Vu Duc Thai, ICTU, Vietnam

Organizing Co-chairs

Tran Thanh Van, TNU, Vietnam
Bui Thu Lam, FISU, Vietnam
Nguyen Duc Dung, IoIT, Vietnam
Do Dinh Cuong, ICTU, Vietnam
Nguyen Van Huan, ICTU, Vietnam
Nguyen Tuan Minh, TNUT, Vietnam

Publication Co-chairs

Nguyen Gia Nhu, DTU, Vietnam
Duc-Binh Nguyen, ICTU, Vietnam
Nguyen Hai Minh, ICTU, Vietnam
Quach Xuan Truong, ICTU, Vietnam
Nguyen Van Hau, FISU, Vietnam

Local Organizing Co-chairs

Nguyen Van Nui, ICTU, Vietnam
Nguyen Duy Minh, ICTU, Vietnam
Do Thi Bac, ICTU, Vietnam
Truong Ha Hai, ICTU, Vietnam
Vu Chien Thang, ICTU, Vietnam
Nguyen Von Dim, ICTU, Vietnam
Vu Xuan Nam, ICTU, Vietnam

Technical Program Committee Members

Dang Quang A., Vietnam
Duong Huu Ai, Vietnam
Dao Nam Anh, Vietnam
Mai The Anh, Vietnam
Nguyen Thi Hai Anh, Vietnam
Nguyen Tuan Anh, Vietnam
Nguyen Viet Anh, Vietnam
Pham The Anh, Vietnam
Do Thi Bac, Vietnam
Ngo Xuan Bach, Vietnam
Doan Van Ban, Vietnam
Ho Tu Bao, Vietnam
Rodrigo C. Belleza, Jr., Philippines
Huynh Thi Thanh Binh, Vietnam
Duc-Binh Nguyen, Vietnam
Nguyen Thanh Binh, Vietnam
Jeong-Hyun Chang, Korea
Nguyen Van Chi, Vietnam
Nguyen Dinh Chien, Vietnam
Ta Quang Chieu, Vietnam
Tien-Yin Chou, Taiwan
Pham Minh Chuan, Vietnam
Mai Van Chung, Vietnam
Nguyen Dinh Cong, Vietnam
Nguyen Huu Cong, Vietnam
Do Dinh Cuong, Vietnam
Duong Chinh Cuong, Vietnam
Nguyen Duy Cuong, Vietnam
Nguyen Hung Cuong, Vietnam
Pham Van Cuong, Vietnam
Tran Van Dai, Vietnam
Ha Manh Dao, Vietnam
Vuong Cong Dat, Vietnam

Nguyen Thi Thu Den, Vietnam
Nguyen Von Dim, Vietnam
Gia Thi Dinh, Vietnam
Luong Khac Dinh, Vietnam
Nguyen Van Doan, Vietnam
Dao Huy Du, Vietnam
Hoang Huu Duc, Vietnam
Le Thi Minh Duc, Vietnam
Ngo Minh Duc, Vietnam
Phan Thanh Duc, Vietnam
Cao Tien Dung, Vietnam
Le Ba Dung, Vietnam
Nguyen Dinh Dung, Vietnam
Nguyen Duc Dung, Vietnam
Nguyen Tien Dung, Vietnam
Tran Tien Dung, Vietnam
Yared Abera Ergu, Taiwan
Nguyen Long Giang, Vietnam
Tran Thi Thu Ha, Vietnam
Doan Thanh Hai, Vietnam
Trinh Thanh Hai, Vietnam
Truong Ha Hai, Vietnam
Nguyen Thi Hang, Vietnam
Le Duc Hau, Vietnam
Nguyen Van Hau, Vietnam
Dao Anh Hien, Vietnam
Le Trung Hieu, Vietnam
Nguyen Huu Hoa, Vietnam
Nguyen Thanh Hoai, Vietnam
Le Hoan, Vietnam
Pham Minh Hoan, Vietnam
Bui The Hong, Vietnam
Pham Duc Hong, Vietnam
Dang Trong Hop, Vietnam
Nguyen Van Huan, Vietnam
Phung The Huan, Vietnam
Hoang Van Hung, Vietnam
Le Van Hung, Vietnam
Tran Dang Hung, Vietnam
Dao Thi Huong, Vietnam
Le Thi Thu Huong, Vietnam
Luu Thi Bich Huong, Vietnam
Ngo Huu Huy, Vietnam
Nguyen Phuong Huy, Vietnam
Pham Thanh Huyen, Vietnam

Do Thuy Ninh, Vietnam
Nguyen Van Nui, Vietnam
Kazuhiro Ogata, Japan
Mizuhito Ogawa, Japan
Huynh Cong Phap, Vietnam
Ho Van Phi, Vietnam
Nguyen Duy Phuong, Vietnam
Nguyen Thi Mai Phuong, Vietnam
Truong Thi Viet Phuong, Vietnam
Dang Minh Quan, Vietnam
Dao Anh Quan, Vietnam
Nguyen Vu Anh Quang, Vietnam
Pham Hong Quang, Vietnam
Vu Vinh Quang, Vietnam
Tran Quang Quy, Vietnam
Vu Khanh Quy, Vietnam
Nguyen Van Quyet, Vietnam
Hoang Duc Quynh, Vietnam
Dang Thai Son, Vietnam
Le Hoang Son, Vietnam
Trinh Hoai Son, Vietnam
Kyounghee Sun, Korea
Tong Thi Hao Tam, Vietnam
Nguyen Duy Tan, Vietnam
Nguyen Thi Thanh Tan, Vietnam
Jose B. Tan, Jr., Philippines
Nguyen Van Tao, Vietnam
Vu Duc Thai, Vietnam
Huynh Quyet Thang, Vietnam
Nguyen Toan Thang, Vietnam
Nguyen Truong Thang, Vietnam
Pham Chien Thang, Vietnam
Tran Anh Thang, Vietnam
Vu Chien Thang, Vietnam
Ha Thi Thanh, Vietnam
Le Nguyen Tuan Thanh, Vietnam
Ngo Ngoc Thanh, Vietnam
Nguyen Tuan Thanh, Vietnam
Vo Thi Thanh Thao, Vietnam
Vu Huy The, Vietnam
Nguyen Si Thin, Vietnam
Dang Dai Tho, Vietnam
Huynh Ngoc Tho, Vietnam
Mai Duc Tho, Vietnam
Pham Duc Tho, Vietnam

Nguyen Thanh Thuy, Vietnam
Nguyen Vinh Thuy, Vietnam
Nguyen Minh Tien, Vietnam
Do Nang Toan, Vietnam
Tran Song Toan, Vietnam
Nguyen Van Toi, Vietnam
Thien-Vu Tran, Vietnam
Nguyen Thi Kieu Trang, Vietnam
Van Hung Trong, Vietnam
Hoang Quang Trung, Vietnam
Nguyen Thanh Trung, Vietnam
Quach Xuan Truong, Vietnam
Le Anh Tu, Vietnam
Duong Quoc Tuan, Vietnam
Luu Minh Tuan, Vietnam
Nguyen Huu Tuan, Vietnam
Nguyen Trung Tuan, Vietnam
Tran Manh Tuan, Vietnam
Bui Van Tung, Vietnam
Tran Minh Tuoc, Vietnam
Nguyen Thi Bach Tuyet, Vietnam
Masashi Unoki, Japan
Tran Thanh Van, Vietnam
Hoang Quoc Viet, Vietnam
Nguyen The Vinh, Vietnam
Nguyen Quang Vu, Vietnam
Vu Viet Vu, Vietnam
Tran Thi Xuan, Vietnam
Tran Quang Yen, Vietnam
Hyun Yoo, Korea

Keynotes

Exploitation of Electronic Medical Records

Ho Tu Bao
Data Science Lab at Vietnam Institute for Advanced Study in Mathematics, Hanoi, Vietnam
bao@viasm.edu.vn

Abstract Exploitation of electronic medical records (EMR) is a new, important, but challenging issue in healthcare in every country. It was started in Vietnam nearly 10 years ago, attracting and requiring much effort from the ICT community. This talk is about the story of EMR exploitation in Vietnam. It has been a long road in which our project has tried to identify the key issues to pursue and to develop the initial solutions along the way. I also point out that mathematics and informatics are essential in those solutions.

Advancing Slope Land Disaster Monitoring and Alert Systems through Synergistic IoT and AI Integration

Tien-Yin Chou
GIS Research Center, Feng Chia University, Taipei City, Taiwan
jimmy@gis.tw

Abstract The escalating global vulnerabilities to natural disasters driven by climate change and geological instabilities have heightened the susceptibility of numerous nations. This presentation expounds upon the deployment of a comprehensive Internet of Things (IoT) framework seamlessly augmented with artificial intelligence (AI) capabilities to revolutionize slope land disaster monitoring and alerting. Within this framework, an assortment of sensor devices facilitates real-time data transmission to a centralized backend infrastructure. Through meticulous integration, an intelligent decision-making model is synthesized, empowering governmental bodies with a robust foundation for informed monitoring and dissemination, distinguished by green, yellow, and red alert levels. The adoption of these sophisticated disaster mitigation technologies adheres to internationally recognized geospatial standards. The incorporation of AI-driven disaster prevention systems refines the precision of essential information conveyance, empowering policy makers to promptly comprehend real-time disaster dynamics. This, in turn, facilitates the timely delivery of preemptive notifications to the populace, consequently mitigating the impact of impending disasters. The convergence of AI and IoT emerges as an indispensable paradigm, meticulously poised to fortify contemporary disaster management strategies.

Audio/Speech Information Hiding Based on Human Auditory Characteristics

Masashi Unoki
School of Information Science, Japan Advanced Institute of Science and Technology, Ishikawa, Japan
unoki@jaist.ac.jp

Abstract Audio information hiding (AIH) has recently been focused on as a state-of-the-art technique enabling copyrights to be protected and defended against attacks and tampering of audio/speech content. This technique has aimed at embedding codes as watermarks to protect copyrights in audio/speech content, which are inaudible to and inseparable by users, and at detecting embedded codes from watermarked signals. It has also aimed at verifying whether it can robustly detect embedded codes from watermarked signals (robust or fragile), whether it can blindly detect embedded codes from watermarked signals (blind or non-blind), whether it can completely restore watermarked signals to the originals by removing embedded codes from them (reversible or irreversible), and whether it can be secure against the publicity of algorithms employed in public or private methods. AIH methods, therefore, must satisfy some of the five following requirements to provide a useful and reliable form of watermarking: (a) inaudibility (inaudible to humans with no sound distortion caused by the embedded data), (b) robustness (not affected when subjected to techniques such as data compression and malicious attacks), (c) blind detectability (high possibility of detecting the embedded data without using the original or reference signal), (d) confidentiality (secure and undetectable concealment of embedded data), and (e) reversibility (removable embedded data from the watermarked signal and/or enable watermarking to be re-edited). In this talk, historical and typical AIH methods (including speech information hiding) are introduced and their drawbacks are pointed out. Then our proposed methods based on human auditory characteristics (cochlear delay, adaptive phase modulation, singular spectrum analysis with psychoacoustic model, formant enhancement, spread-spectrum with LP residue) are introduced. In addition, current research issues such as speech spoofing and deepfake detection will also be introduced.

Research on Intrusion Detection on Host Devices

Ren-Hung Hwang
College of Artificial Intelligence, National Yang Ming Chiao Tung University, Taipei City, Taiwan
rhhwang@nycu.edu.tw

Abstract As networking applications prevail, network security threats also become worsen. Although deep learning-based intrusion detection systems (IDSs) are very promising, there are still unsolved issues: (1) Most of the IDSs are network-based IDSs; (2) Most of the IDSs are based on a single data source (mostly are network traffic); (3) Most of the IDSs do not differentiate different attack stages; (4) Training and testing data sets used by most of the deep learning-based IDSs are offline data sets that are from the public domain or collected by researchers through private experiments. Most IDSs do not consider decentralized learning by exploring the trust relationship between hosts. Our research aims to resolve these issues by developing a deep learning host-based intrusion detection system utilizing multiple data sources. A host-based IDS can be deployed on a personal computer, an application server, or a gateway of IoT devices. Compared to a network-based IDS, a host-based IDS can collect more critical security information, such as system logs and host statistics. These multiple data sources collected at a host benefit the intrusion detection process and preserve privacy. This talk will include three parts. The first part aims to develop a deep learning host-based intrusion detection system utilizing multiple data sources. It also aims to detect different stages of network attacks. In the second part, the trust relationship between hosts is utilized to develop a decentralized learning mechanism such that unknown attacks can be learned through decentralized learning. In the last part, semi-supervised mechanisms will be developed to conduct online learning, which could continuously re-train the deep-learning-based model to detect new network attacks.

Acknowledgements

We would like to express our appreciation to all the members of ICTA 2023' Program Committee for their support and cooperation in this publication. We would like to thank Prof. Prasanna Kumar (Production Editor) and Dr. Thomas Ditzinger (Executive Editor, Interdisciplinary and Applied Science, Engineering, Springer) for their support and cooperation in this publication. We also wish to thank all the authors and participants for their contributions and fruitful discussions that made this conference a success.

About This Book

Technological changes and digital transformation that have taken place over the past decade have had significant impacts on all economic and social sectors. Information and Communication Technology (ICT) in general and artificial intelligence (AI) in particular have driven socio-economic growth.

This book contains four keynote abstracts and 83 best peer-reviewed papers selected from the 179 submissions at the 2nd International Conference on Advances in ICT (ICTA 2023), which share research results and practical applications in ICT research and education. The topics cover all ICT-related areas and their contributions to socio-economic development, focusing on the most advanced technologies, such as AI. Researchers and practitioners in academia and industry can use the books as a valuable reference for their research activities, teaching, learning and advancing current technologies.

The Conference is hosted by the Thai Nguyen University of Information and Communication Technology (ICTU).

December 2023

Nguyen Huu Cong
Dongkyun Kim
Van-Nam Huynh
Le Hoang Son
Nguyen Long Giang
Vu Duc Thai
ICTA 2023

Contents

About the Editors

Assoc. Prof. Phung Trung Nghia received his Engineering degree in Electronics and Telecommunications from Hanoi University of Science and Technology (HUST) in 2002. He completed his Master of Science in Telecommunications from Vietnam National University, Hanoi (VNUH) in 2007 and his Ph.D. degree in Information Science from Japan Advanced Institute of Science and Technology (JAIST) in 2013. He was Dean of the Faculty of Electronics and Telecommunications and Head of Academic Affairs. He has been ICTU's Rector. He has been the Vice President of the Vietnam Club of Faculties-Institutes-Schools-Universities of ICT (FISU) and President of the FISU Branch in the Northern Midlands, Mountains and Coastal Region of Vietnam. He received the Award for Excellent Young Researcher (Golden Globe Award) from Vietnam's Ministry of Science and Technology (MOST) in 2008. His main research interest lies in the interaction between signal processing and machine learning, and he has published more than 70 research papers related to this field. He serves as a technical committee program member, organizing co-chair, program co-chair, track chair, section chair, editorial board member and reviewer of several conferences, journals and books. He is an Associate Editor of Thai Nguyen University Journal of Science and Technology (ICT section).

Dr. Vu Duc Thai got a Master of Science in Computer Science from Thai Nguyen University and a Ph.D. degree in Mathematical Foundations for Computers and Computing Systems from the Institute of Information Technology—Vietnam Academy of Science and Technology in 2012. He has been the ICTU's Vice Rector. His research relates to databases, information systems and embedded systems (Cellular Neural Networks, FPGA). He has published more than 40 articles, books and book chapters. He works as an editor and reviewer for some international journals.

Prof. Nguyen Thanh Thuy received an Engineering degree in Computing in 1982 and a Ph.D. degree in Computer Science from HUST in 1987. He was a professor of computer science at HUST until 2011 and then at the VNU University of Engineering and Technology. Prof. Nguyen Thanh Thuy is now Head of the key laboratory for AI at the Faculty of Information Technology under VNU University of Engineering and Technology. His AI research interests are mainly in knowledge systems, soft-computing, data mining, deep learning, hybrid intelligent systems, general AI and responsible AI. Prof. Nguyen Thanh Thuy is now the President of the National Committee for Professorships of Vietnam in Computer Science (upon Ministry of Education and Training), period 2018–2023. He is also the Vice Director General of the National R&D Program KC-4.0/2019–2025 (upon MOST). He serves as Deputy Chief of Vietnam's National AI Strategy Council in 2020–2030 (established by MOST and other ministries). He keeps the role of the President of the Vietnam Union of AI Institutions and Industry (VINA3I) and the President of FISU, terms 2023–2028.

Assoc. Prof. Le Hoang Son is the Vice Director of VNU Information Technology Institute under Vietnam National University, Hanoi. His main research areas include soft computing foundation, knowledge-based systems, integrated machine learning for multimedia intelligence and multi-modal and environmental AI, with an interest in healthcare and telecommunications network applications. Some applications have been deployed recently with software copyrights, such as a medical imaging support system with a deep learning model and knowledge graph, semi-supervised systems in object detection on remote sensing images, large recommender system (Mega RS) with techniques to extract information on electronic medical records, the software analyzes facial expressions and supports dynamic decision-making. Dr. Son is an Associate Editor of Soft Computing (SCIE), Journal of Intelligent & Fuzzy Systems (SCIE), IEEE Access (SCIE), Data Technologies and Applications (SCIE), Neutrosophic Sets and Systems (ESCI), Vietnam Journal of Computer Science and Cybernetics and Vietnam Research and Development on Information and Communication Technology (2018–2020), VNU Journal of Science: Computer Science and Communication Engineering, Frontiers in Artificial Intelligence. He serves on the Editorial Board of Applied Soft Computing (SCIE), PLOS ONE (SCIE), International Journal of Data Warehousing and Mining (SCIE), International Journal of Web and Grid Services (SCIE) and International Journal of Ambient Computing and Intelligence (ESCI). Dr. Son has been a Vietnam Government Office Consultant (http://vpcp.chinhphu.vn/) since 2019. He received "VNU Young Scientific Research Award 2014", "VNU Annual Research Award 2015", "Vietnam Mathematics Award 2015" and a Certificate of Merit from the President of VNU for outstanding achievements in international publications and intellectual property in 2019 and 2020. He is on the list of 03 Vietnamese scientists in the list of the top 100,000 leading scientists in the world with the most significant influence of the five consecutive years 2019, 2020, 2021, 2022 and 2023.

Prof. Dr. Van-Nam Huynh received a Ph.D. degree in Mathematics from the Vietnam Academy of Science and Technology in 1999. He is a Professor at the School of Knowledge Science, JAIST. Also, he serves as a Research Fellow at The University of Economics in Ho Chi Minh City, Vietnam. He was a visiting professor at the National Electronics and Computer Technology Center (NECTEC), Thailand, in 2019, an adjunct professor at Chiang Mai University, Thailand, from 2015 to 2017, and a part-time lecturer at Tsukuba University, Japan, from 2011 to 2015. His research interests include AI and machine learning, decision analysis and management science, modeling and reasoning with uncertain knowledge, argumentation and multi-agent systems and Kansei information processing and applications. He serves as an Area Editor of the International Journal of Approximate Reasoning, Editor-in-chief of the International Journal of Knowledge and Systems Science and an Area Editor of the journal Array.

Artificial Intelligence

A Combination of Active Learning and Deep Learning for Improving Breast Cancer Prediction

Huong Hoang Luong[1,2], Hai Thanh Nguyen[2], and Nguyen Thai-Nghe[2(✉)]

[1] FPT University, Can Tho, Vietnam
huonghoangluong@gmail.com
[2] Can Tho University, Can Tho, Vietnam
nthai.cit@ctu.edu.vn, ntnghe@cit.ctu.edu.vn

Abstract. In the field of modern medicine, biomedical informatics is crucial for enhancing illness detection, management, and treatment. A strong basis for applications that can enhance people's health and quality of life has been established thanks to the convergence of rich medical data with potent information technology. The detection and diagnosis of breast cancer is one of the key fields where biomedical data is used and labeled data plays an important role in supervised learning models. However, labeling data can costly and time consuming. In the field of breast cancer, there is a similar problem. So that, an approach will be proposed for automatic data labeling using active learning to improve breast cancer prediction by combining least confidence sampling with fine-tuned MobileNet model called BCPDAL. Experimental research on a number of fine-tuned deep learning models combined with uncertainty sampling methods and with the dataset used as BUSI. The results have shown that proposed approach BCPDAL can achieve 90% accuracy with only 50% of the labeled data samples in the entire original dataset.

Keywords: Active learning · Breast cancer · Classification · Uncertainty sampling · Labelling

1 Introduction

Biomedical informatics [1] has become an important field contributing to the improvement of diagnosis, treatment and management of diseases in the field of modern medicine. The combination of rich medical data and powerful information technology has created a solid foundation for applications that can improve people's health and quality of life. There have been many researches and projects on biomedical informatics in life such as: Human Genome Project, Medical Image Analysis, Disease Prediction and Epidemiological Analysis, and so on. One of the important applications of biomedical information is in the field of breast cancer detection and diagnosis [2].

P. T. Nghia et al. (Eds.): ICTA 2023, LNNS 847, pp. 3–10, 2023.
https://doi.org/10.1007/978-3-031-49529-8_1

As reported in [3], Breast cancer is one of the most dangerous and common diseases, especially in women. According to statistics from [4], the number of new cases was 2,261,419 with the number of deaths being 684,996 in 2020. Because of that, early detection of breast cancer will make treatment more effective [5]. One of the ways to detect that is based on images such as: x-ray, ultrasound, etc [6]. From these image data, there have been many studies in building supervised machine learning models to detect and diagnose of the disease based on these breast cancer images [7]. However, the application of supervised machine learning models [8] requires a labeled dataset. The construction of these labeled datasets is very expensive in terms of money and time. And much of the current breast cancer imaging data remains unlabelled. Because of that, in this study, the authors propose an approach in automatic data labeling based on active learning [9] to improve breast cancer prediction by using a combination of fine-tune MobileNet and least confidence sampling. It is called BCPDAL (Breast cancer prediction based on deep active learning). Experimental research on proposed approach and other models combined with uncertainty sampling method [10]. The BUSI dataset [11] is used for experiment and the results have shown that proposed approach will result in 90% accuracy with only 50% of the data in the entire original dataset. The main contributions of this study include:

- Propose an approach using active learning to the breast cancer prediction results in case of very rare data. Find good labeling strategy through evaluation and comparison between different strategies.
- Combine active learning with deep learning, find a suitable model, hyperparameters and fine-tune the deep learning model to help bring the best results to the problem of data labeling with the data set in use.

This article is divided into 5 sections. The first section is an introduction. The next section is related works. The third section is the proposed method. The fourth section is the experiments and final is the conclusion.

2 Related Works

The problem of missing data is one of the most difficult problems in model training with supervised learning. And the same goes for breast cancer. In our knowledge, the application of active learning in the automatic labeling problem to enhance the data and improve the accuracy of the model is very little. The paper [12] is one of the few studies in this area.

In [12], the authors proposed a method called Reversed Active Learning (RAL) to train the CNN model in the breast cancer classification problem. CNN combines with RAL to remove mislabeled data. This study experimented on the ICIAR 2018 Breast Cancer Dataset (IBCD) with an accuracy of 93.75–96.25%.

Authors in [13] proposed an active learning framework on breast cancer detection using an ensemble learning. In this framework, the authors have combined many machine learning models such as: KNN [14], SVM [15], Naive Bayes [16], logistic regression [17], random forest [18] and gradient boost [19]. The experiments were performed on the The Cancer Genome Atlas (TCGA) dataset [20]. The results obtained from the association framework are very satisfactory with an accuracy of 83–98% depending on the classified class.

In [21], a strategy called Batch Active learning by Diverse Gradient Embeddings (BADGE) is proposed based on kMeans++ [22]. This method selected samples based on hallucinated gradients. Since the usage data do not have the true labels available, this hallucinated gradient should be used. Importantly, BADGE trades off uncertainty and diversity without requiring any manually adjusted hyperparameters. While other methods were sometimes successful for specific batch sizes or structures, BADGE always performs as well or better, making it a useful choice for active learning problems in the real world.

3 Proposed Approach

Figure 1 depicts BCPDAL to data labeling and improve breast cancer prediction based on active learning technique [9] for breast cancer imaging data. In which, the combination between the uncertainty sampling methods [10] with other fine-tuned deep models in turn to compare with proposed approach. In addition, the data before being input into the deep learning model will undergo a pre-processing step to increase the image contrast.

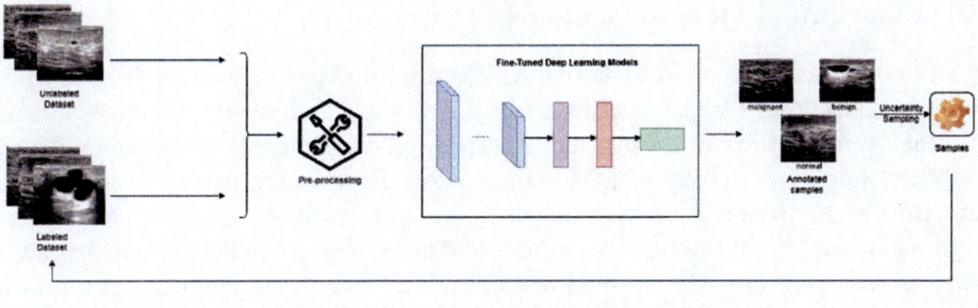

Fig. 1. The proposed approach—BCPDAL.

3.1 Sampling Methods

Uncertainty Sampling [10] is a technique for selecting data samples so that these data samples make the predictive model the least confident to label the data. The goal of this technique is to find data samples that can overcome the weaknesses of the current model. In this paper, three sampling methods will be used includings least confidence, margin sampling and entropy-based sampling.

Least confidence method [10] is focus on selecting the samples that the machine learning model has the highest probability of predicting but is not confident enough in its predictions. Let L be labeled dataset, U be unlabeled dataset, $D = L \cup U$ be all data includings labeled and unlabeled dataset, x be sample in D, $S(x)$ be least confidence score of sample x, $\hat{y}_{max}$ be the model's highest confidence prediction, and $P(\hat{y}_{argmax}|x)$ be predicted probability of the class with the highest probability by the model for sample x. The least confidence score is calculated by formula (1).

$$S_{LC}(x) = 1 - P(\hat{y}_{argmax}|x) \tag{1}$$

Formula (2) represents the margin sampling method [10]. In a classification problem, choosing samples where the model has difficulty distinguishing between two classes can improve the model's performance on new data. In which, $S_M(x)$ is margin score, $\hat{y}_{max}$ is the model's highest confidence prediction and $P(\hat{y}_2|x)$ is the predicted probability of the second lowest probability class by the model for the sample x.

$$S_M(x) = P(\hat{y}_{argmax}|x) - P(\hat{y}_2|x) \tag{2}$$

The final sampling method used in this paper is entropy-based sampling [10]. Let $S_{EB}(x)$ be entropy-based score of sample x. The entropy-based score is calculated according to the formula (3).

$$S_{EB}(x) = -\Sigma_{y \in Y} P(y|x) log_2 P(y|x) \tag{3}$$

3.2 Fine-Tuned Deep Learning Models

To increase the accuracy of feature extraction and classification in deep learning models, a sufficiently large amount of data is required. However, for diseases such as breast cancer, the data analyzed is very limited. Transfer learning method is one of the solutions to overcome this limitation. Fine-tuned pre-train models can result in higher performance for the problem to be solved.

In this paper, different fine-tuned learning models will be performed to improve the results for the applied problem. The weights of the pre-train models still on the ImageNet set are still used in this study. Figure 2 shows an illustration for the fine-tuned deep learning model.

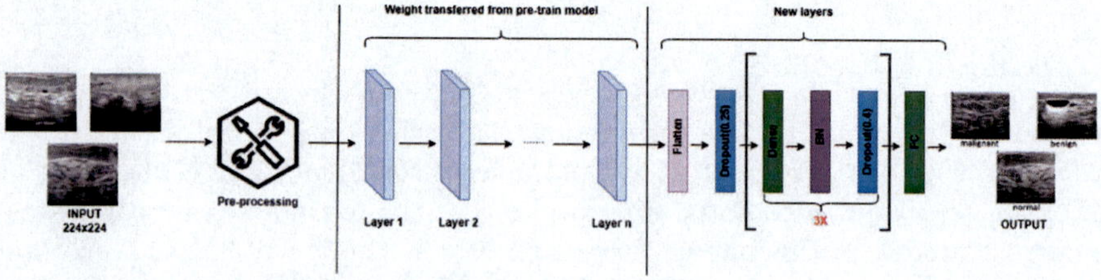

Fig. 2. An illustration of deep model for improving breast cancer prediction.

In the proposed fine-tuning model as Fig. 2, following the layers of the pre-train model is a flatten layer for the purpose of reshape a tensor and dropout layer to skip some units during training in the model, this helps to reduce the overfit rate of the model. This is followed by a set of 3 layers includings dense, batch normalization and dropout that they are repeated 3 times. Final is fully-connected layer. In this study, the fine-tune techniques will be applied for the MobileNet model and other models with purpose for comparison in the experimental section.

4 Experiments

4.1 Dataset and Hyperparameters

Breast Ultrasound Images (BUSI) is a dataset that is published in [11]. This dataset was collected from women aged 25–75 with a total of 600 patients. The included dataset was classified into 3 classes: 210 images for malignant, 437 images for benign, and 133 images of people without disease. All images are saved in PNG format with an average size of 500x500. From the original BUSI dataset, data will be splited into three parts at a ratio of 2.5–77.5-20 consisting of 2.5% for labeled data and 77.5% for unlabeled data and 20% of data used for testing. This labeled data will be the first training set for the fine-tuned models and the unlabeled set to apply the sampling method as described above.

In the experiment, hyperparameters will be used as epoch = 10, 20, 30, learning rate = 0.1, 0.01, 0.001, hidden unit = [512, 256, 128], [256, 256, 64]. The results will be shown in Sect. 4.2 that collect from best hyperparameter set includings epoch = 20, learning rate = 0.001, hidden unit = [512, 256, 128], [256, 256, 64].

To train the models and evaluate the proposed method, Python language is used with two main supporting libraries, tensorflow and keras. The experiment was carried out on a computer with a configuration including Windows 10 64-bit; Intel Core i5-12400F CPU @ 2.50 GHz × 12; and 32 GB of RAM, and NVIDIA GeForce RTX 3060 12 GB VRAM graphics card.

4.2 Evaluate the Experiment Results

The purpose of this experiment is to evaluate BCPDAL and compare with combination between other deep models and strategy sampling in uncertainty sampling.

Figure 3 shows the results of the labeling process with different models using *hidden unit* = {*[512, 256, 128]*}. From above result, BCPDAL just assigning labels to 340 image data samples (At the 17th step, this equates to 56.6% of the dataset) can achieve an accuracy of 87%.

Figure 4 shows the results of the labeling process with another hyperparameter *hidden unit* = {*[256, 256, 64]*}. With the least confidence method, the proposed approach achieves 90% accuracy with only 320 labeled data samples

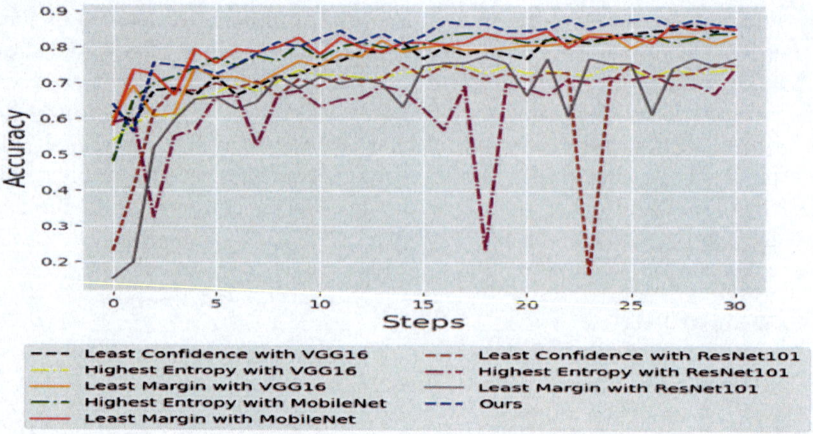

Fig. 3. Labeling results with different methods and models with hyperparameters: epoch = 20, learning rate = 0.001 and hidden units = [512, 256, 128].

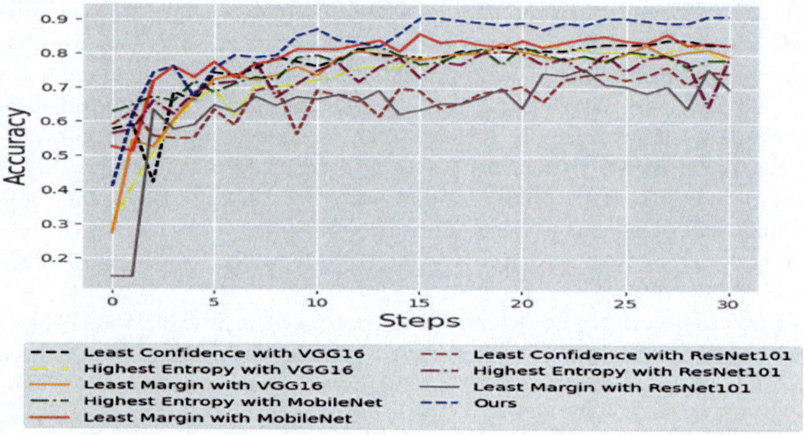

Fig. 4. Labeling results with different methods and models with hyperparameters: epoch = 20, learning rate = 0.001 and hidden units = [256, 256, 64].

(At the 15th step, this equates to 50% of the dataset), which is higher than the results achieved in the above set of hyperparameters.

The Fig. 5 compares the results of applying and not applying BCPDAL. This approach helps to solve the problem of severe data shortage in breast cancer classification based on ultrasound images. Besides, the study can not only perform data augmentation by labeling for unlabeled data, but also improve the accuracy of breast cancer classification.

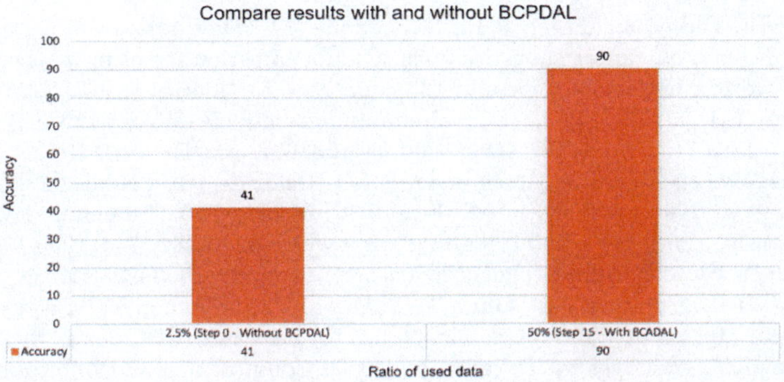

Fig. 5. Performance comparison results with and without using BCPDAL

5 Conclusion

Clearly labeled data is crucial to supervised learning algorithms. Data tagging will cost money and take time, though. Similar issues exist in the field of breast cancer, where the current dearth of labeled data has a significant impact on the outcomes of disease classification models built using breast cancer images. Because of this, this study is provided a method for automatically classifying data and improving breast cancer prediction based on active learning in this work and called BCPDAL. According to experimental research, a good labeling strategy combine with a good model (least confidence with fine-tuned MobileNet) will produce 90% accuracy with only 50% of labeled data samples

In future studies, new techniques or approach will be researched to improve the accuracy of labeling using active learning technique and build a more efficient labeling process for the purpose of using these labels for data enhancement in supervised learning models. Such data enhancement will help the models have enough training data and bring high efficiency to the breast cancer classification problem.

References

1. Sable, J., Hales, J., Bopp, K.: Medical informatics in healthcare organizations: a survey of healthcare information managers. In: Proceedings of the AMIA Symposium, pp. 744–748 (2000). https://europepmc.org/articles/PMC2243871
2. Potamias, G., Analyti, A., Kafetzopoulos, D., Kafousi, M., Margaritis, T., Plexousakis, D., Poirazi, P., Reczko, M., Tollis, I., Sanidas, E., Stathopoulos, E., Tsiknakis, M., Vassilaros, S.: Breast cancer and biomedical informatics: The prognochip project, Sept. 2004
3. Harbeck, N., Penault-Llorca, F., Cortes, J., Gnant, M., Houssami, N., Poortmans, P., Ruddy, K., Tsang, J., Cardoso, F.: Breast cancer. Nat. Rev. Dis. Primers **5**(1), 66 (2019). Sep

4. Sung, H., Ferlay, J., Siegel, R.L., Laversanne, M., Soerjomataram, I., Jemal, A., Bray, F.: Global cancer statistics 2020: Globocan estimates of incidence and mortality worldwide for 36 cancers in 185 countries. A Cancer J. Clin. CA (2021)
5. Pisano, E.D., Yaffe, M.J.: Digital mammography. Radiology **234**(2), 353–362 (2005). https://doi.org/10.1148/radiol.2342030897, pMID: 15670993
6. Pisano, E.D., Yaffe, M.J., Hemminger, B.M., Hendrick, R.E., Niklason, L.T., Maidment, A.D., Kimme-Smith, C.M., Feig, S.A., Sickles, E.A., Braeuning, M.P.: Current status of full-field digital mammography. Acad. Radiol. **7**(4), 266–280 (2000). https://www.sciencedirect.com/science/article/pii/S107663320080478X
7. Luong, H.H., Phan, N.T.L., Dinh, T.C., Dang, T.M., Duong, T.T., Nguyen, T.D., Nguyen, H.T.: Fine-tuning mobilenet for breast cancer diagnosis. In: Smys, S., Kamel, K.A., Palanisamy, R. (eds.) Inventive Computation and Information Technologies, pp. 841–856. Springer Nature Singapore, Singapore (2023)
8. Cunningham, P., Cord, M., Delany, S.J.: Supervised Learning, pp. 21–49. Springer, Berlin Heidelberg (2008). https://doi.org/10.1007/978-3-540-75171-7_2
9. Ranganathan, H., Venkateswara, H., Chakraborty, S., Panchanathan, S.: Deep active learning for image classification. In: 2017 IEEE International Conference on Image Processing (ICIP), pp. 3934–3938 (2017)
10. Liu, S., Li, X.: Understanding uncertainty sampling (2023)
11. Al-Dhabyani, W., Gomaa, M., Khaled, H., Fahmy, A.: Dataset of breast ultrasound images. Data Brief **28**, 104863 (2020). https://www.sciencedirect.com/science/article/pii/S2352340919312181
12. Xie, X., Li, Y., Shen, L.: Active learning for breast cancer identification (2018)
13. Lee, S., Amgad, M., Masoud, M., Subramanian, R., Gutman, D., Cooper, L.: An ensemble-based active learning for breast cancer classification. In: 2019 IEEE International Conference on Bioinformatics and Biomedicine (BIBM), pp. 2549–2553 (2019)
14. Mucherino, A., Papajorgji, P.J., Pardalos, P.M.: k-Nearest Neighbor Classification, pp. 83–106. Springer, New York, New York, NY (2009). https://doi.org/10.1007/978-0-387-88615-2_4
15. Cristianini, N., Ricci, E.: Support Vector Machines, pp. 928–932. Springer, US, Boston, MA (2008). https://doi.org/10.1007/978-0-387-30162-4_415
16. Webb, G.I.: Naïve Bayes, pp. 713–714. Springer, US, Boston, MA (2010). https://doi.org/10.1007/978-0-387-30164-8_576
17. Chung, M.K.: Introduction to logistic regression (2020)
18. Ho, T.K.: Random decision forests. In: Proceedings of 3rd international conference on document analysis and recognition, vol. 1, pp. 278–282. IEEE (1995)
19. Friedman, J.H.: Greedy function approximation: a gradient boosting machine. Ann. Stat. 1189–1232 (2001)
20. Hutter, C., Zenklusen, J.C.: The cancer genome atlas: Creating lasting value beyond its data. Cell **173**(2), 283–285 (2018). https://www.sciencedirect.com/science/article/pii/S009286741830374X
21. Ash, J.T., Zhang, C., Krishnamurthy, A., Langford, J., Agarwal, A.: Deep batch active learning by diverse, uncertain gradient lower bounds (2020)
22. Arthur, D., Vassilvitskii, S.: K-means++: the advantages of careful seeding. In: Proceedings of the Eighteenth Annual ACM-SIAM Symposium on Discrete Algorithms, pp. 1027–1035. SODA '07, Society for Industrial and Applied Mathematics, USA (2007)

A Novel Arm Bone Fracture Detection Using Deep Learning

Huy Hoang Nguyen(✉), Tuan Khoi Nghiem, and Ngoc Minh Dang

School of Electrical and Electronic Engineering, Hanoi University of Science and Technology, Hanoi, Vietnam
hoang.nguyenhuy@hust.edu.vn,
{khoi.nt192941,minh.dn203505}@sis.hust.edu.vn

Abstract. Fracture detection in medical imaging is a critical aspect of diagnostic radiology, with accurate and timely identification being paramount for improved patient outcomes. In this paper, we propose an innovative and comprehensive deep learning-based methodology for humeral fracture detection in X-ray images, utilizing the MURA dataset. Our approach is multi-faceted, integrating cutting-edge techniques across various stages of image processing and analysis. First, we preprocess the X-ray images using the YOLOv7-mask to segment the images. Then, Histogram Equalization and Contrast Limited Adaptive Histogram Equalization algorithms are used to enhance the image's contrast. We propose an image blending method based on the above two algorithms. After that, Super-Resolution Generative Adversarial Networks SRGAN is applied to increase the image resolution. Finally, YOLOv7-E6E was trained on a small dataset with data enhancement techniques to identify and localize fractures on X-ray images. The highest result obtained was 84.03% using our proposed method. When applied to the same dataset, the experimental findings further demonstrate the superiority of our approach over those proposed by other approaches.

Keywords: Deep learning · Fracture detection · X-ray image

1 Introduction

Fracture detection is a crucial problem in medical imaging, particularly in radiology. A timely and correct fracture diagnosis can result in proper care. X-ray imaging is one of the most popular diagnostic techniques for detecting fractures since it is non-invasive, reasonably priced, and widely accessible in most healthcare settings. However, manually identifying fractures in X-ray images is a complex and time-consuming task requiring extensive knowledge and experience. Additionally, subjective elements like observer fatigue, human error, and inter-observer variability may affect a manual diagnosis's accuracy.

To overcome these difficulties, there is growing interest in creating automated defect detection systems employing cutting-edge machine learning techniques.

P. T. Nghia et al. (Eds.): ICTA 2023, LNNS 847, pp. 11–19, 2023.
https://doi.org/10.1007/978-3-031-49529-8_2

Deep learning, a subfield of machine learning, uses multiple-layered artificial neural networks to recognize and learn attributes from the dataset due to its advantages in various applications, including healthcare systems [1,2], security surveillance [3–5], industrial safety [6], etc. Many studies using deep learning to detect fractures have achieved promising results. These include the hip fracture detection models of Kitamura et al. [7], Cheng et al.[8], Yamada et al. [9], and Kim et al. [10] built a model to detect vertebral fractures. In addition, Guan et al. [11] and Choi et al. [12] developed models for detecting arm fractures. The models are built on different datasets. Although there are many studies using deep learning in bone fracture detection, studies focusing on arm fracture detection have still been lacking. To the best of our knowledge, the latest study related to this topic is the work of Nguyen et al. [13], and it still can be improved.

With the above analysis, we will base on the work of Nguyen et al. to propose an advanced arm bone fracture detection using deep learning. This is also our major contribution to this study. To avoid a repetitive writing style, the remainder of this paper follows this structure: Sect. 2 delves into the methodology, Sect. 3 explores the findings and makes comparisons, while Sect. 4 contains the paper's conclusion.

2 Methodology

2.1 Proposed approach

The X-ray image undergoes segmentation with the YOLOv7-mask model, blackening the background pixels. The mask region's contrast is enhanced with CLAHE and EH algorithms before reintegration. The image is then upscaled using an SRGAN model. Finally, it's fed into the YOLOv7-E6E model, producing the raw image with a predicted arm fracture. Figure 1 depicts an overview of the model.

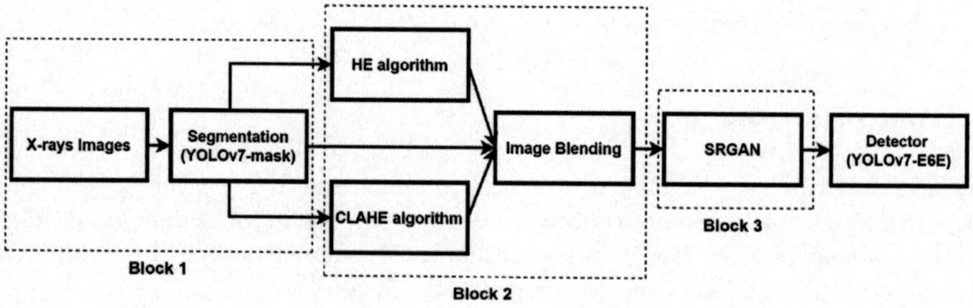

Fig. 1. Method overview.

2.2 YOLOv7-mask Instance Segmentation

YOLOv7-mask: BlendMask [14] and YOLOv7 [15] were combined for instance segmentation. After YOLOv7 object detection has identified the object, BlendMask, a deep learning technique of Hao Chen et al., is utilized to segment the detected object. To do this, the YOLOv7 object identification model was trained for 30 epochs and fine-tuned using the MS COCO instance segmentation dataset. It produces cutting-edge solutions for real-time instance segmentation.

2.3 Image Processing Algorithms

HE algorithm: The Histogram Equalization (HE) algorithm redistributes pixel values in an input image to create a more evenly distributed histogram. This enhances the visibility of structures in X-ray images and reveals more detail in over- or underexposed images by spreading pixel values across the entire dynamic range.

CLAHE algorithm: The Contrast Limited Adaptive Histogram Equalization (CLAHE) algorithm enhances local contrast in image regions while preventing noise amplification. CLAHE limits gain by clipping the histogram to a specified value before computing the Cumulative Distribution Function. It relies on two key parameters, Block Size (BS) and Clip Limit (CL), to improve image quality.

Image blending: The masks for each image are generated using the YOLOv7-mask segmentation model. Subsequently, the HE and CLAHE algorithms are applied sequentially to these masks, with the background converted to black. This process results in three distinct datasets, each equivalent in size, with every image having three versions, including the original. The method for exporting these images is achieved through a blending process as follows:

$$P_{F_{ij}} = \alpha P_{HE_{ij}} + \beta P_{CLAHE_{ij}} + (1 - \alpha - \beta) P_{ij} \tag{1}$$

with $P_{F_{ij}}$ is the pixel value (i, j) of the last image and $P_{HE_{ij}}$, $P_{CLAHE_{ij}}$, P_{ij} are the pixel value (i, j) of the HE's dataset, CLAHE's dataset and original dataset. The alpha and beta weights are adjusted to suit the requirements of each purpose and $\alpha = \beta = \frac{1}{3}$ in this experiment.

2.4 Super Resolution Generative Adversarial Networks

Generator network: The input image for the SRGAN [16] generator will be I_{LR} (low resolution), created from I_{HR} (high resolution). The size of I_{LR} is $W \times H \times C$. The generator takes the input image I_{LR} and produces an image I_{SR} (Super-resolution). $rW \times rH \times C$, where r is the upsample coefficient, will represent I_{SR} and I_{HR}, respectively. The batch normalization and ParametricReLU (PReLU) activation function will come after the B residual block, which comprises two kernels 3×3 with 64 channels in the generator network design. The author employs two PixelShuffle layers, each doubling the image size to upsample the image size.

Discriminator network: The SRGAN discriminator distinguishes between the original and generated super-resolution images. Similar to the VGG network, it comprises 8 convolution layers (3 × 3) with channel counts increasing from 64 to 512 in multiples of 2. Strided convolution reduces image size after doubling feature counts. The 512 feature maps are passed through two dense layers and a sigmoid layer to obtain the image's discriminant probability. Figure 2 illustrates the summary of image preprocessing methods.

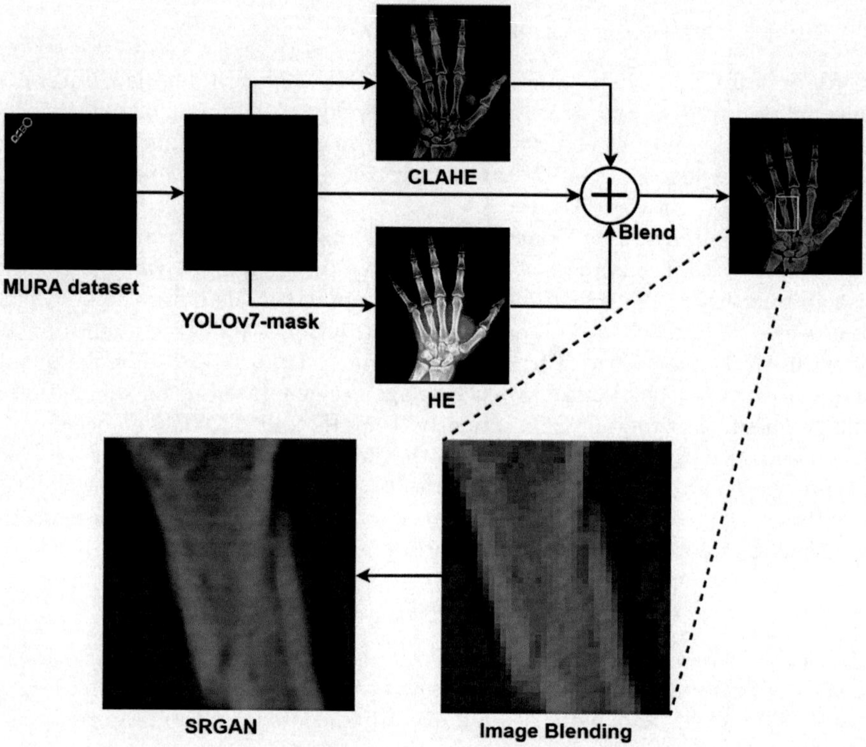

Fig. 2. Preprocessing technique.

2.5 YOLOv7-E6E Object Detection

YOLOv7 based single object detection: ELAN (Efficient Layer Aggregation Network) is the foundation for YOLOv7, incorporating three key components: Cross Stage Partial, Computation Block, and PointWiseConv. ELAN integrates concepts from earlier designs VoVNet and CSPVoVNet. Data passes through the Stem Block to the first ELAN Block in the backbone. Transition Blocks link ELAN boxes, reducing feature map size by half. YOLOv7 utilizes necks like

SPPCSPC, (CSP-OSA)PANet, and RepConv, with the YOLOR head. YOLOv7 variants include YOLOv7-P5 groups trained at 640×640 and 1280×1280 image sizes, as well as YOLOv7-E6E.

YOLOv7-E6E architecture: Our research utilizes YOLOv7-E6E for arm fracture identification, with its foundation being the Extended-ELAN (E-ELAN) architecture. This model is trained on 1280×1280 images from the YOLOv7-P6 family, which employs a distinct scaling method compared to E6E. E-ELAN employs techniques like expansion, shuffling, and cardinality merging within the computation block to enhance learning without disrupting the gradient path. Architecturally, E-ELAN modifies only the compute block design while keeping the transition layer unchanged. It increases the number of channels and computation blocks by maintaining grouping parameters and channel coefficients across all calculation blocks. Feature maps are divided into groups (parameter g) and concatenated, each having the same channels. This cardinality merging enables E-ELAN to instruct different compute block groups to learn diverse characteristics while preserving the original ELAN design architecture.

3 Experimental Results

3.1 Environment Setup

YOLOv7-segment and YOLOv7-detection were trained on an RTX 2080Ti GPU with 11GB VRAM and an Intel i7 CPU. YOLOv7-segment used a batch size of 16 for 500 iterations, with momentum and weight decay rates at 0.937 and 0.0005, respectively. YOLOv7-detection resized input images to 1280×1280, adjusted color channels, saturation, and brightness. It was trained for 250 loops with a batch size of 1 and a learning rate of 0.001, utilizing weight decay rates and momentum set at 0.0005 and 0.949.

Additionally, the SRGAN architecture was employed natively from [16], with a batch size of 16 and a learning rate of 0.0001, trained with ImageNet pre-trained weight initialization. The model underwent training for 5000 update iterations at a learning rate of 0.001, followed by another 5000 at a lower rate of 0.0001. Optimization utilized Adam with $\beta_1 = 0.9$, and both the generator and the discriminant network were updated, equivalent to $k = 1$.

3.2 YOLOv7-mask Training Results

Similar to the work of Nguyen et al. [13], our study uses the same MURA dataset, a sizable public collection of musculoskeletal radiographs gathered from Stanford Hospital. The primary dataset includes 4000 images from five parts, including finger, forearm, wrist, hand, and humerus selected from the MURA dataset. Of these, 3199 images are used for training and 801 for evaluation. Results of training YOLOv7-mask after 500 rounds obtained mAP respectively 98.12%. The error value obtained is 0.001. Then, the masks of each image are taken.

3.3 SRGAN Training Results

Data preparation: This involves collecting a dataset of low-resolution images and their corresponding high-resolution images. The high-resolution image dataset is derived from the x-ray dataset obtained after the image blending step. We obtained the low-resolution images by down-sampling the high-resolution images $(BGR, C = 3)$ using a bicubic kernel with a down-sampling factor $r = 4$. **SRGAN training results**: Peak Signal-to-Noise Ratio (PSNR) and Structural Similarity Index Measurement (SSIM) metrics were calculated on each image. The average PSNR value compared between high-resolution dataset images and images generated from low-resolution dataset images is $35.51dB$. The average SSIM value compared between low-resolution dataset images and images generated from low-resolution dataset images is 0.9660, respectively. Therefore, the dataset used for training YOLOv7-E6E and subsequent in-process tests was generated from high-resolution dataset images passing through SRGAN. Some results are shown in Fig. 3.

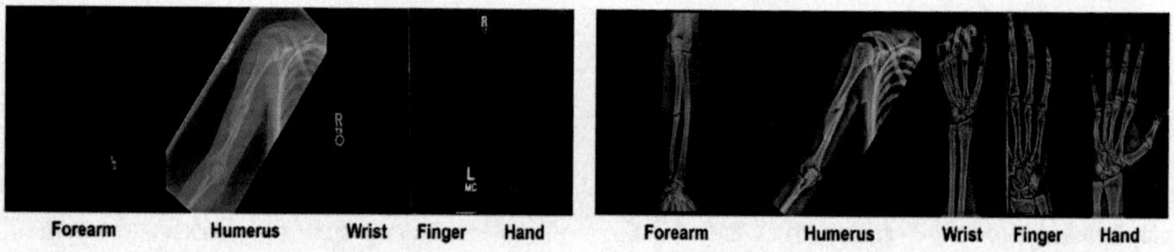

(a) MURA original dataset. (b) Dataset for YOLOv7-E6E.

Fig. 3. Data set for bone fraction detection.

3.4 YOLOv7-E6E Training Results

Table 1. Results of various methodologies and models' evaluation.

Methods			YOLOv7			YOLOv7-E6E		
HE	CLAHE	SRGAN (%)	mAP (%)	Precision (%)	Recall (%)	mAP (%)	Precision (%)	Recall (%)
X			73.02	68.67	72.48	76.75	75.68	74.31
	X		74.16	67.17	75.08	75.57	75.20	75.23
X	X		77.32	82.98	71.56	79.66	84.37	74.29
X		X	75.11	75.20	70.11	76.31	78.92	72.78
	X	X	77.46	80.56	74.83	79.94	84.35	71.20
X	X	X	**81.31**	**85.14**	**78.84**	**84.04**	**91.04**	**75.23**

The dataset has 303 images with 5 categories similar to the segmentation model above, which are selected from the MURA dataset for the YOLOv7-E6E training process, and 102 images are employed throughout the testing process. The size of the training set dramatically affects the performance of the model. Although the training results are excellent, the testing on the test set needs to be more rigorous due to the model's overfitting caused by the lack of data generalization. With 303 images in the dataset, there is a significant chance of overfitting. The data augmentation approach is thus employed to address this issue. Through a flip, a 45° rotation, and a mosaic. Table 1 summarizes the performance of different models with different methods. Models include YOLOv7 and YOLOv7-E6E evaluations using mAP, Precision, and Recall. From the results table, both models have good performance using SRGAN. The YOLOv7-E6E model gave the best results with mAP 84.04%, Precision 91.04%, and Recall 75.23%. These results prove that using SRGAN to create new data gives good results. In addition to using data preprocessing methods, enhancement methods mosaic, brightness, flip, and rotate 45°. Using SRGAN resulted in an approximate 5% increase in mAP compared to the conventional method. In addition, using the YOLOv7-E6E model with an input image size of 1280 × 1280 also results in an approximate 3% increase compared to the YOLOv7 model with an input image size of 640 × 640. In addition to the mapped value, SRGAN's Precision significantly increases compared to conventional methods and smaller models.

3.5 Compare with another method

In this experiment, the YOLOv4 darknet model-based approach of Nguyen et al. is currently the state-of-the-art arm fracture detection method on the dataset used by this study. The best performance of the two methods is summarized in Table 2. The results show that, from the segmentation step, the result is much better, with a 5.46% difference. The result is also better in the fracture detection model, with 2.14%. Better results by using the YOLOv7-mask segmentation model with low loss and very high mAP, resulting in precise, noise-free segmentation. Besides, using the SRGAN model to increase the resolution helps prevent the image from losing features when passing through the YOLOv7-E6E model with an input image size of 1280 × 1280. Therefore, the results we obtained are the best, with 84.04% mAP.

Table 2. Comparison of our approach to the Nguyen's approach.

Matric	Nguyen's methods		Our appoarch	
mAP	Segmentation(YOLACT++)	Detection(YOLOv4 darknet)	Segmentation(YOLOv7-mask)	Detection(YOLOv7-E6E)
	92,66%	81.9%	**98.12%**	**84.04%**

4 Conclusion

In this study, we introduced a radiograph-based approach for arm fracture identification. Our solution incorporates image preprocessing techniques, SRGAN network, and fracture detection via YOLOv7-E6E. To address the limited dataset, we employed various data augmentation methods during training to enhance the model's overall performance. Test results demonstrate that combining CLAHE and HE contrast enhancement, YOLOv7-mask segmentation, and resolution enhancement with SRGAN outperforms traditional contrast enhancement methods and the best result achieved 84.04% mAP for arm fracture detection. Furthermore, this study highlights the effectiveness of our arm fracture detection approach with the current best dataset. Future plans include increasing dataset size and quality, exploring new approaches to expand data applicability, and improving model performance.

References

1. Doan, N.P., Pham, N.D.A., Pham, H.M., Nguyen, H.T., Nguyen, T.A., Nguyen, H.H.: Real-time Sleeping Posture Recognition for Smart Hospital Beds, International Conference on Multimedia Analysis and Pattern Recognition (MAPR), pp. 1–6 (2021). https://doi.org/10.1109/MAPR53640.2021.9585289
2. Sabor, N., Gendy, G., Mohammed, H., Wang, G., Lian, Y.: Robust arrhythmia classification based on QRS detection and a compact 1D-CNN for wearable ECG devices. IEEE J. Biomed. Health Inform. **26**(12), 5918–5929 (2022). https://doi.org/10.1109/JBHI.2022.3207456
3. Nguyen, H.H., Ta, T.N., Nguyen, N.C., Bui, V.T., Pham, H.M., Nguyen, D.M.: YOLO based real-time human detection for smart video surveillance at the edge. In: IEEE Eighth International Conference on Communications and Electronics (ICCE), Vietnam, pp. 439–444 (2021). https://doi.org/10.1109/ICCE48956.2021.9352144
4. Nguyen, H.H., Vu, B.H., Le, H.H., Nguyen, M., Nguyen, V.P.: Description based person retrieval using outfit characteristics and human gender. In: International Conference on Communication, Circuits, and Systems (IC3S), India, pp. 1–6 (2023). https://doi.org/10.1109/IC3S57698.2023.10169676
5. Nguyen, H.H., Trung Le, Q., Nghiem, V.Q., Son Hoang, M., Pham, D.A.: A novel violence detection for drone surveillance system. In: International Conference on Communication, Circuits, and Systems (IC3S), India, pp. 1–6 (2023). https://doi.org/10.1109/IC3S57698.2023.10169405
6. Wu, P.-S., Lin, C.-Y., Cheng, T.-Y., Yao, W.-S.: Analysis and design of industrial safety automatic identification system based on Tiny-YOLOv3. IEIFEEC (2021)
7. Kitamura, G.: Deep learning evaluation of pelvic radiographs for position, hardware presence, and fracture detection. Eur. J. Radiol. (2020). https://doi.org/10.1016/j.ejrad.2020.109139
8. Cheng, C.-T., Chen, C.-C., Cheng, F.-J. et al.: A Human-Algorithm Integration System for Hip Fracture Detection on Plain Radiography: System Development and Validation Study, JMIR Medical Informatics. https://doi.org/10.2196/19416

9. Yamada, Y., Maki, S., Kishida, S., et al.: Automated classification of hip fractures using deep convolutional neural networks with orthopedic surgeon-level accuracy: ensemble decision-making with antero-posterior and lateral radiographs. Acta Orthop. (2020). https://doi.org/10.1080/17453674.2020.1803664
10. Kim, K.C., Cho, H.C., Jang, T.J. et al.: Automatic detection and segmentation of lumbar vertebrae from X-ray images for compression fracture evaluation. Comput. Methods Program. Biomed. (2020)
11. Guan, B. et al.: Arm fracture detection in X-rays based on improved deep convolutional neural network. Comput. Electr. Eng. (2020)
12. Choi, J.W., et al.: Using a dual-input convolutional neural network for automated detection of pediatric supracondylar fracture on conventional radiography. Invest. Radiol. **55**, 101–110 (2020)
13. Nguyen, H.P. et al.: A deep learning based fracture detection in arm bone X-ray images. In: International Conference on Multimedia Analysis and Pattern Recognition (MAPR), pp. 1–6 (2021). https://doi.org/10.1109/MAPR53640.2021.9585292
14. Chen, H., Sun, K., Tian, Z., Shen, C., Huang, Y., Yan, Y.: BlendMask: Top-Down Meets Bottom-Up for Instance Segmentation (2020). arXiv:2001.00309 [cs]
15. Wang, C.-Y., Bochkovskiy, A., Liao, H.-Y.M.: YOLOv7: trainable bag-of-freebies sets new state-of-the-art for real-time object detectors (2022)
16. Ledig, C. et al.: Photo-Realistic Single Image Super-Resolution Using a Generative Adversarial Network. https://arxiv.org/abs/1609.04802

A Systematic Review of Artificial Intelligence in Geographic Information Systems

Son Nguyen-Kim, Vinh Nguyen-The, and Duc-Binh Nguyen(✉)

Thai Nguyen University of Information and Communication Technology, Thai Nguyen, Vietnam
{nkson,vinhnt,binhnd}@ictu.edu.vn

Abstract. This study aims to comprehensively evaluate the use of artificial intelligence (AI) in geographic information systems (GIS) in Scopus database. The PRISMA guideline was adopted to aid article selection. A total of 40 articles were analyzed using the VOSviewer software to identify trends, patterns and related keyword phrases. Results revealed that the number of published articles on the topic has increased significantly between 2021 and 2023, with China and Italy being the countries that rank most positively in publishing papers about the application of AI in GIS. The majority of articles have been published in the journal Sustainability, suggesting that the journal is a popular outlet for research in the field. The three clusters of related keywords identified in this study provide a comprehensive overview of various aspects and applications of AI in GIS, including "Geographic information system", "machine learning", and "big data". This study provided a valuable background for future research in this area and can inform regulators, policymakers, and researchers about the capabilities of AI in GIS for many different fields.

Keywords: Artificial intelligence (AI) · Geographic information system (GIS) · Systematic reviews · Big data · PRISMA

1 Introduction

In the age of digitization and the rapid development of geographic data, Geographic Information Systems (GIS) have become an important tool for collecting, organizing and analyzing spatial data. Meanwhile, Artificial Intelligence (AI) has made significant progress in processing complex information and making intelligent decisions [28]. The combination of AI and GIS has opened up new potential in enhancing the ability to analyze and exploit geographic data.

The application of AI in GIS offers many benefits as it provides automation and intelligence for geographic data processing, including classification, identification, prediction, and mining. The application of AI in GIS enhances the ability to process big data, analyze data in real time, and create accurate predictive

P. T. Nghia et al. (Eds.): ICTA 2023, LNNS 847, pp. 20–31, 2023.
https://doi.org/10.1007/978-3-031-49529-8_3

models [6,13,17,20,35]. This contributes to understanding human-environment interactions, decision support, and solving complex problems in the field of geography. Traditional rule-based geographic information systems and algorithms have been used to process geographic data. However, these methods often have limitations in handling big data and modeling complex relationships in geographic data. The combination of AI and GIS helps to overcome these limitations by using machine learning algorithms, artificial neural networks and data mining methods [7,8,18,32] to create a predictive model real-time prediction and analysis. There have been many previous studies that have sought to develop AI methods and techniques to improve the overall analysis capabilities in the GIS field [10,14,15,35]. In addition, many scientists have stated and presented the potentials, opportunities, difficulties and challenges for the integration of AI in GIS [3,16,19,27,31,34]. However, most of the overview research articles on AI in GIS only focus on research and content classification while other foundation information, such as active authors in the field, influential articles, countries, journals, keywords, etc. rarely investigated in detail.

Therefore, this paper aims to solve the above problem by conducting bibliometric analysis of AI in GIS. By employing this approach, the study intends to provide a comprehensive overview of the prior efforts in the literature, identify research trends, highlight influential works and authors, and uncover emerging themes. This study will address the following research questions:

- RQ1: What is the trend of peer-reviewed papers on the application of AI in GIS from 2014 to 2023, and how has the number of publications changed over time?
- RQ2: What countries are most active in researching AI in GIS and is there any collaboration between them?
- RQ3: Which journals are the most influential in the field of AI in GIS? and
- RQ4: What emerging themes and trends can be synthesized from recent research of AI in GIS?

Answering the research questions above will help researchers gain basic perspectives to approach AI in GIS. New researchers can find future research directions through emerging themes, or relevant journals.

2 Materials and Methods

The Preferred Reporting Items for Systematic Reviews and Meta-Analyses (PRISMA) [5] standards were followed for conducting the systematic review on the use of AI in GIS. First of all, PRISMA is a generally acknowledged and approved scientific standard. It has been supported and accepted as the normative reporting guideline for systematic reviews by various publications, groups, and research institutes [23,25].

2.1 Data Selection

The corpus was created using the Scopus database, which included contained titles, abstracts, and keywords. This database is recognized as one of the most important and reliable sources of excellent publications in the domains of engineering and computer science [24]. To find pertinent papers linked to the study issue, the TITLE-ABS-KEY ("AI" OR "Artificial Intelligence" AND "GIS" OR "Geographic Information Systems") was the search term utilized in this investigation. To ensure eligibility, any journal article addressing the application of AI and GIS needed to meet the following criteria: the topic should pertain to the utilization of AI and GIS. (ii) the article should be published in English. (iii) the article should have been published between 2014 and 2023. Papers that do not meet the previous criteria will be excluded from the corpus. The selection process involved two stages of review: full-text screening and title/abstract screening. During the title/abstract screening, the studies were assessed based on the information provided in their titles and abstracts, and any irrelevant research was excluded. The remaining papers then underwent a more rigorous evaluation during the full-text screening to determine their eligibility for inclusion. After finalizing the selection of relevant research, a standardized data extraction form was used to gather essential information. Key details such as study characteristics (e.g., author and publication year) were entered into the form consistently and comprehensively. The process of data collection and screening, as well as the utilization of the PRISMA technique for conducting the systematic literature review, are depicted in Fig. 1. Following a thorough screening process, 40 publications were selected for inclusion in the research.

2.2 Data Analysis and Reporting

In order to generate visual representations of the data, we utilized a widely recognized computer tool called VOSviewer. This powerful tool allowed us to create visualizations encompassing various aspects of information, such as authors, keywords, nations, and organizations. By examining the frequency of co-occurrence among these elements within the literature data, we established connections between them.

3 Results

3.1 RQ1: What Is the Trend of Peer-Reviewed Papers on the Application of AI in GIS from 2014 to 2023, and How Has the Number of Publications Changed Over Time?

Figure 2 illustrates the distribution of assessment literature focused on the utilization of AI in GIS from 2014 to 2023. The information highlights the growing interest in the current topics being investigated by researchers. Notably, the number of review papers published annually has experienced significant growth starting from 2021 compared to previous years. The information also showed that the number of publications examining the use of AI in GIS grew gradually throughout the period from 2014 to 2023. The year 2022 witnessed the highest number of peer-reviewed publications, with 15 papers, indicating a heightened interest in the integration of AI in GIS. Additionally, 2023 has also seen a notable number of peer-reviewed papers published, with 7 articles.

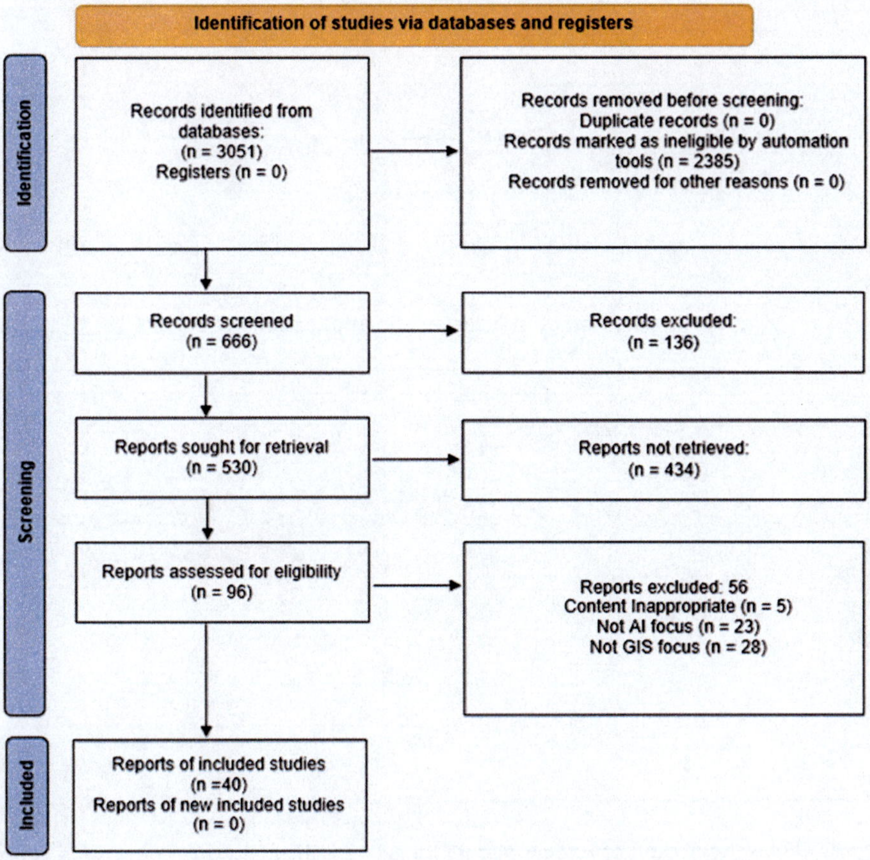

Fig. 1. The PRISMA flow chart depicts the selection of research articles.

3.2 RQ2: What Countries Are Most Active in Researching AI in GIS and Is There Any Collaboration Between Them?

Figure 3 shows the widespread interest in the use of AI in GIS by the number of publications published on the subject, with primary authors hailing from 20 different nations. China is the nation that contributes the most articles, with 8, followed by Italy with 6, and Saudi Arabia and Vietnam with 3 articles.

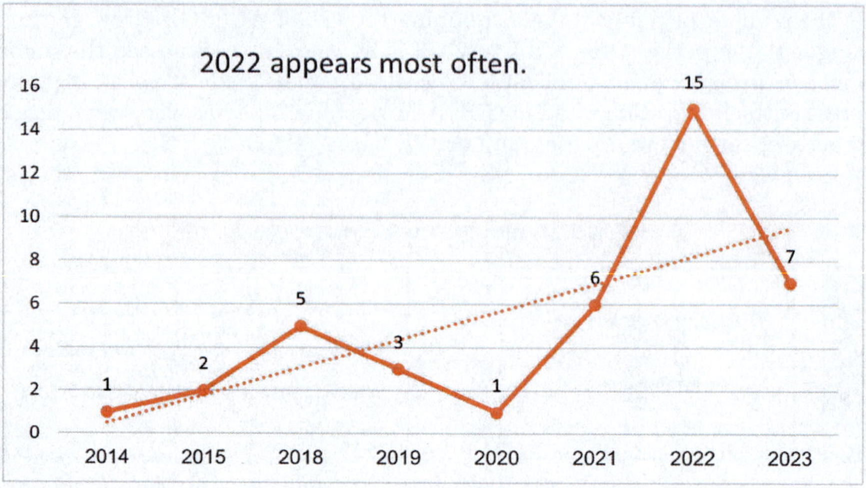

Fig. 2. Distribution of articles on the applications of AI in GIS.

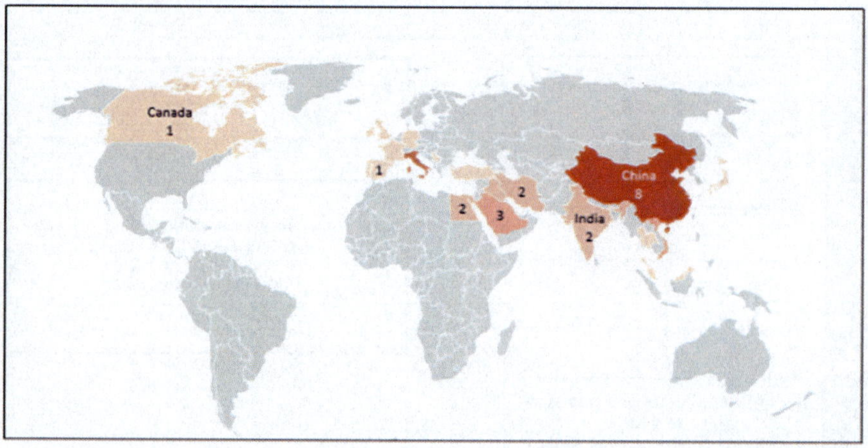

Fig. 3. Global landscape of review papers on AI in GIS: analyzing countries's contributions from 2014 to 2023

The patterns in publication volume differ from country to country and year to year. On the subject of AI in GIS, Portugal just one work was published in 2014. There were two pieces from Malaysia and France the next year. Italy is the leading nation in 2018 with two publications; Serbia, Vietnam, and the United Kingdom each have one publication. Vietnam is the only one of the three countries - the others being Canada and Iraq - that will continue to have a 2019 publication in 2020. China contributes to two publications in 2021, and Iraq, Thailand, India, and Japan each contribute to one publication. The majority of articles on this topic have been published by 2022 in China, Italy and Saudi Arabia [1,7], with a total of 9 publications. Spain [21], Germany [4], Egypt [33], Turkey [30], India [22] and Iran [26] are the other countries that have published a study on the topic. China [10,12] continues to be the most published country on this topic in 2023. In addition, articles on this topic have also been published in several other countries, including Iran [20], Ireland [29], Saudi Arabia [2], Egypt [9] and Italy [11].

Indicating an increasing interest in this field of study, the number of nations publishing on the subject of AI in GIS climbed from one in 2014 to 15 in 2022 and 7 publications as of July 2023. The distribution of publications across several nations demonstrates the interest and collaboration in research on the use of artificial intelligence in geographic information systems that exists on a worldwide scale.

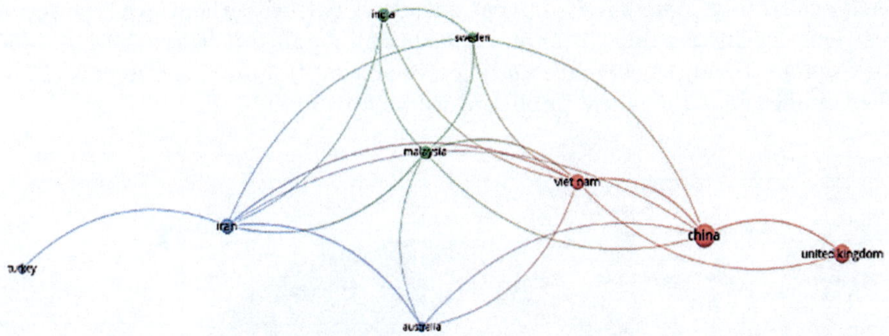

Fig. 4. Countries collaborate to explore the application of AI in GIS: an analysis of review papers from 2014 to 2023

Figure 4 demonstrates that international partnerships were used in certain articles on the use of AI in GIS. In 2018, papers by Viet Nam, Iran, Australia, South Korea, China, and Malaysia were published, a paper was published by Serbia and Italy, and another was published by Italy and France. In 2019, Viet Nam, Sweden, Malaysia, Iran, and India all published a paper, and another paper was published by Canada and Italy. In 2020, a paper was published by Viet Nam and United Kingdom. In 2021, a paper was published by Thailand and United Kingdom, and another paper was published by China and United Kingdom. In 2022, a paper was published by Iran, Turkey, and Netherlands, a paper was published by Spain and Chile, a paper was published by China and United Kingdom, a paper was published by Saudi Arabia and Cyprus, and another paper was published by China and Viet Nam. In 2023, a paper was published by Ireland and United Kingdom, a paper was published by Saudi Arabia and Egypt, and another was published by Iran and Australia. These partnerships imply that nations are collaborating to research the use of AI in GIS and exchange knowledge in this area. Collaborations across nations can result in a more varied variety of viewpoints and ideas, which can ultimately help to develop more practical methods for utilizing AI in GIS. While partnerships are valuable, it is crucial to note that the bulk of the papers were written by individual nations, showing that these nations are also following their own research agendas and examining the use of AI in GIS in their own particular ways.

3.3 RQ3: Which Journals Are the Most Influential in the Field of AI in GIS?

There were a total of 40 publications published between 2014 and 2023 across various journals, according to data on journal articles relevant to the use of AI in GIS in Fig. 5. The bulk of the papers (6) were published in Sustainability (Switzerland): GIS, demonstrating the popularity of this journal as a venue for this type of study. With three publications, Sensors (Switzerland) was the second most journal. Sustainable Cities and Society (2), Applied Sciences (Switzerland) (2), Computational Intelligence and Neuroscience (2), and IEEE Access (2) are other publications that have published works on this subject.

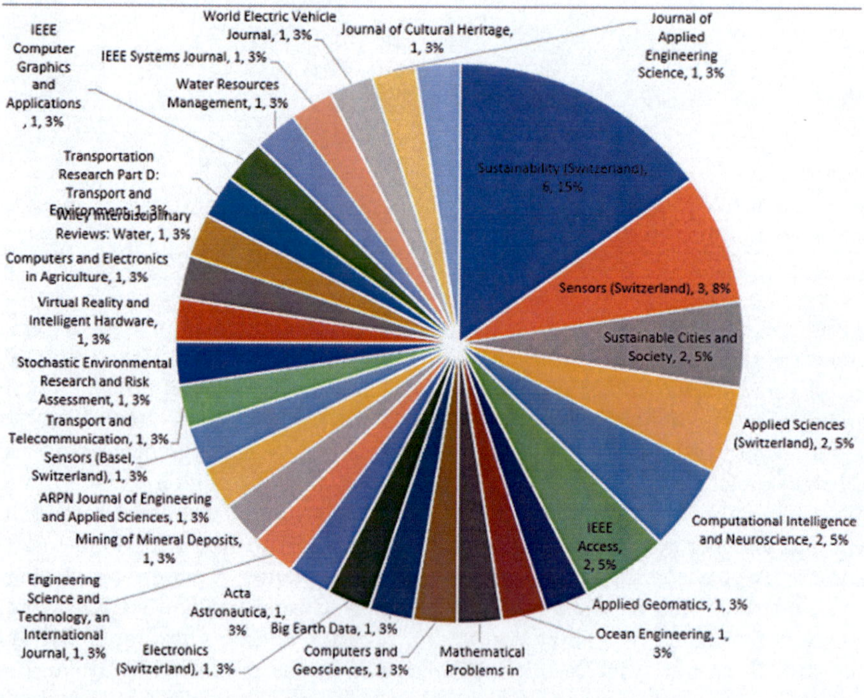

Fig. 5. An overview of published journal articles on the application of AI in GIS: a critical analysis of the distribution and trends from 2014 to 2023

The remaining journals published one article each, including Applied Geomatics, Ocean Engineering, Mathematical Problems in Engineering, Computers and Geosciences, Big Earth Data, Electronics (Switzerland), Acta Astronautica, Engineering Science and Technology, an International Journal, Mining of Mineral Deposits, ARPN Journal of Engineering and Applied Sciences, Sensors (Basel, Switzerland), Transport and Telecommunication, Stochastic Environmental Research and Risk Assessment, Virtual Reality and Intelligent Hardware, Computers and Electronics in Agriculture, Wiley Interdisciplinary Reviews: Water, Transportation Research Part D: Transport and Environment, IEEE Computer Graphics and Applications, Water Resources Management, IEEE Systems Journal, World Electric Vehicle Journal, Journal of Applied Engineering Science, and Journal of Cultural Heritage. Every year, the number of publications in this area has been gradually rising. This shows that interest in the use of AI in GIS is expanding, as seen by the rise in publications on the subject in recent years. The dispersion of articles across many journals, including contributions from disciplines like computer science and engineering, emphasizes the multidisciplinary character of research on the use of AI in GIS.

3.4 RQ4: What Emerging Themes and Trends Can Be Synthesized from Recent Research of AI in GIS?

Using a keyword co-occurrence network (KCN) analysis, the study looked at the structure and expertise in the area of AI in GIS. Figure 6 shows the 9 keywords most frequently used by authors the in the literature, which include “geographic information systems”, “artificial intelligence”, “machine learning”, “computer vision”, “deep learning”, “image classification”, “remote sensing”, “big data”, and “bim”.

Keyword	Occurrences
geographic information systems	20
artificial intelligence	13
machine learning	3
computer vision	2
deep learning	2
image classification	2
remote sensing	2
big data	2
bim	2

Fig. 6. Keyword analysis

Figure 7 highlights the KCN to help the reader better comprehend the links between these keywords. From the analysis, three clusters of AI in GIS were revealed, with cluster 1 consisting of four items, cluster 2 consisting of three items and cluster 3 consisting of two items. The largest label and circle, which is for "Geographic information systems", reveals that the keyword "Geographic information systems" has the largest number of occurrences in the Scopus database (20 times). The second cluster, which has the second-largest font, represents "machine learning". The third cluster represents "big data".

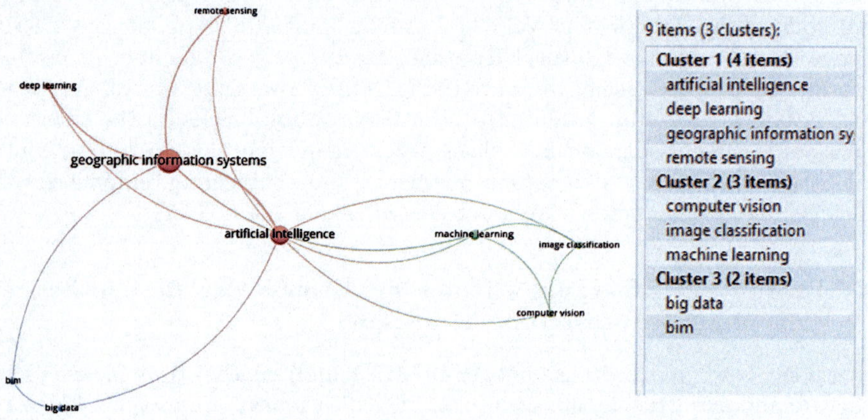

Fig. 7. Keyword co-occurrence network analysis

4 Discussion

In recent years, the integration of Artificial Intelligence (AI) techniques with Geographic Information Systems (GIS) has gained significant attention in the research community. The combination of AI and GIS has the potential to revolutionize spatial analysis, decision-making, and resource management across various domains. Several thematic reviews of the system have highlighted the potential benefits of AI in GIS. AI techniques, including machine learning, deep learning, and natural language processing, have been widely applied to GIS tasks, enabling improved data analysis, feature extraction, classification, and prediction. Machine learning algorithms, such as support vector machines, random forests, and neural networks, have been successfully employed for spatial modeling, pattern recognition, and predictive analysis in GIS. Deep learning methods, particularly convolutional neural networks, have demonstrated remarkable performance in image-based GIS tasks, such as land cover classification and object detection. Natural language processing techniques have facilitated the interpretation of unstructured spatial data, such as textual descriptions and social media feeds, contributing to enhanced situational awareness and decision-making. The integration of AI techniques with GIS has found applications in various domains, including urban planning, environmental monitoring, transportation, and disaster management. In urban planning, AI-based GIS models have been utilized for land use prediction, urban growth simulation, and transportation network optimization. Environmental monitoring has benefited from AI in GIS through the analysis of remote sensing data for forest cover change detection, water quality assessment, and species distribution modeling. AI techniques have also been employed in transportation planning to optimize route planning, traffic prediction, and intelligent transportation systems. In disaster management, AI in GIS has played a crucial role in early warning systems, damage assessment, and post-disaster recovery planning.

5 Conclusion and Future Work

In conclusion, this bibliometric analysis of systematic reviews of AI in GIS provides valuable insights into the trends, patterns, and clusters of keywords related to this field of research. The use of the PRISMA guideline as a tool for selecting relevant papers ensures the reliability and relevance of the 40 papers in this study. The information highlights the significant increase in the number of review papers published on this topic from 2014 to 2023, indicating the growing interest in the use of AI in GIS. The involvement of China and Italy in publishing review papers on this topic reflects their active participation in the development and application of AI in GIS. The identification of three clusters of related keywords provides a comprehensive overview of the different aspects and applications of AI in GIS, including AI Techniques in GIS, Image Classification uses Machine Learning algorithms in Computer vision, and the impact of big data from BIM and GIS on smart urban futures. This study provides a valuable foundation for future research in this field, and can inform policymakers, educators, and researchers about the potential of AI to enhance the ability to analyze and exploit geographic data.

References

1. Abalkhail, A.A.A., Al Amri, S.M.A.: Saudi Arabia's management of the Hajj season through artificial intelligence and sustainability. Sustainability **14**(21), 14142 (2022)
2. Alamgeer, M., Alkahtani, H.K., Maashi, M., Othman, M., Hilal, A.M., Alsaid, M.I., Osman, A.E., Alneil, A.A.: Optimal fuzzy wavelet neural network based road damage detection. IEEE Access (2023)
3. Alhasan, M., Hasaneen, M.: Digital imaging, technologies and artificial intelligence applications during covid-19 pandemic. Comput. Med. Imaging Graph. **91**, 101933 (2021). https://doi.org/10.1016/j.compmedimag.2021.101933
4. Annanias, Y., Zeckzer, D., Scheuermann, G., Wiegreffe, D.: An interactive decision support system for land reuse tasks. IEEE Comput. Graphics Appl. **42**(6), 72–83 (2022)

5. Arya, S., Kaji, A.H., Boermeester, M.A.: Prisma reporting guidelines for meta-analyses and systematic reviews. JAMA Surg. **156**(8), 789–790 (2021). https://doi.org/10.1001/jamasurg.2021.0546
6. Bayoudh, M., Roux, E., Richard, G., Nock, R.: Structural knowledge learning from maps for supervised land cover/use classification: application to the monitoring of land cover/use maps in french guiana. Comput. Geosci. **76**, 31–40 (2015)
7. Benaafi, M., Yassin, M.A., Usman, A., Abba, S.: Neurocomputing modelling of hydrochemical and physical properties of groundwater coupled with spatial clustering, gis, and statistical techniques. Sustainability **14**(4), 2250 (2022). https://doi.org/10.3390/su14042250
8. Chun, P.J., Yamane, T., Tsuzuki, Y.: Automatic detection of cracks in asphalt pavement using deep learning to overcome weaknesses in images and GIS visualization. Appl. Sci. **11**(3), 892 (2021). https://doi.org/10.3390/app11030892
9. Embaby, A., Ismael, A., Ali, F.A., Farag, H., Mousa, B., Gomaa, S., Elwageeh, M.: An approach based on machine learning algorithms, geostatistical technique, and GIS analysis to estimate phosphate ore grade at the Abu Tartur Mine, Western Desert, Egypt. Min. Min. Depos. **17**(1) (2023)
10. Fang, H., Shao, Y., Xie, C., Tian, B., Shen, C., Zhu, Y., Guo, Y., Yang, Y., Chen, G., Zhang, M.: A new approach to spatial landslide susceptibility prediction in karst mining areas based on explainable artificial intelligence. Sustainability **15**(4), 3094 (2023). https://doi.org/10.3390/su15043094
11. Guarino, S., Oliva, G., Di Pietro, A., Pollino, M., Rosato, V.: A spatial decision support system for prioritizing repair interventions on power networks. IEEE Access (2023)
12. Hu, J., Lv, Z., Yuan, D., He, B., Chen, W., Ye, X., Li, D., Yang, G.: A spatiotemporal intelligent framework and experimental platform for urban digital twins. Virtual Real. Intell. Hardw. **5**(3), 213–231 (2023)
13. Iadanza, E., Benincasa, G., Ventisette, I., Gherardelli, M.: Automatic classification of hospital settings through artificial intelligence. Electronics **11**(11), 1697 (2022). https://doi.org/10.3390/electronics11111697
14. Jung, K., Nguyen, V.T., Lee, J.: Blocklyxr: an interactive extended reality toolkit for digital storytelling. Appl. Sci. **11**(3), 1073 (2021). https://doi.org/10.3390/app11031073
15. Jung, K., Nguyen, V.T., Yoo, S.C., Kim, S., Park, S., Currie, M.: Palmitoar: the last battle of the us civil war reenacted using augmented reality. ISPRS Int. J. Geo Inf. **9**(2), 75 (2020). https://doi.org/10.3390/ijgi9020075
16. Kamel Boulos, M.N., Peng, G., VoPham, T.: An overview of GeoAI applications in health and healthcare. Int. J. Health Geogr. **18**, 1–9 (2019). https://doi.org/10.1186/s12942-019-0171-2
17. Leung, C.K., Braun, P., Cuzzocrea, A.: AI-based sensor information fusion for supporting deep supervised learning. Sensors **19**(6), 1345 (2019)
18. Liu, J., Dong, F., Li, Z.: Application of convolutional neural network to GIS and physics. Comput. Intell. Neurosci. **2022** (2022). https://doi.org/10.1155/2022/8559343
19. MacEachren, A.M., Gahegan, M., Pike, W., Brewer, I., Cai, G., Lengerich, E., Hardistry, F.: Geovisualization for knowledge construction and decision support. IEEE Comput. Graphics Appl. **24**(1), 13–17 (2004). https://doi.org/10.1109/MCG.2004.1255801
20. Masoumi, H., Shirowzhan, S., Eskandarpour, P., Pettit, C.J.: City digital twins: their maturity level and differentiation from 3D city models. Big Earth Data **7**(1), 1–36 (2023). https://doi.org/10.1080/20964471.2022.2160156

21. Moreno, M., Ortiz, R., Cagigas-Muñiz, D., Becerra, J., Martin, J., Prieto, A., Garrido-Vizuete, M.d.l.A., Macías-Bernal, J.M., Chávez, M., Ortiz, P.: Art-risk 3.0 a fuzzy-based platform that combine GIS and expert assessments for conservation strategies in cultural heritage. J. Cult. Herit. **55**, 263–276 (2022)
22. Rahla Rabia, M.P., et al.: BIM and GIS integrated utility supply station location optimization and possibilities. J. Appl. Eng. Sci. **20**(4), 1384–1394 (2022). https://doi.org/10.5937/jaes0-40600
23. Nguyen, V.T., Jung, K., Gupta, V.: Examining data visualization pitfalls in scientific publications. Vis. Comput. Ind. Biomed. Art **4**(1), 1–15 (2021). https://doi.org/10.1186/s42492-021-00092-y
24. Nguyen, V.T., Nguyen, C.T., Yoo, S.C., Jung, K.: Unveiling augmented reality applications: exploring influential factors through comprehensive review. SN Comput. Sci. **4**(5), 694 (2023). https://doi.org/10.1007/s42979-023-02175-4
25. Nguyen, V.T., Nguyen, C.T.: A systematic review of structural equation modeling in augmented reality applications. Indones. J. Electr. Eng. Comput. Sci. 328–338 (2022). https://doi.org/10.11591/ijeecs.v28.i1.pp328-338
26. Nikoobakht, S., Azarafza, M., Akgün, H., Derakhshani, R.: Landslide susceptibility assessment by using convolutional neural network. Appl. Sci. **12**(12), 5992 (2022)
27. Rathnasiri, P., Adeniyi, O., Thurairajah, N.: Data-driven approaches to built environment flood resilience: a scientometric and critical review. Adv. Eng. Inform. **57**, 102085 (2023). https://doi.org/10.1016/j.aei.2023.102085
28. Rejeb, A., Rejeb, K., Abdollahi, A., Zailani, S., Iranmanesh, M., Ghobakhloo, M.: Digitalization in food supply chains: a bibliometric review and key-route main path analysis. Sustainability **14**(1), 83 (2021)
29. Saddiqi, M.M., Zhao, W., Cotterill, S., Dereli, R.K.: Smart management of combined sewer overflows: from an ancient technology to artificial intelligence. Wiley Interdiscip. Rev. Water **10**(3), e1635 (2023)
30. Senocak, A.A., Goren, H.G.: Forecasting the biomass-based energy potential using artificial intelligence and geographic information systems: a case study. Eng. Sci. Technol. Int. J. **26**, 100992 (2022)
31. Tekouabou, S.C.K., Diop, E.B., Azmi, R., Jaligot, R., Chenal, J.: Reviewing the application of machine learning methods to model urban form indicators in planning decision support systems: potential, issues and challenges. J. King Saud Univ.-Comput. Inf. Sci. **34**(8), 5943–5967 (2022). https://doi.org/10.1016/j.jksuci.2021.08.007
32. Tengtrairat, N., Woo, W.L., Parathai, P., Aryupong, C., Jitsangiam, P., Rinchumphu, D.: Automated landslide-risk prediction using web GIS and machine learning models. Sensors **21**(13), 4620 (2021)
33. Wael, S., Elshater, A., Afifi, S.: Mapping user experiences around transit stops using computer vision technology: action priorities from Cairo. Sustainability **14**(17), 11008 (2022)
34. Wang, M., Wang, C.C., Sepasgozar, S., Zlatanova, S.: A systematic review of digital technology adoption in off-site construction: Current status and future direction towards industry 4.0. Buildings **10**(11), 204 (2020). https://doi.org/10.3390/buildings10110204
35. Wang, Y., Wang, Z., Ma, T., Li, G., Tie, H.: Research on the realization path of railway intelligent construction based on system engineering. Sustainability **14**(11), 6945 (2022). https://doi.org/10.3390/su14116945

Build an Effective System for Container Code Recognition

Nong Thi Hoa[1](✉) and Nguyen Van Tao[2]

[1] East Asia University of Technology, Ha Noi, Vietnam
hoant1@eaut.edu.vn

[2] Thai Nguyen University of Information and Communication Technology, Nguyen, Vietnam
nvtao@ictu.edu.vn

Abstract. An automatic system for container code recognition supports to monitor containers at ports. It helps to decrease time for collecting information of containers. Therefore, it is essential for many logistics enterprises. Previous studies take a long time by using many complex computing for extracting features and recognizing characters. In this paper, we propose a system for container code recognition with an improved method for code location and code segmentation. Our system consists of three phrases including code location, code segmentation, code recognition. Preprocessing methods are applied to enhance input images. Scan line analysis and high difference of colors on an interested region are used to locate a code region in an input image. Next, the code region is segmented into a letter region and a digit region based on obtained information of code location. Then, CRAFT, Character-Region Awareness for Text detection - a neural network with deep learning, is used to recognize a text from the letter region as well as a text from the digit region. Post processing is applied to improve results of code recognition based on knowledge of similar characters and two parts of a container code. Experiments are conducted on various images from ports of China and Singapore. Results show our approach effectively recognizes container codes from distorted and lossless container codes.

Keywords: container code · container code recognition · scan line analysis · neural network · CRAFT · image analysis

1 Introduction

Transporting on sea has been developed to provide goods. Automatic container code recognition helps monitor containers at ports and warehouses. From a container code, we infer information including owner, product group, registration number, and check digit. Container code recognition usually takes much time because of the difference of font and size, and distortion of characters. Moreover, closeness of characters, illumination and invalidation of color also cause to decrease of recognition's accuracy.

Previous studies can be divided into two categories including a system with three phrases (code location, character segmentation and character recognition), and a system

P. T. Nghia et al. (Eds.): ICTA 2023, LNNS 847, pp. 32–39, 2023.
https://doi.org/10.1007/978-3-031-49529-8_4

with one phrase using neural networks with deep learning. In the first category, code regions are located based on color and geometrical features. Popular methods include change of the luminance values [1, 2], Maximally Stable Extremal Regions (MSER) [4, 6, 17], a horizontal high-pass filter and scan line analysis [14, 16]. Next, character segmentation are done by sliding window [1], spatial structure template matching [4], Connected Component Analysis (CCA) [10, 13, 14], K-Means algorithm [15]. Then, characters are recognized by neural networks with deep learning [2, 4, 6–9, 17], and Support Vector Machine (SVM) [1, 3, 16]. These studies use complex computing to extract each character and some studies use neural networks with deep learning for character recognition. As a result, consume both CPU time and memory. In the second one, many input images are provided to train neural networks with deep learning. Then, they use these neural networks to recognize a container code from an input image without locating code and segmenting characters. Neural networks with deep learning for character recognition include Convolutional Neural Network (CNN) [11, 12], Convolutional Recurrent Neural Network (RCNN) [8, 17], Faster-CNN [5]. Neural networks with deep learning usually have many layers and many neurons on each layer. Therefore, they take a long time to output results and train models. In this paper, we propose an improved system for container code recognition that decreases the time for code location and increases the accuracy of code recognition. Code location, code segmentation, code recognition are three steps of our system. Preprocessing methods are applied to enhance input images. Locating code is done by scan line analysis and high difference of colors on an interested region. A code region is divided into two parts including a letter one and a digit one. Characters in each part are recognized by a CRAFT (a neural network with deep learning) to output a text. A result is obtained by combining texts of the letter part and the digit part. Knowledge of similar characters and two parts of a container code are used for post processing. Results of code recognition are improved by post processing. Experiments are conducted on various images from ports of China and Singapore. Results show that our approach recognizes effectively on distorted and lossless container codes.

This paper is organized as follows. The next section is the proposed system. Experiment results are shown in Sect. 3. Finally, conclusions are written in Sect. 4.

2 Our Approach

The proposed system is a pipeline with three phrases including code location, code segmentation and code recognition. Figure 1 shows phrases of our system. An input image is an image showing the rear of a container.

2.1 Code Location

First, input images are converted to grey-scale images and resize to a specific size. Preprocessing methods are done to enhance input images. Methods of preprocessing include increasing contrast, normalization, and increasing brightness when background is dark. A code region is detected based on the high difference between background colors and code colors.

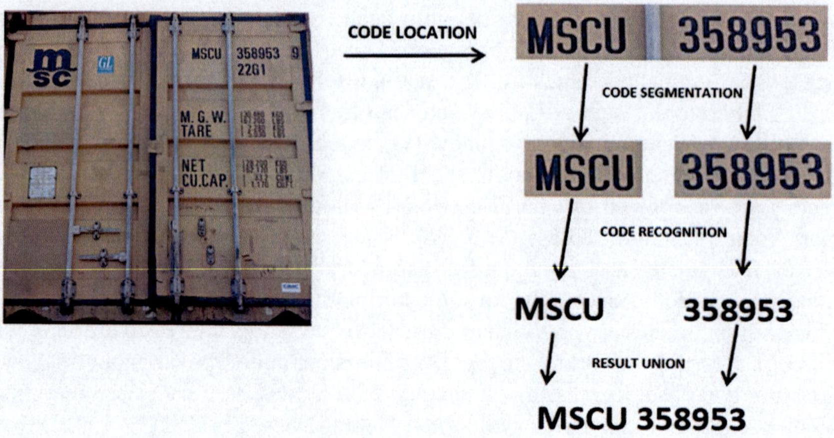

Fig. 1. Phrases of the proposed system for container code recognition.

We observe many input images with many various colors of background to find knowledge for code location. We obtained following information:

- A container code is on a door on the rear of a container.
- Popular size of characters of a container code in images is summarized.
- Color of a container code is white or black. Color of a background (after converting to grey-scale) is usually light grey. We find popular colors of a background in grey scale.
- A container code includes 10 characters in a text line and on at the right top corner of an input image.
- A code region contains about 10 sub-regions whose colors are white or black. Distance between two sub-regions is usually similar.

We extract a rectangle from the right top corner of image. The rectangle's high is chosen based on the popular size of container codes in input images. Its width is estimated based on the width of a door on the rear of a container. We test rectangles from top to bottom until finding a rectangle containing a container code. Figure 2 presents position and size of a rectangle in an input image.

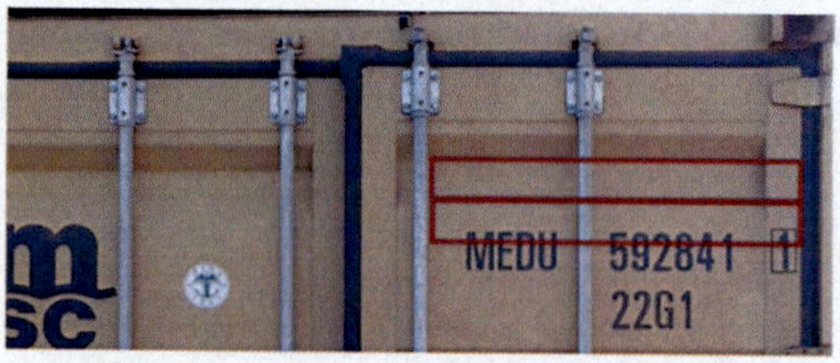

Fig. 2. Red rectangles show size and position of regions to locate code.

Scan line analysis is applied in each column of an interested region (the current rectangle) to find columns whose pixels present colors of background. Other columns contain pixels presenting a container code. From position of columns showing background, we estimate number of characters in the current rectangle. Figure 3 presents columns containing colors of a background by yellow lines. If a rectangle contains a container code, number of yellow regions is usually greater than 10 because of 10 characters in a container code. A yellow region contains several columns showing background. If a container code is distorted or overlapped, number of yellow regions decreases. We count number of yellow regions to decide the current rectangle is a code region. Code location is failed for images with dark background because background's colors are considered as code's colors when code's color is black. Therefore, we increase brightness of background's colors to solve this problem.

Fig. 3. Yellow lines show columns containing background.

We formulate height and position of characters in the letter part of the current rectangle. If the height is too small, the current line is moved up or is moved down on the current rectangle to obtain a region containing a full container code. As a result, a rectangle containing a full container code is extracted. Figure 4 presents an example of a region containing a full container code.

Fig. 4. An example of a region containing a full container code.

2.2 Code Segmentation

Code segmentation is done based on distance between a letter region and a digit region. We see that distance between a letter region and a digit region is sharply greater than distance between two sub-regions containing a character of a container code. The yellow line in Fig. 5 shows distance between a letter region and a digit region of an input image. Therefore, results of using scan line in code location are used to isolate two parts (a letter region and a digit region) of a container code. Figure 6 presents the results of code segmentation of an image in Fig. 5.

Fig. 5. Distance between a letter region and a digit region of an input image.

Fig. 6. Results of code segmentation of an image in Fig. 5.

2.3 Code Recognition

We use a CRAFT neural network for recognizing each part of a container code. A CRAFT network is an effective neural network for text recognition in sophisticated environments [8]. CRAFT is a correct choice for code recognition because characters of a container code may be distorted and/or faded.

CRAFT is a new scene text detection method to effectively detect text area by exploring each character and affinity between characters. To overcome the lack of individual character level annotations, CRAFT exploits both the given character-level annotations for synthetic images and the estimated character-level ground-truths for real images acquired by the learned interim model. In order to estimate affinity between characters, the network is trained with the newly proposed representation for affinity. This procedure includes 5 steps: (1) crop the word-level image; (2) predict the region score; (3) apply the watershed algorithm; (4) get the character bounding boxes; (5) unwarp the character bounding boxes.

3 Experiment Results

Experiments are conducted on 1000 images collecting from ports in China and Singapore. Codes are run in Matlab.

Images are scaled to 600 × 800. Background of input images are removed by the background subtraction in frame difference method. A top half of images is analyzed to locate a code region. Code location is effectively done. The accuracy of code location is 99.8%. Figure 7 shows some results of code location. The accuracy of code segmentation is 100%. We use the *ocr* function of Matlab to character recognition. Post processing is done for errors based on knowledge of two parts of a container code and similar characters. Some results of experiments are shown in Table 1. Typical errors are corrected in Table 2.

The accuracy of code recognition without post processing is 69.7% and 78.4% for code recognition with post processing. The accuracy of character recognition with post processing is 98.2%. The *ocr* function is used for recognizing both letters and digits. Therefore, the accuracy of code recognition is decreased. Quality of input images causes to drop the accuracy of code recognition. The accuracy of character recognition is 98.2%. As a result, most characters of container code are correctly recognized.

Advantages of the proposed system include (i) time for code location is short. Time for extracting two parts of a container code is short. As a result, time for container code recognition significantly drops; and (ii) distorted and/or lossless container codes are effectively recognized because of using a CRAFT.

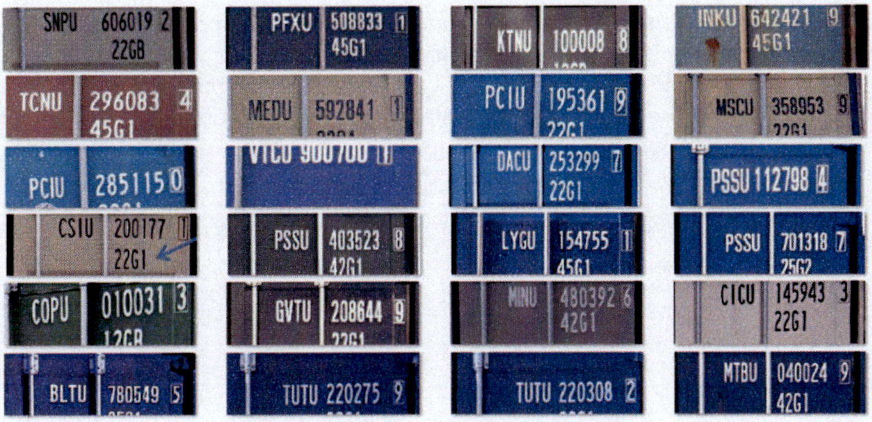

Fig. 7. Some results of code location.

Table 1. Some results of experiments.

Container code	Letters	Digits	Code recognition without post processing	Code recognition without post processing	Number of error characters	Error description
SNPU 606019	SNPU	606019	true	true		
PFXU 508833	PFXU	51]8B33	false	false	2	0 =>l], 8 => B
KTNU 100000	KTNU	100000	false	false	1	8 => 0
INKU E42421	INKU	E42421	false	false	1	6=>E
TCNU 296083	TCNU	296083	true	true		
MEDU 592841	IVIEDU	592841	false	true	1	IVI=>M
PCIU 195361	PCIU	19536]	false	true	1	1=>]
MSCU 358953	MSCU	358953	true	true		
PCIU 285115	PCIU	285115	true	true		

4 Conclusions

We proposed an improve systems for container code recognition. Our system consists of three phrases including code location, code segmentation and code recognition. Scan line analysis and color difference are applied for interested regions to locate a code region and isolate parts of a container code. We present an improved method for locating code and extracting parts of a container code. A CRAFT network is used to output a text of a letter region or a digit region. The text of a letter region and one of a digit region are combined to output a text of a container code. Post processing is applied to improve results of code recognition. Experiments are conducted on images from port of China and Singapore. Results of experiments show our approach improves the accuracy of container code recognition from distorted and lossless container codes.

We will investigate retrain two CRAFTs (one for letter recognition and one for digit recognition) to increase the accuracy of container code recognition, and run experiments on big datasets.

Table 2. Post processing to correct characters after recognizing a container code.

Error	Post processing	Part of code
I]	0	Digits
B	8	Digits
]	1	Digits
II	11	Digits
Z	2	Digits
I	1	Digits
1	4	Digits
IJ	U	Letters
IVI	M	Letters
I	T	Letters
|	I	Letters

References

1. Li, Y., Li, H., Gao, G.: Towards end-to-end container code recognition. Multimed. Tools Appl. **81**, 15901–15918 (2022)
2. Zhang, R., Bahrami, Z., Wang, T., Liu, Z.: An adaptive deep learning framework for shipping container code localization and recognition. IEEE Trans. Instrum. Meas.Instrum. Meas. **70**, 1–13 (2021)
3. Nguyen, D., Nguyen, D., Nguyen, T., Ngo, K., Cao, H., Vuong, T., Quan, T.: Automatic container code recognition using MultiDeep pipeline. In: International Conference on Computational Collective Intelligence, pp. 139–153 (2020)
4. Cao, L., Gai, Z., Liu, E., Gao, H., Li, H., Yang, L., Li, H.: Automatic container code recognition system based on geometrical clustering and spatial structure template matching. In: Lecture Notes in Electrical Engineering, vol. 463, pp. 2198–2204 (2019)
5. Zhiming, W., Wuxi, W., Yuxiang, X.: Automatic container code recognition via faster-RCNN. In: International Conference on Control, Automation and Robotics, pp. 870–874 (2019)
6. Jiwei, W., Jianming, G., Xueqi, L.: Key methods of recognizing container number automatically using video stream in intelligent tally. In: Lecture Notes in Electrical Engineering, vol. 593, pp. 10–17 (2019)
7. Li, C., Liu, S., Xia, Q., Wang, H., Chen, H.: Automatic container code localization and recognition via an efficient code detector and sequence recognition. In: IEEE International Conference on Advanced Intelligent Mechatronics, pp. 532–537 (2019)

8. Baek, Y., Lee, B., Han, D., Yun, S., Lee, H.: Character region awareness for text detection. In: Computer Vision and Pattern Recognition (2019). arXiv:1904.01941
9. Liu, Y., Li, T., Jiang, L., Liang, X.: Container-code recognition system based on computer vision and deep neural networks. In: AIP Conference Proceedings (2018)
10. Verma, A., Sharma, M., Hebbalaguppe, R., Hassan, E., Vig, L.: Automatic container code recognition via spatial transformer networks and connected component region proposals. In: IEEE International Conference on Machine Learning and Applications, pp. 728–733 (2016)
11. Wang, Z., Ma, S.: Container code detection by deep convolutional network. In: Global Conference on Mechanics and Civil Engineering, vol. 132, pp. 82–87 (2017)
12. Mei, L., Guo, J., Liu, Q., Lu, P.: A novel framework for container code-character recognition based on deep learning and template matching. In: International Conference on Industrial Informatics—Computing Technology, Intelligent Technology, Industrial Information Integration, pp. 78–82 (2016)
13. Yoon, Y., Ban, K.D., Yoon, H., Kim, J.: Automatic container code recognition from multiple views. ETRI J. **38**, 767–775 (2016)
14. Wu, W., Liu, Z., Chen, M., Liu, Z., Wu, X., He, X.: A new framework for container code recognition by using segmentation-based and HMM-based approaches. Int. J. Pattern Recognit. Artif. Intell. **29** (2015)
15. Koo, K.M., Cha, E.Y.: A novel container ISO-code recognition method using texture clustering with a spatial structure window. Int. J. Softw. Eng. Its Appl. **7**, 51–62 (2012)
16. Wu, W., Liu, Z., Chen, M., Yang, X., He, X.: An automated vision system for container-code recognition. Expert Syst. Appl. **39**, 2842–2855 (2012)
17. Kim, K.B., Woo, Y.W., Yang, H. K.: An intelligent system for container image recognition using ART2-based self-organizing supervised learning algorithm. In: Asia-Pacific Conference on Simulated Evolution and Learning. LNCS, vol. 4247, pp. 897–904 (2006)

Building an Efficient Retriever System with Limited Resources

Quang Nhat Nguyen and Huong Thanh Le(✉)

Hanoi University of Science and Technology, Hanoi, Vietnam
quang.nn204925@sis.hust.edu.vn, huonglt@soict.hust.edu.vn

Abstract. Despite significant advancements in Question-Answering (QA) systems based on Large Languge Models (LLMs), the issue of generating imprecise answers leading to less informative responses persists. To develop effective QA systems for open-domain datasets, particularly content-specific datasets, dense passage retrieval and the two-stage retriever-reader model remain a rational choice. However, when being applied in real-world systems, these approaches encounter challenges posed by the limitation of computational resources and training data. To address the scarcity of training data, we propose fine-tuning the pretrained BERT-based encoder using masked language modeling before employing a dual-encoder architecture—an established and efficient technique. Additionally, we introduce a modified loss function for dual-encoder training that reduces memory usage during training without compromising system performance. The new loss function is employed in a multi-stage training strategy, yielding enhanced retriever performance at each training stage. To further augment the system's capabilities, we train a cross-encoder to construct a robust retriever for domain-specific datasets. The effectiveness of these proposed techniques is validated by experiments with significant increases in performance compared to the baseline models, underscoring their potential to advance the state-of-the-art in open-domain question-answering systems.

Keywords: Question answering · Information retrieval · Dual-encoder · Vietnamese legal texts

1 Introduction

Nowadays, with the development of large language models (LLMs), question-answering systems can use pre-trained knowledge to generate answers to input questions without the need for a retrieval step. However, due to the vast amount of training data, these systems might provide inaccurate results when answering questions in a domain-specific area. Therefore, when building domain-oriented question-answering applications, Information Retrieval (IR) remains a critical step that determines the accuracy of the system.

There are two primary approaches to constructing a retrieval model: word matching and semantic searching. Semantic retriever is more optimal because of its ability to truly understand the question and documents to deal with acronyms and paraphrases

P. T. Nghia et al. (Eds.): ICTA 2023, LNNS 847, pp. 40–50, 2023.
https://doi.org/10.1007/978-3-031-49529-8_5

that remain as the weakness of the word matching strategy. A typical semantic retriever uses a dual-encoder, which has been widely explored [8, 10, 13, 14, 18], to retrieve the most relevant documents for a query and then a cross-encoder [3, 10, 12, 13] to rank them further.

However, training an effective dual-encoder usually requires a large and diverse dataset, while real-world question-answering systems often focus on specific domains with limited data. This can lead to overfitting, where the model performs poorly because it's too specialized. Moreover, training a dual-encoder requires significant computing resources, which may not be readily available, making resource-intensive approaches impractical. Using limited computing resources results in smaller training batch sizes, reducing model performance. Additionally, the inappropriate percentage of hard negative passages in a batch can also affect the model's performance negatively.

This paper addresses these challenges faced by real-world question-answering systems. To handle limited data, we fine-tune the pre-trained encoder on domain-specific text using masked language modeling (MLM) [2]. We employ a multi-stage training strategy for the dual-encoder to enhance its accuracy and integrate a cross-encoder re-ranker to optimize the retriever's performance. To effectively navigate the limitations of parallel computing memory, we introduce a modified loss function in fine-tuning pre-trained encoders for dense passage retrieval, in order to reduce the impact of hard negatives, allowing the model to converge better. We identify and address the shortcomings of the existing dataset by expanding it through targeted web crawling and appending corresponding titles to the beginnings of all passages. Through comprehensive experimentation with these new passages, we achieve notable performance improvements. These enhancements highlight the importance of titles and reinforce the effectiveness of their innovative methods, offering valuable insights for real-world question-answering systems.

2 Background and Related Works

Since this paper concentrates on improving the performance of the retriever in the situation of lacking training data and computing resources, this section discusses researches involving the dual-encoder and the cross-encoder.

2.1 The Dual-Encoder Passages Retriever

The semantic search uses pre-trained language models to acquire semantic representations of queries and passages. Robust pre-trained models with extensive knowledge and learnability for every domain can efficiently encode questions and contexts with dense embedding vectors and assess their relevancy using similarity functions like cosine or inner product. The actual use of this strategy necessitates the saving of passage vectors as well as the use of an efficient index to retrieve relevant contexts rapidly. The dual-encoder [7], which utilizes two distinct deep encoders for questions and passages, is the most commonly investigated architecture for the semantic-encoding technique and provides the most stunning results. However, it often requires relatively large amounts of training data.

The dual-encoder's training processes include certain notable characteristics:

- There are two main approaches: self-supervised pre-training [1, 4, 5, 8] and fine-tuning on labeled question-passage pairs dataset. We employ the second approach, which saves parallel training resources.
- Hard negative passages (hard negatives) are contexts that do not contain the answers but have some levels of semantic or lexical relevance to the question. Utilizing an appropriate quantity of hard negatives in training helps the model better recognize the incorrect results, thus enhancing the model's performance [7, 10, 13, 18, 19].
- Updating hard negatives during training (dynamic hard negatives) [19] or between training stages (multi-stage training) [13] has been demonstrated to be an optimal technique. As word-occurrence vectors utilizing TF-IDF [17], BM25 [16] and variants are frequently used to choose hard negatives, updating hard negatives using the recently-trained model between the training steps is a powerful design that helps the model address their shortcoming without costing too much resources.

2.2 The Cross-Encoder Passages Re-ranker

Besides dual-encoder, cross-encoder is also shown as an outstanding method for information retrieval and question-answering problems [3, 10, 12, 13]. Unlike dual-encoder, cross-encoder only needs to use one pre-trained BERT-based encoder model in both training and inference procedures. The questions will be attached to paragraphs, separated by a special character and used as input to the model. Then the vector representation of the first token (e.g. [CLS] symbol in BERT) will continue to be put into a classifier to decide if this is a question-paragraph pair containing the answer or not. This is a classic method associated with the advent of pre-trained Transformer-based encoders but so far, it still gives good results, even better than dual-encoder [10]. However, using only a cross-encoder is an expensive and infeasible method in practice. Dual-encoder model is always preferred in information retrieval problems because of its high retrieving speed despite poorer accuracy. Cross-encoder architecture is currently applied in building a re-ranker - a model after the dual-encoder retriever to re-evaluate the top passages returned by the dual-encoder. We also built a cross-encoder re-ranker in our model to improve the performance of the whole retrieval system on the dataset.

3 Our Proposed Approach

In the context of a Question Answering (QA) system, the optimal objective is to provide precise answers for given queries. In this pursuit, it is better for the retriever to return passages instead of documents. To this end, our retriever utilizes a collection of passages, rather than entire documents, within the corpus.

The retrieval problem is formulated as follows: Given a question q and a corpus of passages C, the system needs to retrieve the passages that may contain the answer to the input question. Our proposed system is applied to the Vietnamese legal dataset.

Our retrieval system architecture is shown in Fig. 1, which includes two main components: the dual-encoder and the cross-encoder. The dual-encoder contains two separate pre-trained encoders: one for the question and another for the passage. PhoBERT

[11], a pre-trained encoder for Vietnamese language is applied to encode the input text. PhoBERT was trained based on RoBERTa [9] which optimized the BERT pre-training procedure for more robust performance. The encoded passages are stored and indexed using the FAISS indexer [6]—a proficient algorithm for approximate similarity search, renowned for its scalability to billions of vectors. The similarity of the question and context is calculated as the dot product of their pair of embedding vectors. The high dot product score means that this is more likely a pair of a question and a passage containing the answer you are looking for. In this paper, we retrieve top k passages with k equal to 30. This ensures both the retrieval speed and the accuracy of the model. These passages are fed into the cross-encoder.

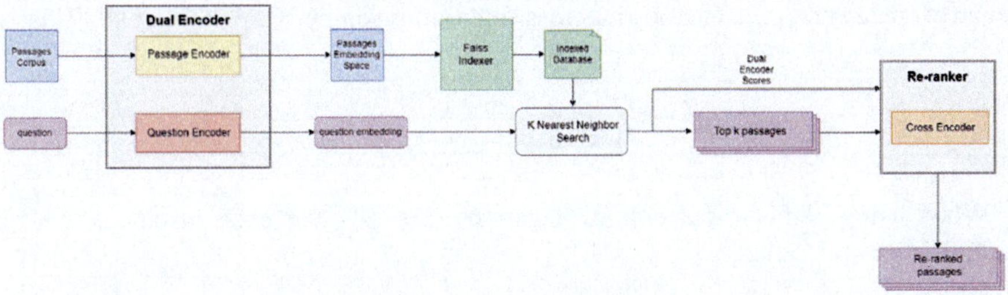

Fig. 1. Our retrieval system.

The cross-encoder includes the PhoBERT pre-trained encoder and a classifier which consists of a dropout, a fully connected and a softmax layer. The model will be trained to return 0(YES) if the passage attached to the question contains the answer, and 1(NO) otherwise.

The scores of the dual-encoder and cross-encoder are combined to form a final re-rank score for a passage:

$$score_{rerank} = \frac{score_{dual}}{100} + score_{cross} \tag{1}$$

After the k passages are re-ranked, we can choose a number of passages or set a score threshold to return the best paragraphs. The results of the retriever can be passed directly to the user (in case k is small) or further fed into other models (reader, generator…) to return the most concise answer to the question. The rest of this section will introduce our proposed strategies to improve the retriever's performance.

3.1 Fine-Tuning PhoBERT for Domain-based QA Systems

PhoBERT was pre-trained with 20GB of Wikipedia and News texts. It can well understand daily-life Vietnamese text. However, when working with data in specific domains such as law or bioinformatics, its performance is getting worse since each specific domain has a different vocabulary. Given a small closed-domain QA dataset, it is not good for just fine-tuning the PhoBERT with this dataset, since it is not enough for the pre-trained encoders to understand the domain problem. To solve this problem, we fine-tune the

PhoBERT with the task masked-language modeling (MLM) on content passages of the dataset before training the dual-encoder. This training process helps the model learn closed-domain information for better semantic understanding in the encoding step. The fine-tuned PhoBERT is used to encode the input text of the QA system.

3.2 The Dual-Encoder

Improve the loss function. Let us consider a training dataset $D = \left\{\langle q_i, p_i^+, p_{i,1}^-, \ldots, p_{i,n}^-\rangle\right\}_{i=1}^{m}$ comprising m instances, each instance contains a unique question, q_i, and a pertinent (positive) passage, p_i^+, alongside n inconsequential (negative) passages, $p_{i,j}^-$. The original loss function given by Karpukhin et al. [7] is:

$$L_1 = -\log \frac{e^{sim(q_i, p_i^+)}}{e^{sim(q_i, p_i^+)} + \sum_{j=1}^{n} e^{sim(q_i, p_{i,j}^-)}} \quad (2)$$

When using some in-batch hard negatives (i.e., $p_{i,k}^*$ for each instance), the loss function is now calculated as:

$$L_2 = -\log \frac{e^{sim(q_i, p_i^+)}}{e^{sim(q_i, p_i^+)} + \sum_{k=1}^{m} e^{sim(q_i, p_{i,k}^*)} + \sum_{j=1}^{n} e^{sim(q_i, p_{i,j}^-)}} \quad (3)$$

According to Karpukhin et al. [7] and some other studies [10, 13, 18], utilizing at least one hard negative per instance increases the model's ability to distinguish between positive and negative passages, resulting in better retrieval results. Nevertheless, when dealing with the Vietnamese legal dataset, it has come to our attention that the model's performance deteriorates when one or a few hard negatives generated by BM25+ [15] are added. Upon an exhaustive investigation into the potential factors contributing to this regression, we have identified a potential source of this issue—discrepancies in batch size between our implementation and the original paper [7]. Karpukhin et al. [7] employed large batch sizes, such as 128 or more, facilitated by modern computing resources. Conversely, due to constraints related to parallel resources, our batch size configurations are limited at 64 and 32, particularly when incorporating an additional hard negative. This discrepancy leads to an imbalance in the ratio of hard negative passages to the overall count of negative passages within our implementation. Consequently, this skewed ratio results in overfitting of the model, causing it to excessively prioritize hard negatives at the expense of generalization capabilities.

So, how to still use hard negatives and achieve better results when working with limited parallel resources? We propose a newly modified loss function to control the influence of hard negative segments during training:

$$L' = -\alpha.L_1 + -(1-\alpha).L_2 \quad (4)$$

The new loss function will comprise a weighted combination of two negative log-likelihood loss components: one incorporating hard negatives and the other exclusively accounting for random negatives. To balance these two loss functions, we propose an α hyper-parameter between 0 and 1. A good α value will sufficiently control the influence of hard negatives in the training process. $\alpha = 0$ means we are not using hard negatives; $\alpha = 1$ means we are actually just using a loss function that contains hard negatives. Different values of α are tested in our experiments in order to pick the optimal one. This will be presented in the experimental results section.

Choosing Efficient Training Samples. In addition to positive and negative samples, the integration of hard negatives proves instrumental in augmenting system performance. Methods such as TF-IDF [17], BM25 [16], and their variants are often used to select hard negatives for training. During the training process, to a certain point, the model was able to learn those hard negatives well but at the same time, could not distinguish some negatives with complex semantics. Therefore, it is very important to update the hard negatives. The new hard negatives will be the negatives that are mistaken by the current model as positives. Because of resource consumption, the update can only happen when we stop the training procedure.

3.3 Cross-Encoder

The top k results of the dual-encoder are reranked by the cross-encoder, which works as a classifier. Positive samples of the classifier's training dataset are the passages corresponding to the question from the dataset. Negative samples are taken from the top negative passages returned by the dual-encoder. Passages that are longer than the maximum input length that PhoBERT can accept will be truncated to a reasonable length. Since one positive sample accompanies with several negative samples, we repeat the positive samples many times to balance the training data.

4 Experiments

4.1 Dataset

We utilize the Vietnamese legal dataset in the text retrieval task from the Zalo AI Competition 2021[1] for evaluating the system's performance in text retrieval. This dataset consists of a collection of legal documents organized into passages, accompanied by a training dataset containing individual questions along with corresponding answers, some of which may have multiple responses. Since only the training dataset is public by the Zalo AI Competition, we randomly divide 3196 samples in that set into 3 smaller train-val-test sets with the number of samples of 2400, 350, and 446 respectively. For the corpus of law passages, after filtering out duplicated ones, we get 60830 law passages from 3263 law documents.

1 https://challenge.zalo.ai/portal/legal-text-retrieval.

4.2 Experimental Setup

Since the PhoBERT encoders are used in both the dual-encoder and cross-encoder, preprocessed questions and passages need to go through the word segmentation process before being fed into the models. *Pyvi*[2] - a word segmenter - is used for this task.

To get the hard negative passages for each question in the training process, we selected passages returned by BM25+ in $rank - bm25$[3] library, which are not the correct passages for each input question.

Dual-encoder two-stage training. After getting the initial hard negatives with BM25+, we train the dual-encoder in the first stage on 80 epochs. The last model is then saved and used to update hard negatives, which are used in the next 10-epoch-training stage. In the training procedure of dual-encoder, we set the learning rate for both stages as $1e^{-5}$ and use the Adam optimization with linear scheduling. We experiment with batch sizes from 16 to 64.

Cross-encoder training. We truncate long passages into smaller ones and let the trained dual-encoder prepare the training dataset for the cross-encoder as described before. We also use Adam optimization with linear scheduling and a dropout rate of 0.1. We set the batch size as 16 and train the model for 10 epochs, evaluating the model on the validation set after each 1000 training steps. The checkpoint of the model having the highest results on the validation set is saved. We restate the importance of data balancing through duplicating positive samples. Both the training procedures of cross-encoder and dual-encoder are conducted on Kaggle[4] with GPU P100 or two GPUs T4.

Evaluation metric. One feature of the Zalo legal question-answering dataset is that out of 3196 samples, there are 3103 with only one positive passage (accounting for more than 97%). Questions with two or three positive passages make up the remaining 3%, a rather modest number. Therefore, instead of using the popular recall score like other information retriever studies, we use the accuracy to measure the proportion of samples with at least one answering passage returned in the top k passages scored by the models. The formula for the measure is as follows:

$$\boldsymbol{acc}_k = \frac{n_{p,k}}{n}.100\% \tag{5}$$

where n is the number of samples in the evaluating dataset, $n_{p,k}$ is the number of samples having at least one positive passage retrieved in the top-k articles. We evaluate the dual-encoder with the top 1, 5, 10, 30, and 100 retrieved passages. With the cross-encoder re-ranker, we only consider the improvement in performance in the top 1, 5, and 10 after re-ranking the top 30 articles returned by the dual-encoder. We also recorded the results when the trained models inferred with the texts that were truncated to match the maximum input length of PhoBERT.

2 https://pypi.org/project/pyvi/.

3 https://pypi.org/project/rank-bm25/.

4 https://www.kaggle.com/.

4.3 Experimental Results

In this section, we denote the experimented dual-encoder versions as $s - p - b - h - \alpha$ where $s \in \{1, 2\}$ is the ordinal number of the training stage, $p \in \{0, 3\}$ is the number of epochs that the PhoBERT encoders were fine-tuned with the MLM task on the law corpus before fine-tuning on the question-passage dataset in the first stage of dual-encoder training (in the second stage or $s = 2$, we start from the resulted encoder in the first stage, so $p = 0$), $b \in \{16, 32, 64\}$ is the batch size, $h \in \{0, 1, 3\}$ is the number of hard negatives used for each training sample, and $\alpha \in [0, 1]$ is the hyperparameter used in the loss function.

Table 1. Dual-encoder: Hard negatives experiments.

$s - p - b - h - \alpha$	acc_1	acc_5	acc_{10}	acc_{30}	acc_{100}
1 – 0 – 16 – 0 – 0	56.5	83.1	88.5	93.4	**97.0**
1 – 0 – 32 – 0 – 0	54.2	80.7	87.2	**94.1**	96.1
1 – 0 – 64 – 0 – 0	53.6	81.6	86.7	**94.1**	96.6
1 – 0 – 16 – 1 – 1	56.5	79.8	85.2	92.8	96.1
1 – 0 – 32 – 1 – 1	56.5	82.0	87.4	**94.1**	96.6
1 – 0 – 16 – 3 – 1	60.0	80.0	85.4	92.6	96.4
1 – 0 – 16 – 3 – 0.1	**63.9**	**84.0**	**89.6**	**94.1**	**97.0**

Table 1 represents our first experimental results in training a dual-encoder model, with different usages of hard-negative. Table 1 shows that the use of hard negatives does not improve and even worsens the performances, which is contrary to previous studies [7, 13, 18]. After proposing a new loss function, we try it with different values of α. With $\alpha = 0.1$, we obtain the best performance that surpasses all previous results.

Table 2 represents our extremely impressive performance after fine-tuning PhoBERT for 3 epochs with the MLM task. This performance is even higher than when the new modified loss is applied. Using both fine-tuning and the new loss (with $\alpha = 0.1$) in the first training stage brings us the best result for the dual-encoder in the first stage of training.

Training the dual-encoder in the second stage does benefit the model performance, with higher results in the top 1, 5, 10, 30 (Table 2). Besides, using the new loss function continues to show its effect that when $\alpha = 0.3$, we obtain equal or higher accuracy on top 1, 10, 30, 100 retrieved passages than using the old loss function ($\alpha = 1$).

Table 3 shows the results of the whole retriever model including the dual-encoder and cross-encoder re-ranker. Truncating long articles into smaller sub-passages that help PhoBERT encode the entire content can improve the dual-encoder performance. When using a cross-encoder and re-rank the top 30 retrieved passages with the combined score introduced in Sect. 3, the results increase significantly, stating the importance of a re-ranker after the dual-encoder in a retrieval system.

Table 2. Dual-encoder: Multi-stage training results.

$s-p-b-h-\alpha$	acc_1	acc_5	acc_{10}	acc_{30}	acc_{100}
$1-0-16-3-1$	60.0	80.0	85.4	92.6	96.4
$1-0-16-3-0.1$	63.9	84.0	89.6	94.1	97.0
$1-3-16-3-1$	64.3	84.7	89.4	94.8	**98.4**
$1-3-16-3-0.1$	65.4	86.7	91.9	96.4	**98.4**
$2-0-16-3-1$	70.6	**91.0**	**93.7**	97.0	97.9
$2-0-16-3-0.3$	**73.0**	88.7	**93.7**	**97.3**	**98.4**

Table 3. Results when using cross-encoder re-ranker.

$s-p-b-h-\alpha$	trunc	re	acc_1	acc_5	acc_{10}	acc_{30}	acc_{100}
$2-0-16-3-0.3$	No	No	73.0	88.7	93.7	**97.3**	**98.4**
$2-0-16-3-0.3$	Yes	No	73.5	90.1	92.6	96.8	**98.4**
$2-0-16-3-0.3$	Yes	Yes	**83.4**	**95.2**	**96.4**	96.8	–

Dataset Limitation. After analyzing the cases where the retriever model failed, we found that the main reason can be mentioned is the questions that require high inference. However, at the same time, there is a problem caused by the lack of relevant information on the positive passages to the answer. The corpus only includes legal passages taken from a large legal document and omits the title of the whole legal document. We find that the title of a legal document can also be an important source of information that improves the quality of the model. Therefore, we crawl[5] the titles of legal documents in the corpus to attach to each passage and re-train both the dual-encoder and cross-encoder. For documents with too short titles, we have also added the content of *Article 1: Scope of Regulation* to increase the amount of information for the passages. The results showed that adding text titles to the rules improved the model's results significantly.

Table 4. Dual-encoder: Stage 1 results with titles appending.

$s-p-b-h-\alpha$	title	acc_1	acc_5	acc_{10}	acc_{30}	acc_{100}
$1-3-16-3-1$	No	64.3	84.7	89.4	94.8	**98.4**
$1-3-16-3-0.1$	No	65.4	86.7	91.9	96.4	**98.4**
$1-3-16-3-1$	Yes	66.3	86.9	91.2	**96.6**	98.2
$1-3-16-3-0.1$	Yes	**69.2**	**89.0**	**93.2**	96.4	**98.4**

[5] From the website: https://thuvienphapluat.vn/.

According to Tables 4 and 5, appending law titles does improve the performance of the dual-encoder in both training stages. When we use the old loss function ($\alpha = 1$), the results with attached passages can approximate the highest performance with the modified loss and original articles. Combining the modified loss and titles-appending brings us very comparative results. After two-stage training, the dual-encoder can retrieve around 74.4% of the questions with one returned passage.

Table 5. Dual-encoder: Stage 2 results with titles appending.

$s-p-b-h-\alpha$	title	acc_1	acc_5	acc_{10}	acc_{30}	acc_{100}
$2-0-16-3-1$	No	70.6	91.0	93.7	97.0	97.5
$2-0-16-3-0.3$	No	73.0	88.7	93.7	97.3	97.9
$2-0-16-3-1$	Yes	73.5	**91.9**	93.4	96.8	98.2
$2-0-16-3-0.3$	Yes	**74.4**	91.4	**94.1**	**97.3**	**98.4**

After truncating and re-ranking, our new retriever surpasses the previous on all top 1, 5, 10 re-ranked retrieved passages (see Table 6).

Table 6. Re-ranker results with titles appending.

$s-p-b-h-\alpha$	title	trunc	re	acc_1	acc_5	acc_{10}	acc_{30}
$2-0-16-3-0.3$	No	Yes	Yes	83.4	95.2	96.4	96.8
$2-0-16-3-0.3$	Yes	Yes	Yes	**84.3**	**95.7**	**96.6**	**97.3**

5 Conclusion

In this paper, we introduce a training strategy to improve the performance of the retriever on closed-domain datasets despite facing limited parallel computing resources for the training procedure. By demonstrating the effectiveness of the techniques through experiments, we conclude that fine-tuning the pre-trained encoder on the passages corpus and applying a modified loss in a multi-stage training procedure is the right approach for the dual-encoder. The dual-encoder should be combined with a cross-encoder to form a final robust retriever, which shows potential to advance the state-of-the-art in question-answering systems, especially on specific domains.

References

1. Chang, W.C., Yu, F.X., Chang, Y.W., Yang, Y., Kumar, S.: Pre-training tasks for embedding-based large-scale retrieval (2020)
2. Devlin, J., Chang, M.W., Lee, K., Toutanova, K.: Bert: pre-training of deep bidirectional transformers for language understanding (2019)
3. Fajcik, M., Docekal, M., Ondrej, K., Smrz, P.: R2-d2: a modular baseline for open-domain question answering (2021)
4. Gao, L., Callan, J.: Condenser: a pre-training architecture for dense retrieval (2021)
5. Guu, K., Lee, K., Tung, Z., Pasupat, P., Chang, M.W.: Realm: retrieval-augmented language model pre-training (2020)
6. Johnson, J., Douze, M., Jégou, H.: Billion-scale similarity search with gpus (2017)
7. Karpukhin, V., Oğuz, B., Min, S., Lewis, P., Wu, L., Edunov, S., Chen, D., Yih, W.: Dense passage retrieval for open-domain question answering (2020)
8. Lee, K., Chang, M.W., Toutanova, K.: Latent retrieval for weakly supervised open domain question answering (2019)
9. Liu, Y., Ott, M., Goyal, N., Du, J., Joshi, M., Chen, D., Levy, O., Lewis, M., Zettlemoyer, L., Stoyanov, V.: Roberta: a robustly optimized bert pretraining approach (2019)
10. Luan, Y., Eisenstein, J., Toutanova, K., Collins, M.: Sparse, dense, and attentional representations for text retrieval (2021)
11. Nguyen, D.Q., Nguyen, A.T.: Phobert: pre-trained language models for Vietnamese (2020)
12. Nogueira, R., Cho, K.: Passage re-ranking with BERT (2020)
13. Qu, Y., Ding, Y., Liu, J., Liu, K., Ren, R., Zhao, W.X., Dong, D., Wu, H., Wang, H.: RocketQA: an optimized training approach to dense passage retrieval for open-domain question answering (2021)
14. Reimers, N., Gurevych, I.: Sentence-bert: sentence embeddings using Siamese bert-networks (2019)
15. Robertson, S., Walker, S.: Some simple effective approximations to the 2-poisson model for probabilistic weighted retrieval, pp. 232–241 (Jan. 1994). https://doi.org/10.1007/978-1-4471-2099-524
16. Robertson, S.E., Zaragoza, H.: The probabilistic relevance framework: Bm25 and beyond. Found. Trends Inf. Retr.Retr. **3**, 333–389 (2009)
17. Salton, G., Buckley, C.: Term-weighting approaches in automatic text retrieval. Inf. Process. Manag.Manag. **24**, 513–523 (1988)
18. Xiong, L., Xiong, C., Li, Y., Tang, K.F., Liu, J., Bennett, P., Ahmed, J., Over-Wijk, A.: Approximate nearest neighbor negative contrastive learning for dense text retrieval (2020)
19. Zhan, J., Mao, J., Liu, Y., Guo, J., Zhang, M., Ma, S.: Optimizing dense retrieval model training with hard negatives (2021)

Deep Learning Based Coconut Fruit Maturity Classification

Nhut-Lam Nguyen(✉), Ngoc Van Khanh Duong, and Hoang Duy Thien Nguyen

School of Engineering and Technology, Tra Vinh University, Tra Vinh, Vietnam
{lamnn,vankhanh,thiennhd}@tvu.edu.vn

Abstract. Coconut trees are perennial crops that bring high economic benefits to farmers in Southeast Asian countries including Vietnam. Coconut fruits are used for different purposes depending on the type of coconut and the purpose of harvesting such as for food, drink, and handicrafts. Coconut fruits are often harvested and classified according to the color of its shell. Harvesting coconut fruits at the right time will improve the quality of products made by them. In this paper, we are interested in harvesting coconut fruits for commercial production by applying deep learning techniques to classify coconut fruits into two major maturity stages: mature and overly mature. We collected images of coconut fruits and trained three popular deep learning models including VGG19, Inception-V3, and ResNet101 to classify their maturity. Experimental results show that our approach can classify coconut fruit maturity with 95% accuracy using the ResNet101 model.

Keywords: Coconut · Deep learning · Maturity detection · Transfer learning

1 Introduction

Recently, Vietnam and the World are in the Fourth Industrial Revolution. In agriculture, we not only focus on preserving traditional agriculture but also consider applying new technology to enhance the quality of agricultural products. The development of information technology, especially Artificial Intelligence (AI), has become a trend toward the development of smart, efficient, and sustainable agriculture.

Coconut trees are important perennial crops that bring high economic efficiency to people in the Mekong Delta. Products from Vietnam's coconuts have been presented in over 90 countries worldwide [1]. The economic efficiency that coconuts bring to farmers is considerably high compared to other crops. Most of the ingredients of coconuts are multipurpose. For example, coconut water is used to process beverages, handicraft products are made from coconut shells, and coconut fibers are used for production of floor mats, mattresses, etc. In particular, the quality of products made from coconut fruits depends on the maturity of the coconut fruits.

According to [2], the time for pollination of coconut flowers to mature lasts 12 months. Copra is formed from the 5th month and the coconut water used for drinking is usually harvested in the 7th and 8th months. The copra for commercial production is

P. T. Nghia et al. (Eds.): ICTA 2023, LNNS 847, pp. 51–59, 2023.
https://doi.org/10.1007/978-3-031-49529-8_6

harvested in the 12th month. Harvesting coconut fruits after the 12th month results in degrading copra quality.

For the above-mentioned reasons, harvesting coconut fruits at the right time, depending on harvesting purposes, is crucial. Currently, most farmers harvest coconut fruits based on their experiences. Therefore, determining the maturity of coconut fruits could be inaccurate, especially for those who are inexperienced. Inaccurate determination of coconut fruit maturity will lead to a decrease in quality coconut copra as well as the post-harvest processing requirements. In addition, improving the accuracy in determining the maturity of coconut fruits will help farmers optimize their profits as well as the quality of coconut fruit components. Therefore, in this study, we applied deep learning techniques to determine the maturity of coconut fruits. The goal of the study is to help farmers determine the maturity of coconut fruits automatically with high accuracy.

2 Related Works

Determining the maturity of coconut fruits has attracted attention from researchers in recent years, especially authors from Southeast Asian countries. In 2021, Nguyen et al. [3] developed a system to determine the wax percentage of waxed coconuts grown in Tar Vinh province, Vietnam, using the Support Vector Machine (SVM) and K-nearest Neighbors machine learning algorithms. The proposed system has proven the effectiveness of applying machine learning to this research area. Caladcad et al. [4] proposed a model to classify the maturity of Philippine coconut fruits based on the acoustic signal obtained by shaking the coconut fruits. Several machine learning models have been evaluated in the paper. Their experimental results show that the Random Forest model achieved the best performance compared to the Artificial Neural Network (ANN) and SVM models.

Javel et al. [5] proposed an approach to determine the maturity of coconut fruits based on fuzzy logic. This fuzzy logic-based classification model takes the color of the coconut fruits and the sound recorded while shaking the coconut as inputs. However, the accuracy of their system was not mentioned. Fadchar et al. [6] applied ANN to build a model to classify the maturity of coconut fruits based on the acoustic signal emitted when shaking the coconut fruits. Although the accuracy of their proposed model is better than the manual classification method, current deep learning-based learning models can be promising solutions to this problem.

In another study, Noppon and Nipon [7] introduced a model to classify coconut fruit maturity using a convolutional neural network (CNN). The accuracy of their proposed model is quite high (97.27%), however, the dataset used for model training is relatively small (50 samples per class). Madue et al. [8] proposed an approach to classify fresh aromatic coconut polynomial regression. However, the detection accuracy was not good.

Overall, studies on determining the maturity of coconut fruits are still challenging. Therefore, the purpose of this paper is to propose an efficient deep learning-based model for coconut fruit maturity recognition.

3 Methodology

Our main contributions are to create a coconut fruit maturity dataset and to propose an efficient framework for coconut fruit maturity classification based on transfer learning. We first acquired the dataset and performed preprocessing steps. The dataset was then labeled by experts. Different deep transfer learning models were deployed and evaluated for coconut maturity classification. The best model was finally selected as the coconut maturity classification model. Our framework is shown in Fig. 1.

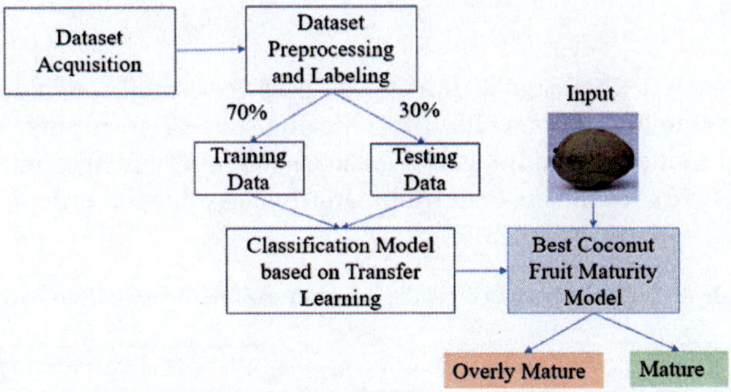

Fig. 1. The proposed coconut fruit maturity classification framework.

3.1 Dataset Acquisition

Our dataset was acquired using the Canon EOS 6D with lens Canon EF 16-35mm f/4L IS USM. The original images were cropped and resized to 640 × 640 pixels. The coconut fruit maturation levels can be grouped into three categories: immature coconut (120–200 days), mature coconut (220–300 days), and overly mature coconut (320–380 days) [9]. However, in this paper, we only considered two mature stages including the mature stage and overly mature stage as we did not focus on harvesting coconut for drink but copra production. Harvesting coconut at the overly mature stage can lead to a decrease in copra quality [2]. The coconut fruit images were then labeled by experienced farmers who are experts in mature coconut fruit harvesting. Finally, we obtained a dataset of a total of 500 images with 250 samples for each class. Figure 2 illustrates examples of the acquired images in our dataset.

3.2 Classification Models

In this paper, we deployed three popular deep transfer learning models for coconut fruit maturity classification. These models are VGG19, Inception-V3, and ResNet101, which are briefly introduced in the following sections.

VGG19 model: VGG19 is a deep convolutional neural network proposed by Simonyan and Zisserman at the Visual Geometry Group, University of Oxford for image classification tasks [10]. VGG19 placed the first and second in the localization and classification

Fig. 2. Samples of collected images coconut fruits (640 × 640 pixels).

track of the ImageNet Challenge 2014. VGG19 uses very small convolution filters (3 × 3) instead of having big filters while still achieving state-of-art results. VGG19 has 19 layers deep and trained on the ImageNet database to classify images into 1000 classes. The VGG19 network for the coconut fruit maturity classification is depicted in Table 1.

Table 1. VGG19 network for the coconut fruit maturity classification.

Layer name	Size/stride	Filter
input	224 × 224 × 3	3 × 3
conv1_1, conv1_2	224 × 224 × 64/1	3 × 3
max pooling	112 × 112 × 64/2	3 × 3
conv2_1, conv2_2	112 × 112 × 128/1	3 × 3
Max pooling	56 × 56 × 128/2	3 × 3
conv3_1, conv3_2	56 × 56 × 256/1	3 × 3
max pooling	28 × 28 × 256/2	3 × 3
conv4_1, conv4_2, conv4_3, conv4_4	28 × 28 × 512/1	3 × 3
max pooling	14 × 14 × 512/2	3 × 3
conv5_1, conv5_2, conv5_3, conv5_4	28 × 28 × 512/1	3 × 3
max pooling	7 × 7 × 512/2	3 × 3
fc	4096	
fc	4096	
softmax	2	output

Inception-V3 model: Inception-V3 [12] is a 48-layers deep convolutional neural network from the Inception family. The Inception-V3 model consists of symmetric and asymmetric building blocks. Each of these blocks comprised of convolutions, average pooling, max pooling, concatenations, dropouts, and fully connected layers. This model won the ImageNet competition in 2014 for its speed and size. The architecture of the Inception-V3 model for coconut fruit maturity classification is shown in Table 2.

Table 2. Inception-V3 network for the coconut fruit maturity classification.

Layer name	Size	Filter/stride
input	299 × 299 × 3	3 × 3/2
conv	149 × 149 × 32	3 × 3/1
conv	147 × 147 × 32	3 × 3/1
max pooling	147 × 147 × 64	3 × 3/2
conv	73 × 73 × 64	3 × 3/1
conv	71 × 71 × 80	3 × 3/2
conv	35 × 35 × 192	3 × 3/1
3 × inception	35 × 35 × 288	1 × 1, 3 × 3
5 × inception	17 × 17 × 768	n × 1, 1 × n, 1 × 1
3 × inception	8 × 8 × 1280	1 × 3, 3 × 1, 1 × 1
max pooling	8 × 8 × 2048	8 × 8
linear	1 × 1 × 2048	logits
softmax	1 × 1 × 2	output

ResNet101 model: The ResNet [11] is a popular deep learning model for image recognition based on the concept of residual learning. There are various versions of ResNet which are named according to the number of its layers. In this paper, we used ResNet101 to implement the coconut fruit maturity detection model. The ResNet101 has 101 layers, and it uses the skip connection technique to solve the problem of information loss during the deep network training. The residual network architecture allows this model to learn complex image features while still achieving high performance. Our proposed ResNet101 network for the coconut fruit maturity classification are shown in Table 3.

Table 3. ResNet101 network for the coconut fruit maturity classification.

Layer name	Size	Filter/stride
input	244 × 244 × 3	3 × 3/2
conv1	112 × 112 × 64	7 × 7/1
max pooling	56 × 56 × 64	3 × 3/2
conv2_x	56 × 56 × 64, 256	1 × 1, 3 × 3
conv3_x	28 × 28 × 128, 512	1 × 1, 3 × 3
conv4_x	14 × 14 × 256, 1024	1 × 1, 3 × 3
conv5_x	7 × 7 × 512, 2048	1 × 1, 3 × 3
average pooling	1 × 1 × 2048	
softmax	1 × 1 × 2	output

3.3 Evaluation Metrics

We used confusion matrix to evaluate the performance of classification models. In the confusion matrix, true positive (TP) is the total number of samples that are correctively classified as mature; false positive (FP) is samples that are overly mature but misclassified as mature; true negative (TN) is the samples that are correctly overly mature; false negative (FN) denotes the samples that are mature but incorrectly classified as overly mature. The standard metrics used in this paper derived from the confusion matrix, are calculated as follows.

$$Accuracy = \frac{TP + TN}{TP + TN + FP + FN}$$

$$Precision = \frac{TP}{TP + FP}$$

$$Recall = \frac{TP}{TP + FN}$$

$$F1 - score = 2 * \frac{Precision * Recall}{Precision + Recall}$$

3.4 Experiments and Results

In this section, we present our experimental results for coconut fruit maturity based on different deep transfer learning models. In our dataset, there were 250 samples for each class, i.e., the mature and overly mature classes. We split the dataset into two parts: the training and testing sets with a ratio of 7:3, which means that there were 350 and 150 images in the training and testing sets, respectively. Twenty percent of the training set was used as the validation set. Keras deep learning framework was used for experiments and the models were trained on the Google Colab using the T4 GPUs. For data argumentation, the ImageDataGenerator class, which is a real-time data augmentation utility in Keras, was used. A summary of parameters and its values used in the experiments are shown in Table 4.

Table 4. Model parameter setting.

Parameter	Value
Optimizer	Adam
Loss function	Categorical cross entropy
Epoch	100
Learning rate	0.00001
Batch size	32

Table 5 shows the accuracy and loss during the training steps. It can be seen from the table that the VGG19 model outperformed the Inception-V3 and ResNet101 models with accuracy and loss rates of 1.000 and 0.012, respectively. The ResNet101 shows its lowest accuracy and highest loss during the training. Figure 3 presents the accuracy and loss during each iteration of the experiments. It can be seen from the figure that the accuracy and loss are approximately stable at iteration 80, which indicates that the three models reach convergence.

Table 5. Model performance for 100 epochs.

Model	Accuracy	Loss
VGG19	1.000	0.012
Inception-V3	0.986	0.037
ResNet101	0.957	0.074

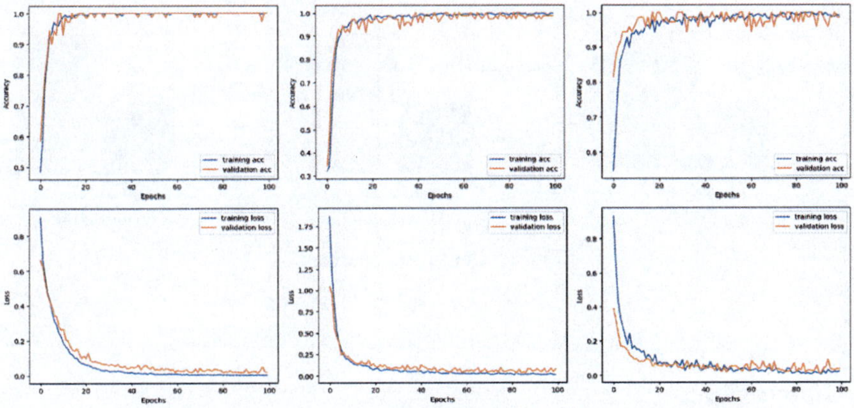

(a) VGG19 (b) Inception-V3 (c) ResNet101

Fig. 3. Training and validation loss during iterations

To verify the performance of various models on unseen data, we performed experiments on the test set (30% of the dataset). The results are shown in Fig. 4. It can be seen from the figure that the ResNet101 archives the best performance compared to other models for all evaluation metrics. Although the ResNet101 model did not perform well compared to other models during the training step, it shows its best classification capability for coconut fruit maturity problems (0.95 for accuracy, 0.96 in terms of precision, recall, and f1-score). In contrast, VGG19 reveals its lowest accuracy for unseen data as its performance was relatively poor. In addition, it can also be observed from Fig. 4 that the Inception-V3 model is also good for coconut fruit mature stage recognition. The accuracy, precision, recall, and f1-score of the Inception-V3 model archives 0.95 for all metrics.

We further verify the performance of the three models in each class based on the confusion matrix shown in Fig. 5. From the figure, we can see that all models are accurate at recognizing overly mature coconuts. However, the number of samples that

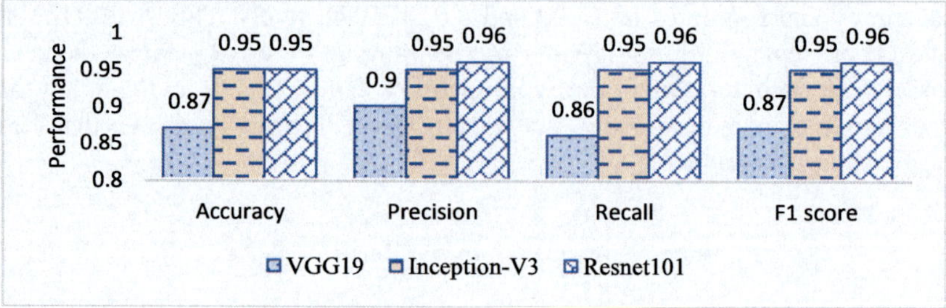

Fig. 4. Model performance on test set.

were mature but misclassified as overly mature was relatively large. The VGG19 models miscategorized 19 out of 75 samples, which shows its weakness in coconut fruit maturity detection.

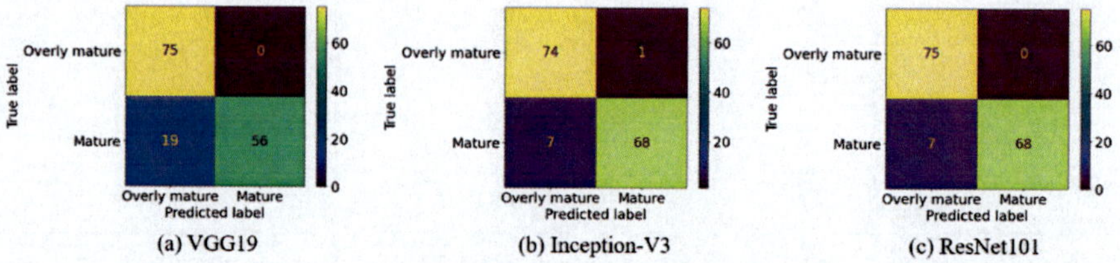

Fig. 5. Confusion matrices.

4 Conclusion

In this paper, we propose a coconut fruit maturity classification using deep transfer learning model. We collected images of coconut fruits and grouped them into two classes: mature and overly mature. We deployed three well-known deep learning models including VGG19, Inception-V3, and ResNet101. Experimental results show that the ResNet101 model can effectively classify the ripeness of coconut fruits with an accuracy of 0.95 and 0.96 for all precision, recall, and F1-score. The results confirm that our proposed model is feasible for classifying the maturity of coconut fruits. As our dataset was relatively small, in future work, we are planning to increase the dataset and try other machine learning models to improve the model's efficiency as well as to deploy it in practice.

Acknowledgement. This study was fully funded by Tra Vinh University under grant contract number 246/2022/HĐ.HĐKH&ĐT-ĐHTV.

References

1. D`ưa Việt Nam "vươn xa" có mặt tại gần 90 quốc gia, https://dangcongsan.vn/y-te/dua-viet-nam-vuon-xa-co-mat-tai-gan-90-quoc-gia-547776.html. Accessed 29 Aug 2023
2. Đặc điểm sinh học cây d`ưa, http://dost-bentre.gov.vn/tin-tuc/933/dac-diem-sinh-hoc-cay-dua. Accessed 29 Aug 2023
3. Hòa, N.M., Tần, N.T., Hùng, D.M., Khanh, N.V.: Application of SVM and KNN algorithms to build a classification model of makapuno coconuts in Vietnam. Tạp chí Khoa học và Công nghệ Đại học Đà N˜ăng. 41–46 (2021)
4. Caladcad, J.A., et al.: Others: determining Philippine coconut maturity level using machine learning algorithms based on acoustic signal. Comput. Electron. Agric.. Electron. Agric. **172**, 105327 (2020)
5. Javel, I.M., Bandala, A.A., Salvador, R.C., Bedruz, R.A.R., Dadios, E.P., Vicerra, R.R.P.: Coconut fruit maturity classification using fuzzy logic. In: 2018 IEEE 10th International Conference on Humanoid, Nanotechnology, Information Technology, Communication and Control, Environment and Management (HNICEM), pp. 1–6 (2018)
6. Fadchar, N.A., Cruz, J.C.D.: Design and development of a neural network—based coconut maturity detector using sound signatures. In: 2020 IEEE 7th International Conference on Industrial Engineering and Applications (ICIEA), pp. 927–931 (2020)
7. Noppon, L., Nipon, P.: The application of convolution neural network for coconut maturity classification. In: 2021 18th International Conference on Electrical Engineering/Electronics, Computer, Telecommunications and Information Technology (ECTI-CON), pp. 112–115 (2021)
8. Madue, S., Khaorapapong, T., Karnjanadecha, M., Limsiroratana, S.: Classification of fresh aromatic coconuts by using polynomial regression. In: Second International Conference on Digital Image Processing, pp. 182–187 (2010)
9. Lazim, M.I.M., Badruzaman, N.A., Peng, K.S., Long, K.: Quantification of cytokinins in coconut water from different maturation stages of Malaysia's coconut (Cocos nucifera L.) varieties. J. Food Process. Technol. **6**, 1 (2015)
10. Simonyan, K., Zisserman, A.: Very deep convolutional networks for large-scale image recognition. arXiv:1409.1556 (2014)
11. He, K., Zhang, X., Ren, S., Sun, J.: Deep residual learning for image recognition. In: Proceedings of the IEEE Conference on Computer Vision and Pattern Recognition, pp. 770–778 (2016)
12. Szegedy, C., Vanhoucke, V., Ioffe, S., Shlens, J., Wojna, Z.: Rethinking the inception architecture for computer vision. In: Proceedings of the IEEE Conference on Computer Vision and Pattern Recognition, pp. 2818–2826 (2016)

Enhanced Prediction of mRNA Subcellular Localization Using a Novel Ensemble Learning and Hybrid Approach

Thi-Tuyen Nguyen[1], Van-Nui Nguyen[1(✉)], Thi-Xuan Tran[2], and Nguyen-Quoc-Khanh Le[3,4(✉)]

[1] University of Information and Communication Technology, Thai Nguyen, Vietnam
{nttuyen,nvnui}@ictu.edu.vn

[2] University of Economics and Business Administration, Thai Nguyen, Vietnam
tranxuantbhd@tueba.edu.vn

[3] Artificial Intelligence in Medicine, College of Medicine, Taipei Medical University, Taipei 110, Taiwan
khanhlee@tmu.edu.tw

[4] AIBioMed Research Group, Taipei Medical University, Taipei 110, Taiwan

Abstract. Unraveling the subcellular localization of mRNA is an imperative aspect in the realm of biotechnology. This resolution can illuminate the inner workings of genetic regulatory mechanisms, gene expression modalities, and the evolution of cellular physiological and developmental processes. However, the experimental delineation of mRNA subcellular localization imposes significant temporal and resource commitments. Despite the development of multiple algorithms and predictive models for mRNA subcellular localization, their performance indexes have not been markedly high. In this paper, we introduce a novel hybrid approach to categorize mRNA into five distinct subcellular locales, including the cytoplasm, endoplasmic reticulum, extracellular region, mitochondria, and nucleus. Our model exploits the strengths of ensemble learning with a hybrid methodology, incorporating multiple biologically pertinent features extracted from the input sequencing data. Additionally, the model dynamically adjusts the weightages of functions and the minority class, through the modulation of the weight ratio of disparate models during their contribution to the principal model. Overall, our model delivers promising results, with an average accuracy of 0.89 in an independent dataset for the classification of mRNA subcellular localizations into five subclasses. This displays a significant performance elevation in contrast to preceding algorithms, particularly in instances where the classes are adequately sampled.

Keywords: mRNA subcellular localization · Sequence analysis · Ensemble learning · Hybird · Machine learning · Bioinformatics

P. T. Nghia et al. (Eds.): ICTA 2023, LNNS 847, pp. 60–68, 2023.
https://doi.org/10.1007/978-3-031-49529-8_7

1 Introduction

Eukaryotic RNA (mRNA) is an essential type of RNA in eukaryotic cells. It is responsible for transferring genetic information from DNA to the ribosome to carry out the process of protein synthesis. mRNA has a unique structure and plays a crucial role in the transmission sequence of genetic information [1]. During genetic information transmission, genes in DNA are copied into mRNA in a process called transcription. mRNA is a copy of a portion of the gene sequence in DNA, but unlike DNA, mRNA only contains a single-stranded sequence of nucleotides instead of a double-stranded sequence. It makes mRNA more flexible and quickly moves out of the cell nucleus [2, 3]. The structure of mRNA comprises various types of nucleotides: adenine (A), cytosine (C), guanine (G), and uracil (U). In mRNA, these nucleotides are arranged into nucleotide sequences called codons. Each codon consists of three nucleotides, and each codon corresponds to a specific amino acid [4, 5]. Therefore, mRNA in eukaryotic organisms ensures accurate protein synthesis and regulates cellular activities [6]. The intracellular localization of mRNA within the cell reveals characteristics of the timing and spatial aspects of gene expression [7]. Inaccurate mRNA localization can lead to various disorders, such as Alzheimer's disease, Parkinson's disease, and different types of cancer [3]. The discovery of the mechanism that regulates subcellular mRNA localization plays an essential role in the discovery of human pathogenesis, through which it is possible to know the causes of related diseases at the molecular level. From there, propose and develop drugs to treat related diseases. However, since these experiments are time- and resource-intensive, it is necessary to find more efficient methods to solve this problem.

Currently, predicting the subcellular localization of mRNA has been extensively studied and is of great interest to scientists. RNATracker is the first prediction tool for subcellular localization of mRNA within cells [8]. In this work, Yan et al. introduce, evaluate and interpret RNATracker, a deep neural network predictor of subcellular localization combining two convolutional layers, a bidirectional LSTM layer and an attention module. Afterwards, several other prediction methods have been proposed. In [9], Zhang et al. proposed the iLoc-mRNA method, in which they employed Support Vector Machine (SVM) as the classifier [9]. In this paper, the authors introduce a new prediction tool called SubLocEP, based on machine learning, to predict the subcellular localization of eukaryotic mRNA. Jing Li et al. [10] developed a predictor, called SubLocEP, utilizing a set of diverse prediction models, including recurrent neural networks, SVM, and KNN, and combining the results from these models to generate the final prediction. In this study, Jing Li et al. proposed a two-layer integrated prediction model for accurately predicting the location of sequence samples. In [12], Shihang Wang et al. proposed a predictor named DeepmRNALoc using a two-stage feature extraction strategy to predict eukaryotic mRNA subcellular localization. Although these proposed models have contributed to the study of mRNA subcellular localization, they have disadvantages and performance is not high enough; therefore it still have a room for us to study to improve the deep understanding of mRNA. This paper introduces a novel hybrid approach to predict mRNA by ensemble learning and hybrid approach.

2 Material and Methods

2.1 Data Collection

In this paper, we utilized data from previously published research [10, 11] to compare the performance with the reported results on the same dataset. The origin of this dataset was extracted from the RNALocate database [13]. While downloading the original dataset, we found that it consisted of two subsets: a training set and an independent validation set. Based on their subcellular locations, the samples in this dataset were divided into five classes: Cytoplasm (3590), Endoplasmic reticulum (1426), Extracellular region (855), Mitochondria (421), and nucleus (5073). Looking at the number of samples in the data classes, we observe that the data classes are imbalanced, with significant variations in the number of samples among them.

2.2 Feature Engineering

In this paper, the authors utilized machine learning to predict the function of mRNA subcellular localization in the nucleus of standard organisms. Data preprocessing and feature selection from amino acid sequences were crucial for classification models. These feature types can be grouped into five categories *(Kmer, NAC - Nucleic Acid Composition; DNC - Di-Nucleotide Composition; CKSNAP - Composition of k-spaced Nucleic Acid Pairs)*, and in this paper, the authors tested the efficiency of five common features. The detail information about these features and features enginering process has been presented in iLearn [14]. After feature selection process, we identified the best feature based on evaluating each feature's effectiveness by calculating the mean and standard deviation of the scores array. Testing each feature individually helps assess its performance in the specific problem. We have identified three out of five features that performed better for our dataset, including Kmer, DNC, and CKSNAP. Then, from these three features, we combined them to create hybrid features to achieve the highest performance on our model, including "DNC + CKSNAP", "CKSNAP + Kmer", "DNC + Kmer", and "DNC + CKSNAP + Kmer".

2.3 Model Construction, Learning and Evaluation

The classification problem is to determine and separate data into different groups or classes based on available features and information. Machine learning provides methods and algorithms for classification models based on known training data. This paper used various traditional machine learning algorithms to construct classifiers. They included support vector machine (SVM), logistic regression (LR), random forest (RF), decision tree, and k-nearest neighbors (KNN), adaptive boosting (AB), gradient boosting (GB), and extreme gradient boosting (XGB). They are commonly used algorithms in classification models [15, 16]. However, since the dataset in the paper is imbalanced, the results obtained after evaluating the models built on the selected feature set could have been more satisfactory. Therefore, the authors continued with ensemble learning and hybridizing various machine learning models to find the model with the best predictive performance for our given dataset. To find the best algorithm on the given dataset, we experimented with each algorithm on individual features of the dataset. Then, using the same algorithm, we ran it on combined features.

To find the best final model, the predictive performance of models using different parameters was evaluated through 10-fold cross-validation. The training dataset was divided into ten approximately equal-sized subgroups. The training process was run ten times, each selecting one subgroup as the testing dataset and the remaining subgroups as the training dataset. The results were combined to produce a single estimation for the 10-fold cross-validation evaluation. 10-fold cross-validation has the advantage of improving the reliability of evaluation because it considers all original data as training and testing datasets, with each data used for validation exactly once [8, 17, 18]. To evaluate the predictive proficiency of the trained models, an array of statistical indices was employed. This approach was necessitated by the fact that most evaluative metrics can only illustrate the predictive capacity of the model from a singular perspective. The following measures were often used: accuracy, recall, precision, and F1-score [19, 20]. The *ACC* indicates the prediction accuracy; however, when the sample size of different classes in the data is uneven, ACC's assessment conclusions may be questionable. *Recall* is the proportion of correctly predicted positive sets out of all positive sets. *Precision* represents the proportion of correctly predicted positive sets out of all predicted positive sets. The *F1-score* is defined as the harmonic mean of the model's *precision* and *recall*.

$$Recall = \frac{TP}{TP + FN};\ Precison = \frac{TP}{TP + FP}$$

$$ACC = \frac{TP + TN}{TP + TN + FP + FN};\ F1_{score} = \frac{2 * Precison * Recall}{Precison + Recall},$$

where: *TP*, *TN*, *FP* and *FN* represent the numbers of true positives, true negatives, false positives and false negatives, respectively.

3 Results and Discussion

3.1 Impact of Single and Hybrid Features in Predictor of Eukaryotic mRNA Subcellular Localization

In this paper, we calculated the frequency of 4 amino acids, A, C, G, and T in five mRNA subcellular compartments to explore the differences between each cellular subtype. As displayed in Fig. 1, the distinctions amongst the subcellular compartments appear minimal; nonetheless, several noteworthy observations merit attention. In the Nucleus compartment, Alanine (A), Cysteine (C), Glycine (G), and Threonine (T) were the most abundant amino acids. On the other hand, these four amino acids appeared less frequently in the Mitochondria compartment. The distribution and frequency of different amino acids across the subcellular compartments certainly play a pivotal role in distinguishing them. Therefore, we propose a method to accurately predict mRNA subcellular compartments by combining these amino acids in feature encoding. Kmer represents the frequency of different kmers in the nucleotide sequence, providing the foundation for analyzing and predicting nucleotide sequence properties. The CKSNAP method generates a feature vector by calculating the frequency of these nucleotide pairs in the nucleotide sequence, thereby providing information about the structure and properties of the nucleotide sequence [14]. DNC is used to learn patterns and important information from the nucleotide sequence by applying convolutional layers to detect structural features and model the relationships between nucleotides. Therefore, these feature extraction methods can contribute to predicting eukaryotic mRNA subcellular localization.

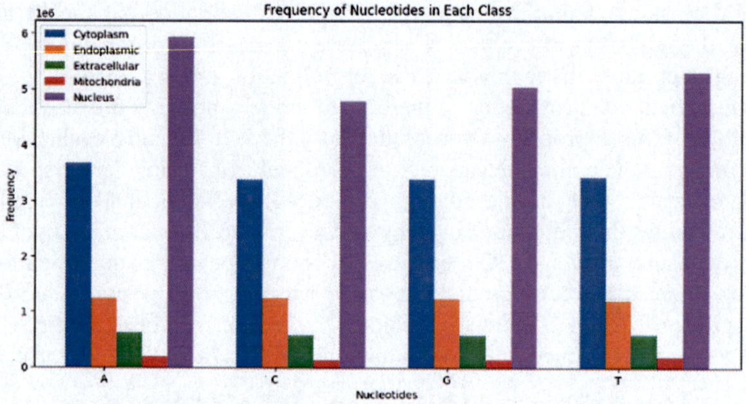

Fig. 1. The distribution of data samples across

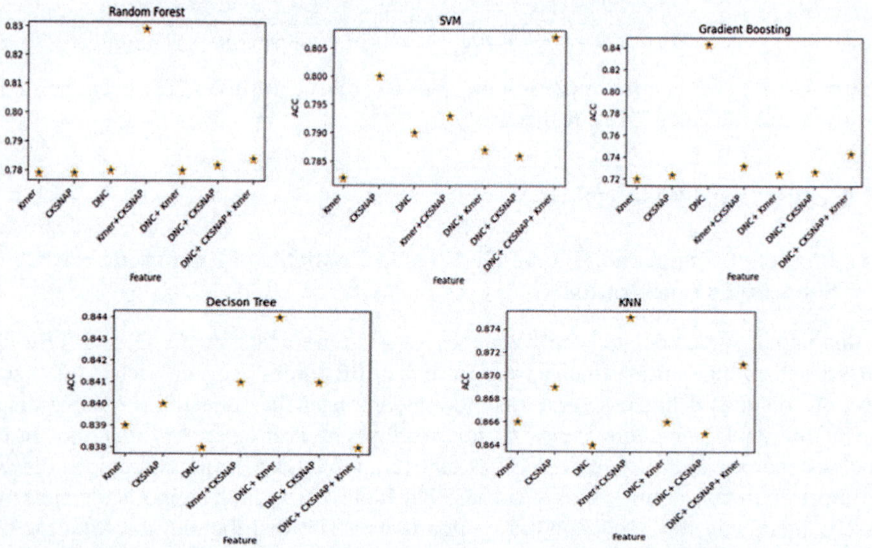

Fig. 2. Performance of Eukaryotic mRNA Subcellular Localization classification using different features and machine learning classifiers

Figure 2 shows the average performance results of the classes on machine learning models based on individual and combined features. Using standalone machine learning methods, whether on individual or combined features, did not yield satisfactory results. Among all the models run on our dataset, the KNN model with optimized parameters achieved the highest performance on the extracted features. This model yielded an average performance of 0.875 in predicting the classes on the combined features Kmer + CKSNAP and the lowest performance of 0.864 on the individual Kmer feature.

3.2 Ensemble Learning of Machine Learning Models

To improve the predictive performance, the authors employed ensemble learning by combining multiple machine learning models. As shown in Fig. 2, the KNN model's results on the feature set Kmer + CKSNAP demonstrated the highest performance. Therefore, to enable a fair comparison between the models, the authors used the Kmer + CKSNAP feature set to train the ensemble learning models. During the experimentation process, the authors conducted ensemble learning of the following model combinations: KNN and Random Forest (RF), KNN and Random Forest and Decision Tree (DT), SVM, Random Forest and Decision Tree, GradientBoosting, AdaBoost, and XGBoost. In the process of ensemble learning, to achieve the highest performance, the authors focused on two issues:

1. Changing parameters for the LogisticRegression() function in the stacking_classifier.
2. Due to the dataset's imbalance, as shown in Fig. 1, where the Mitochondria class (421 samples) is the smallest, comprising only 8.35% compared to the most abundant Nucleus class (5,073 samples), the authors also applied the Synthetic Minority Oversampling Technique (SMOTE) [21] to balance the data before inputting it into the model and adjusted the weights of this class during the model training process (increasing its weight compared to the weights of other classes). It allows the models to focus more on learning from the class with higher weight to balance predictive performance and improve accuracy in the more important classes. Adjusting weights can be beneficial in problems with heavily imbalanced class distribution in the training data.

Table 1. Performance evaluation by Ten-fold cross-validation on *Kmer + CKSNAP* features

Method	Model	Precision	Recall	ACC	F1-score
	KNN	0.614	0.641	0.875	0.611
Ensemble Learning	KNN + RF	0.621	0.651	0.878	0.602
	RF + DT	0.508	0.507	0.860	0.517
	GB + AB + XGB	0.613	0.603	0.868	0.602
	KNN + RF + DT + SVM	**0.671**	**0.662**	**0.880**	**0.671**

However, despite using ensemble learning methods in machine learning, the class with the least data (Mitochondria) still does not have a high prediction rate. Therefore, the authors proceeded to construct a hybrid model from four different machine learning models: KNN + RF + DT + SVM. In this hybrid model, to optimize performance, the authors focused on adjusting the parameters of each machine learning model. Additionally, we continued to use the method of adjusting the weights of classes with limited data using the class_weights ratio and applied the SMOTE technique, similar to the Ensemble Learning of machine learning models method. However, in this hybrid model, the authors also incorporated the voting method. The voting method is a technique in machine learning used to combine predictions from multiple different models to generate a final prediction. In this method, they utilized Weighted Voting technique [22]. Each model was assigned a weight, and the final prediction was calculated based on the weights of the models. This allowed prioritizing models with better performance or deeper expertise in specific predictions. This means that during prediction, the models were used sequentially to make predictions, and the final result was calculated based on the assigned weight ratios. As displayed in Table 2, the performance of both the ensemble learning model and hybrid models are the most optimal models; therefore, our final model generated by incorporating of ensemble learning and hybrid approach is high and promising for the prediction of mRNA subcellular localization.

Table 2. Performance evaluation by Independent testing

Method	Model	Precision	Recall	ACC	F1-score
	KNN	0.614	0.641	0.875	0.611
Ensemble Learning	KNN + RF	0.621	0.651	0.878	0.602
	RF + DT	0.508	0.507	0.860	0.517
	GB + AB + XGB	0.613	0.603	0.868	0.602
	KNN + RF + DT + SVM	**0.671**	**0.662**	**0.880**	**0.671**
Hybird	SVM + RF + DT	0.641	0.640	0.850	0.671
	GB + AB + XGB	0.630	0.611	0.871	0.642
	KNN + RF + DT + SVM	**0.682**	**0.691**	**0.890**	**0.681**

3.3 Independent Testing Performance

As mentioned earlier, the original dataset we downloaded contains two sets of data: a training dataset and an independent validation dataset. We utilized the provided independent testing dataset to assess the trained models' feasibility. In Tables 1 and 2, we present the results of ACC, Precision, Recall, and F1-score as the average values of these metrics for the five classes predicted.

To assess the performance of our model in the real case, we compared it with several existing machine learning models, including iLoc-mRNA [9], mRNALoc [10], SubLocEP [11]. Fortunately, our proposed model obtains the highest performance with 0.89 accuracy and F1-score value of 0.681 (Table 3).

In machine learning with imbalanced datasets, alongside metrics such as ACC, Precision, and Recall, the F1-score is the most accurate metric to evaluate model performance.

Table 3. Comparison of performance among training-based machine learning models

Model	Precision	Recall	ACC	F1-score
iLoc-mRNA	0.183	0.425	0.425	0.256
mRNALoc	0.549	0.491	0.491	0.509
SubLocEP	0.671	0.601	0.601	0.578
Hybird KNN + RF + DT + SVM	**0.682**	**0.691**	**0.890**	**0.681**

As displayed in Table 3, our proposed model outperforms previously published models, reaching the highest F1-score at 0.681. This demonstrates that our proposed model and data processing techniques have brought positive results in cases of data imbalance.

4 Conclusion

In this study, we conducted experiments on mRNA subcellular localization to find the model that achieves the highest performance for that dataset. These tasks often involved feature extraction of nucleotides and finding the best features for the dataset. We then combined these features to input into the model. Next, we combined different machine learning models using ensemble learning and a hybrid of machine learning models. During this process, we focused on changing the weights of the models, adjusting the weights of the class with the least data when inputting into the model, and altering the contribution weight ratio of each model in the overall model and the decision-making process of the final model. Finally, we obtained the hybrid machine learning model (*KNN + RF + DT + SVM,* generated based on features of *CKSNAP + Kmer*) with the highest performance when evaluation on both cross-validation and testing approach. Moreover, when comparing with some existing predictors on mRNA subcellular localization, our proposed model also demonstrated the highest performance and F1-score. This suggests that our proposed model in this study is valuable and stable for predicting mRNA subcellular localization.

References

1. Lodish, H., et al.: Molecular Cell Biology. Macmillan (2008)
2. Alberts, B., et al.: Molecular Biology of the Cell 6th ed. 2014: Garland Science
3. Cooper, T.A., Wan, L., Dreyfuss, G.J.C.: RNA and disease **136**(4), 777–793 (2009)
4. Watson, J.D.: Molecular Biology of the Gene. Pearson Education India (2004)
5. Nelson, D.L., Lehninger, A.L., Cox, M.M.: Lehninger Principles of Biochemistry. Macmillan (2008)
6. Berg, J., et al.: Protein synthesis **8**, 893–924 (2015)
7. Martin, K.C., Ephrussi, A.J.C.: mRNA localization: gene expression in the spatial dimension **136**(4), 719–730 (2009)
8. Yan, Z., Lécuyer, E., Blanchette, M.J.B.: Prediction of mRNA subcellular localization using deep recurrent neural networks **35**(14), i333–i342 (2019)

9. Zhang, Z.-Y., et al.: Design powerful predictor for mRNA subcellular location prediction in Homo sapiens **22**(1), 526–535 (2021)
10. Garg, A., et al.: mRNALoc: a novel machine-learning based in-silico tool to predict mRNA subcellular localization **48**(W1), W239–W243 (2020)
11. Li, J., et al.: SubLocEP: a novel ensemble predictor of subcellular localization of eukaryotic mRNA based on machine learning **22**(5), bbaa401 (2021)
12. Wang, S., et al.: DeepmRNALoc: a novel predictor of eukaryotic mRNA subcellular localization based on deep learning **28**(5), 2284 (2023)
13. Cui, T., et al.: RNALocate v2. 0: an updated resource for RNA subcellular localization with increased coverage and annotation **50**(D1): D333–D339 (2022)
14. Chen, Z., et al.: iLearn: an integrated platform and meta-learner for feature engineering, machine-learning analysis and modeling of DNA, RNA and protein sequence data **21**(3), 1047–1057 (2020)
15. Cai, C., et al.: SVM-Prot: web-based support vector machine software for functional classification of a protein from its primary sequence **31**(13), 3692–3697 (2003)
16. Géron, A.: Hands-on Machine Learning with Scikit-Learn, Keras, and TensorFlow. O'Reilly Media, Inc. (2022)
17. Kermany, D.S., et al.: Identifying medical diagnoses and treatable diseases by image-based deep learning **172**(5), 1122–1131. e9 (2018)
18. Wekesa, J.S., et al.: A deep learning model for plant lncRNA-protein interaction prediction with graph attention **295**, 1091–1102 (2020)
19. Chu, Y., et al.: DTI-CDF: a cascade deep forest model towards the prediction of drug-target interactions based on hybrid features **22**(1), 451–462 (2021)
20. Shan, X., et al.: Prediction of CYP450 enzyme–substrate selectivity based on the network-based label space division method **59**(11), 4577–4586 (2019)
21. Chawla, N.V., et al.: SMOTE: synthetic minority over-sampling technique **16**, 321–357 (2002)
22. Dogan, A., Birant, D.: A weighted majority voting ensemble approach for classification. In: 4th International Conference on Computer Science and Engineering (UBMK) (2019)

Improved Approach for Mapping Anthropological Facial Features Based on A Convolutional Neural Network

Nguyen Thu Huong[1,2,3](✉), Nguyen The Long[1], and Pham Thi Lien[3]

[1] Department of Informatics, Institute of Cybersecurity and Digital Technologies, MIREA – Russian Technological University, 119454 Moscow, Russia
nthuongcntt@ictu.edu.vn

[2] Laboratory of Artificial Intelligence and Machine Learning, Institute of Information Technology and Data Science, Irkutsk National Research Technical University, 664074 Irkutsk, Russia

[3] University of Information and Communication Technology, Thai Nguyen University, 70000 Thai Nguyen, Vietnam
ptlien@ictu.edu.vn

Abstract. Our facial anthropological and actions are non-verbal communication tools that account for 93% of emotions in human communication, 55% of which are facial expressions and gestures. Facial emotions can be easily analyzed from facial images, and computers can interact with people the way people interact with each other. For this reason, automatic classification and analysis of facial emotions is currently an interesting and challenging problem that has great implications for society. In this work, the random forest method combined with a neural network was proposed to improve the classification of emotions by building feature maps. Training and testing models were built to perform the experiments, which achieved very good accuracy in the range of 98.5% to 99.7% for the two datasets.

Keywords: Deep learning · Anthropological face feature · Feature extraction · Anthropological face classification · Convolutional neural network

1 Introduction

Data processed in a classification system are divided into three categories: Training Set, Test Set, and Validation Set. The training set is the facts set used to train the model. System learning algorithms learn model from this training set. The learning process varies relying on the set of rules and utilization pattern. The proposed linear regression version uses [1], the factors in the training set are used to find the fine becoming feature or line that describes the connection between the input and the output of the dataset. The training statistics is trained the use of an optimization technique such as the basis formula or approximated optimization algorithms such as gradient descent [2] or stochastic gradient descent. Inside the k-Nearest neighbors set of rules, the factors inside the training records set are the points that may be friends (are close to every different) learned by using greedy

P. T. Nghia et al. (Eds.): ICTA 2023, LNNS 847, pp. 69–76, 2023.
https://doi.org/10.1007/978-3-031-49529-8_8

strategies. The accuracy or blunders of the trained predictive model is tested by new facts samples. by means of including the common of these kinds of mistakes, we will calculate the prediction error for the take a look at set. These prediction errors are evaluated with the aid of a selection of metrics together with precision, take into account, F1 rating [3, 4], RMSE [5], MAE [6].

The extraction of anthropological face features by traditional methods and methods using deep learning have been fully exploited and have had certain successes. The machine uses a Haar cascade classifier for characteristic extraction [7], and the Fisher face set of rules makes use of a uniform Local Binary pattern. This goal's most important motive is to create a whole some environment for the corporate zone and help patients with Asperger syndrome. This will have huge usage through psychologists as properly for treating their sufferers in therapy.

To recognize the emotion from EEG sign, main components are needed to be taken into account, the first is sizeable function extraction from emotional EEG signals and the second is Emotion class [8]. Having a number of noises and artifacts in EEG signal, extracting the ideal features from more than one channel EEG signals is a tough difficulty. But, some of characteristic extraction strategies have already been proposed through researchers for emotion recognition or type. These features are extensively in time area, frequency area, time–frequency domain, or concurrent time and frequency area technique [9, 10]. Some other trouble is the channel selection from the multichannel EEG indicators due to the fact the reduced channel statistics reduces the characteristic dimension that allows to growth the class accuracies [11].

There are various techniques for function extraction, inclusive of the traditional approach of feature engineering based totally on many signal processing strategies and facts, or automatic feature engineering, which may be immediately extracted by way of neural networks. Many researches were achieved on every conventional and automatic function engineering to advise a powerful EEG-based totally emotion type. Extracted abilities are given as input to powerful preferred gadget studying model like SVM [12], KNN [13], and lots of others. Currently, deep analyzing networks have validated large strength in feature extraction and type duties, and lots of researchers have implemented certainly one of a type neural networks to EEG records [14] to beautify accuracy. The intention of [15] is to broaden a clever method to enhance the accuracy of emotion detection with the useful resource of discrete signal processing techniques and applying optimized useful resource vector system classifier with genetic evolutionary set of rules. An included network, this is able to spotting emotion intensity levels from anthropological face pix in actual time the usage of deep mastering technique is proposed. The cognitive take a look at of anthropological face based on expression intensity stages are beneficial in programs collectively with healthcare proposes to decorate emotion reputation with 2 special vital parameters, valence and emotion depth. This allows in better automatic responses with the useful resource of a device to an emotion.

The main contributions of this work are as follows: In this work, a combination of machine learning algorithm – Random Forest – and deep learning technique CNN-VGG16 is proposed. The proposed method achieves competitive performance on both Oulu-CASIA (Oulu-CASIA NIR & VIS facial expression database) and FER-2013 datasets. This has very positive implications for implementing a combination of machine learning and deep learning approaches in classification applications. Identifying regions of interest in images using CNN-VGG16 to create a feature map that extracts patient's anthropological face. Combining the depth of feature maps with spatial information, focusing on extracting important features and using the joint loss function to give the network a better feature effect. Adjusting the parameters in the random forest algorithm and reducing the difference between same anthropological face within the class, expanding the feature distance between different anthropological face and ensuring the classification accuracy [16, 17].

The article is presented in the following sections: "Introduction" section presents relevant work, "Methodology" section describes the algorithm in this paper, and "Experiments and Results" section presents experiments and experimental results. In the "Conclusion" section, the research results of this work are presented.

2 Methodology

2.1 Creating Data for Data Augmentation in Deep Learning

Currently, the data problem plays a very important role in deep learning projects. Therefore, there are areas where there is limited data, so it is difficult to train the model to give good prediction results. Especially in emotion recognition, the direction of the subject's head is very important because it influences up to 70% of the anthropological face muscles to express emotions. Therefore, an additional step of data-assisted processing is proposed in this study. There are many proposed techniques, such as: Mirroring, random cropping, color shifting, noise/occlusion… Based on the analysis of FER -2013 database for the training/validation/testing processes on the following criteria: Number sample quantity, calculation of balance/imbalance of sample, so the technique used in this article.

Using this image enhancement is very useful for the process of constructing CNN structures to build feature maps because the augmentation for each training/epoch stage can be different. The first epoch at a large learning rate may be different from the last epoch at a small learning rate. The way of Augmentation for each class, the frequency of using Augmentation may be different (see Fig. 1).

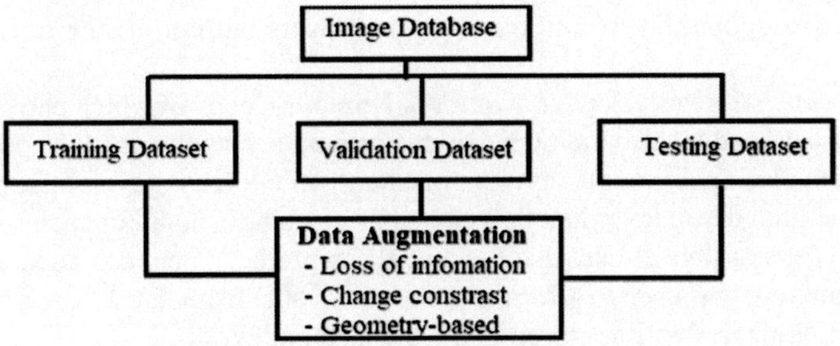

Fig. 1. Description of the data enhancement process for the stages: training/validation/testing

- Loss of information: Part of the image is lost. The obscured case can be illustrated.
- Change contrast: change the contrast of the picture, saturation.
- Geometry-based: Various types can be rotated, flipped, scaled, padded, squashed, and deformed.

2.2 Anthropological Face Feature Detector Using Convolutional Neural Network

CNN uses learned filters to convolve the function maps from the previous layer. Filters are – dimensional weights and these weights have a spatial relationship with each different.

Deep learning creates a feature map for data to which the classifier is applied to solve the problem. Each problem has a unique data set and the strategies applied are different for different problems. The convolutional neural network architecture is implemented with many layers such as convolution, ReLu, and max-pooling. Other layers such as dropout are added to the network to improve performance during training. The dropout layer is activated only during training. The dropout layer randomly a certain number of neurons during the forward pass (input to the function) and remembers the neurons that remained during the forward pass. In the backward pass, only the neurons that were not dropped out are updated. The dropout is a function that provides regularization. The dropout layer ensures that the model learns robust features that are independent of the neurons, which in turn avoids overfitting during the training phase.

2.3 Visualizing Feature Maps Generated in a CNN

Feature maps are generated with the aid of making use of filters or features detectors to the input image or the feature map output of the prior layers. This process will offer insight into the internal representations for particular input for each of the Convolutional layers inside the model.

Visualize the characteristic maps primarily based on these steps:

Step 1: Define a new model: The output of the model might be functions maps, which can be an intermediate representation for all layers after the first layer.

Step 2: Load the input image and use visualization models to track object-specific changes layer by layer.

Step 3: Prepare data for Training and Validation process. In this step will proceed to build the convolutional layers and the max pool layers but not for the fully connected layer.

A CNN consists of a stack of convolutional modules, each of which performs feature extraction. Each module also includes a convolutional layer followed by a pooling layer. The final convolutional module consists of one or more dense layers that perform the layering. The final dense layer in a CNN consists of a single node for each specific layer in the model (all possible classes that the model can predict). For each node, the softmax activation function was used to generate the predict value from 0 to 1. The next step uses the Adam Optimizer algorithm to optimize the CNN network.

3 Experiments and Results

3.1 Experimental Setup and Experimental Data Set

All experiments have been carried out using a Intel core i5 CPU @ 3 GHz and 4 GB RAM. With a purpose to ensure the fairness of the test on the advanced community and the contrast community, the training parameters used in the experiment are precisely the identical. All model training techniques undertake the mastering charge attenuation method. The preliminary gaining knowledge of charge is 0.01 and the attenuation coefficient is 0.001, the batch length is 100. After the training model is completed, all images within the training records set are called an epoch, and a hundred and fifty epochs are set in the experiment. Adam optimization algorithm is used to optimize the network quicker.

The 2 datasets used in this article are FER-2013, Oulu-CASIA (Oulu-CASIA NIR&VIS facial expression database). The FER-2013 dataset was collected from Google images about emotion of human, such as: happy, sad, angry, afraid, surprise, disgust, and neutral, with happy. This database consists of 28,000 image for training process, 3500 for validation process and 3500 for testing process. The Oulu-CASIA NIR&VIS facial expression database are surprise, happiness, sadness, anger, fear and disgust. It almost is males (73,8%), other is female.

3.2 Experimental Results

Model performance was determined based on the accuracy of each model in the FER -2013 and Oulu-CASIA (Oulu-CASIA NIR&VIS facial expression database) datasets. Tables 1 and 2 show the confusion matrix of all models and help in evaluating their respective performances. This matrix also helps in identifying the probable presence of misclassifications, which are also possible in anthropological face identification, since accurate anthropological face identification is always difficult for faces. The model CNN-RF, which uses real images as input, and the model CNN, which uses hybrid images, have approximately the same results.

Table 1. Confusion matrix of CNN – RF for FER-2013 database.

	Angry	Disgust	Fear	Happy	Sad	Surprise	Neutral
Angry	544	2	116	155	151	25	190
Disgust	10	340	11	5	56	8	65
Fear	17	61	760	18	23	109	37
Happy	32	11	68	881	45	3	5
Sad	2	32	71	54	826	8	69
Surprise	17	56	111	230	91	704	2
Neutral	21	81	79	133	25	17	450

Table 2. Confusion matrix of CNN – RF for Oulu-CASIA database.

	Angry	Disgust	Fear	Happy	Sad	Surprise	Neutral
Angry	231	10	36	45	17	8	29
Disgust	47	357	18	26	35	23	44
Fear	61	11	459	19	72	60	100
Happy	19	50	80	781	105	20	17
Sad	10	35	37	105	997	7	52
Surprise	63	19	105	71	8	563	27
Neutral	78	37	96	10	56	67	708

Figure 2 show the results of the training, testing and validation processes of CNN on two datasets (a) FER-2013; (b) Oulu-CASIA. The results have shown the stability of the proposed method, which avoids the problem of overfitting by using the optimal Adam algorithm and the appropriate data partitioning for the two training and testing processes.

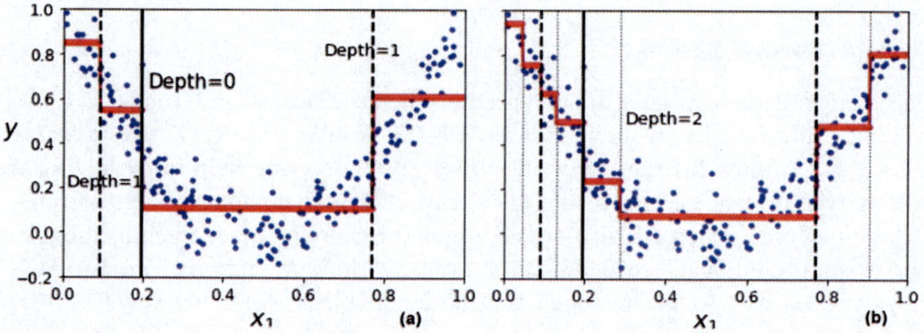

Fig. 2. Build emotion feature maps based on convolution neural network for (a) FER-2013 and (b) Oulu-CASIA databases

The results of anthropological face of two information sets with the standards for assessment of Accuracy, Precision, Recall of the methods from the Table 3 that the CNN has accuracy highest and stable. Overfitting refers to a situation where a version learns statistical regularities unique to the training set. This is certainly one of the largest demanding situations in deep getting to know, as an overfitted model cannot be generalized to records in no way visible before.

A habitual take a look at to stumble on overfitting to the training records is to screen the loss and accuracy of the training and validation units. In building the CNN community, we used pooling, dropout, normalization, and regularization strategies to optimize training time and avoid overfitting. The excellent approach to lessen overfitting is to attain greater schooling information. A version trained with a huge records set is typically simpler to generalize.

Table 3. The results of classification based on ANN, SVM, CNN – RF (proposed method)

Criteria	Dataset	ANN	SVM	CNN -RF
Accuracy (%)	FER – 2013	97.61	98.02	98.97
	Oulu-CASIA	96.2	95.89	99.47
Precision (%)	FER – 2013	95.15	98.1	99.6
	Oulu-CASIA	97.01	97.16	99.24
Recall (%)	FER – 2013	96.05	94.11	98.55
	Oulu-CASIA	95.94	94.76	99.15
MSE	FER – 2013	0.072	0.046	0.015
	Oulu-CASIA	0.105	0.082	0.0059

4 Conclusion

In this work, the main tasks duties have been anthropological face statistics era and feature extraction. The experimental results have proven that the extracted functions play a key role in the prediction and studying section of the classifier. While the extracted features have the capability to uniquely describe the signal styles in exclusive classes, the classifier's potential to are expecting increases and the version works extra correctly. Therefore, function extraction strategies might be taken into consideration as a priority of our research criteria in future work. Data augmentation is more efficient for neural network models than for standard machine learning models. Data augmentation advanced the prediction accuracy of the ANN, SVM, and CNN classifiers. It was clean that the geometry-primarily based accuracy of the three strategies improved due to the lack of facts and the change in contrast. Convolutional neural community model requires extra information than traditional machine learning models. Feature extraction or data augmentation with generative adversarial networks was proposed to solve these problems.

The implemented method had better accuracy for CNN, compared to ANN and SVM classifiers, implying that statistics sparsity is more important for neural network models than for traditional machine learning models.

References

1. Khushbu, K., Suniti, Y.: Linear regression analysis study. J. Pract. Cardiovasc. Sci. **4**(1), 33–36 (2018)
2. Yu, X., Yankang, W.: A self-adaptive gradient descent search algorithm for fully-connected neural networks. Neuro Comput., 70–80 (2022)
3. Nguyen, T. L., Afanasiev, A. D., Nguyen, T.H.: Segmentation of Forest Fire Images Based on Convolutional Neural Networks. Int. J. Artif. Intell., 21–35 (2021)
4. Nguyen, L.T., Nguyen, H.T., Afanasiev, A.D.: Automatic Identification Fingerprint Based on Machine Learning Method. J. Oper. Res. Soc. China, 273–285 (2021)
5. Nguyen, T. H., Nguyen, T.L.: Deep learning methods for classification of road defects. Int. J. Artif. Intell., 178–192 (2021)

6. Nguyen, T.H., Nguyen, T.L., Dreagle, I.A.: Robust approach to detection of bubbles based on images analysis. Int. J. Artif. Intell. **16**, 167–177 (2018)
7. Kaushal, D., Devasheesh, T., Yashvi, S.: Feature extraction technique for emotion detection using machine. SSRG Int. J. Electron. Commun. Eng., 41–45 (2020)
8. Zheng, W.: Multichannel EEG-based emotion recognition via group sparse canonical correlation analysis. IEEE Trans. Cogn. Dev. Syst. **9**(3), 281–290 (2017)
9. Zheng, W.L., Lu, B.L.: Investigating critical frequency bands and channels for EEG-based emotion recognition with deep neural networks. IEEE Trans. Auton. Ment. Dev. **7**(3), 162–175 (2015)
10. Jirayucharoensak, S., Pan-Ngum S., Israsena, P.: EEG-based emotion recognition using deep learning network with principal component based covariate shift adaptation. Sci. World J., 1–10 (2014)
11. Ansari, K., Chanel, G., Pun, T.: A channel selection method for EEG classification in emotion assessment based on synchronization likelihood. In: Proceedings of 15th European Signal Processing Conference, 1241–1245 (2007)
12. Suchitra S., Shikha T.: An intelligent facial expression recognition system with emotion intensity classification. Cogn. Syst. Res., 39–52 (2022)
13. Gurram S. K.: Intelligent deep learning based ethnicity recognition and classification using facial images. Image Vis. Comput., 104–124 (2022)
14. Kamnitsa, K.: Efficient multi-scale 3D CNN with fully connected CRF for accurate brain lesion segmentation. Med. Image Anal., 211–232 (2017)
15. Konstantinidis, D.: A modular CNN-based building detector for remote sensing images. Comput. Netw, 176–185 (2020)
16. Kulikov, A.A.: The model is a reprint of an object in the image. Russ. Technol. J. **8**(3), 7–13 (2020)
17. Kulikov, A.A.: Application of biometric systems in face identification technologies. Russ. Technol. J. **9**(1), 7–14 (2021)

Intrusion Detection System Using Semi-supervised Learning with Hybrid Labeling Techniques

Ren-Hung Hwang[1](✉), Tzu-Hao Tsai[2], and Jia-You Lin[2]

[1] College of Artificial Intelligence, National Yang Ming Chiao Tung University, Tainan 71150, Taiwan
rhhwang@nycu.edu.tw

[2] Department of Computer Science and Information Engineering, National Chung Cheng University, Chiayi 62102, Taiwan

Abstract. With the continuous development of the Internet and a substantial increase in interconnected devices, devices have become more vulnerable to cybersecurity attacks. Deploying Network-based Intrusion Detection Systems (NIDS) can effectively detect malicious traffic. However, a drawback of many existing methods is that once the model is deployed, it is no longer updated, resulting in inadequate predictive capabilities. To make NIDS adapt to new traffic patterns after deployment, some researchers have adopted semi-supervised learning methods, which label the unlabeled data and periodically retrain the model, leading to an improved detection rate. Nevertheless, semi-supervised learning methods often face challenges related to data labeling accuracy, as incorrect labels can negatively impact the model's performance. To address this issue, we propose a hybrid labeling technique to tackle such problems. Experimental results demonstrate that the proposed hybrid labeling method outperforms a single labeling approach, enhancing model accuracy.

Keywords: NIDS · Network traffic · DL-based anomaly detection · Semi-supervised learning

1 Introduction

Recent advances in machine learning-based network intrusion detection systems (NIDS) have demonstrated their superior detecting performance compared to traditional approaches. However, after a ML-based NIDS is deployed in the real environment, it will continuously encounter new traffic patterns that were not seen during its training phase. The existing models may fail to accurately predict these unseen attack flows. To address this challenge, we can collect new data observed over a period of time and periodically retrain the model. However, during this process, the collected data often faces the challenge of being unlabeled or incorrectly labeled. In such scenarios, semi-supervised machine learning techniques are commonly used to tackle these problems. Semi-supervised

P. T. Nghia et al. (Eds.): ICTA 2023, LNNS 847, pp. 77–85, 2023.
https://doi.org/10.1007/978-3-031-49529-8_9

machine learning leverages both labeled and unlabeled data as training samples and applies various algorithms to solve tasks in different environments. In [1], the authors introduced several existing semi-supervised machine learning algorithms, and in our experiments, we adopted the self-training approach to address the problem.

2 Related Work

Currently, there are only a few relevant studies on the application of semi-supervised learning to network intrusion detection systems, and we have summarized and compared them in Table 1.

In semi-supervised learning methods, the task of correctly labeling unlabeled data is crucial and worth researching because incorrect labels can lead to a decrease in the model's detection capabilities. The actual detection performance of the model when deployed online is also a widely discussed topic. For instance, papers [14–18] have proposed methods that maintain excellent detection performance for intrusion detection systems deployed in real online scenarios.

In our research, we are dedicated to addressing the challenge of data labeling and evaluating the accuracy of data labeled as true positives. Additionally, we investigate the impact of deploying the ML-based NIDS in practical scenarios.

Table 1. Related work

Paper	Online mechanism	Dataset	Algorithm
[2]	X	Fsecurify, self-collected	TextCNN
[3]	X	NSL-KDD	ANN, DT, KNN, RF, SVM
[4]	X	CICIDS2017, CTU	CNN
[5]	X	CICIDS2017, KDD CUP99	LN
[6]	X	NSL-KDD	Adversarial Auto-encoder
[7]	X	KDD CUP99	K-means
[9]	X	NSL-KDD	ASVM, FCM
[10]	X	UNSW-NB15	SSAE-DeepCNN
[11]	X	CISIDS2017, NSL-KDD	K-means, RFCNN, LSTM
[12]	X	CICIDS2017	PCA, Isolation-Forest, One-Class SVM, Autoencoder
[13]	X	CICIDS2017, UNSW-NB15	FSMG
[19]	X	CICIDS2017, ISCX 2012	D-PACK (CNN+Autoencoder)
[14]	V	MAWIFlow dataset	RF, Adaboost, DT
[15]	V	Many kinds of public datasets	K-NN
[16]	V	KDD CUP99, Moore, Sperotto	MSOINN
[17]	V	NSL-KDD, KDD CUP99	LAD
[18]	V	CICIDS2017, UNSW-NB15	DT, RF
Ours	V	CICIDS2017	CNN+clustering

3 Hybrid Labeling Techniques for Intrusion Detection Systems

3.1 Overview

The proposed system architecture, as depicted in Fig. 1, comprises two phases: the offline phase and the online phase. During the offline phase, the ML model is trained using a pre-collected dataset containing both training and testing data. During the online phase, traffic data observed in the real world is collected periodically. At the end of each period, the proposed semi-supervised learning mechanism is employed to retrain the ML model.

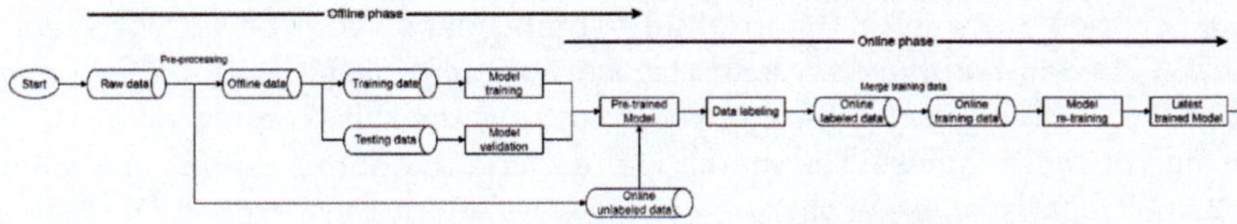

Fig. 1. Overview of system structure

Specifically, during the offline phase, the training data is utilized to train the model, and the testing data is used to validate the model, resulting in a pre-trained model. Moving on to the online phase, the pre-trained model is applied to label the unlabeled data collected from the real world, generating online labeled data. This data is then combined with the original dataset to create a new set of training samples. The model is retrained using this augmented training data. The system continues this process until the ML model is deprecated. At that point, a new ML-based NIDS will be deployed.

3.2 Data Preprocessing

In the current internet environment, there is an enormous amount of network traffic. In most of the literature, the input for the ML-based NIDS is either the statistics of a traffic flow or the raw data of an entire packet. As a result, it is either too late to detect a malicious flow or too time-consuming to process a packet. In this study, we employed the approach proposed in our previous research method [19]. In [19], only the first few bytes of the first few packets from a flow are examined by the ML model, effectively reducing the pre-process complexity and detection delay.

3.3 Deep Learning Model

In addition to excelling in image recognition tasks, CNN (Convolutional Neural Network) models have also found widespread use in recent times for NIDS tasks. A survey and statistical analysis of numerous research papers in this field [20] have shown promising results for CNN models. CNN models utilize automatic feature extraction, enabling them to effectively learn features related to different classes. This ability allows them to achieve excellent recognition capabilities. In this study, we adopt the 1D-CNN model proposed in our previous work [19] for intrusion detection.

3.4 Hybrid Labeling Techniques

We adopt the Self-training method introduced in [1]. The approach involves first using supervised learning for model pre-training. Then, the pre-trained model is utilized to predict unlabeled data, generating pseudo-labels. From these pseudo-labels, we select the most confident predictions as newly labeled data. After merging this new data with the original training data, we proceed with model re-training. As Self-training is a heuristic learning approach, it allows for experimentation with different labeling methods during the self-training process.

During the online phase, the model will undergo model retraining. As shown in Fig. 2, in the early phase of the self-learning, we adopt the "Predict then label" approach to select data with high confidence for data selection. Once a sufficient amount of data is collected, we combine it with the "Clustering then label" method to label and select data. Next, let's introduce the two aforementioned data labeling methods used in the system architecture. The two labeling methods are further elaborated as follows.

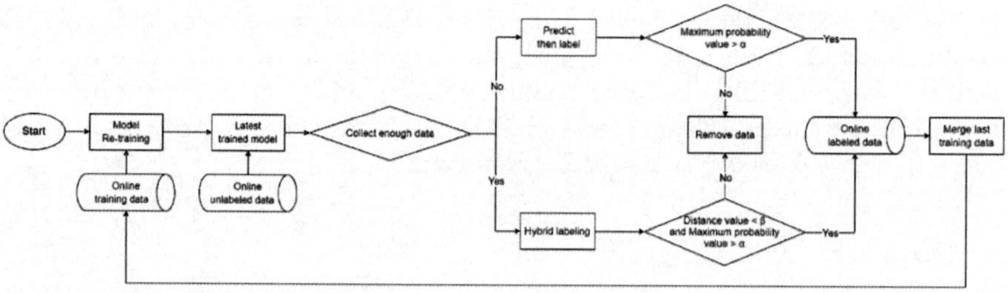

Fig. 2. Proposed architecture for model re-training in the online phase

Predict Then Label: This method utilizes the softmax layer of the CNN model to output floating-point values between 0 and 1, representing probabilities for each class. The data is then labeled with the class that has the maximum probability value, provided it is larger than the pre-defined threshold.

Clustering Then Label: This method begins by extracting class-specific features from the data using the CNN model. Subsequently, it applies K-means clustering to obtain cluster centers for each class, with K representing the number of network attacks. The distances between new data points and the cluster centers are calculated to determine the shortest distance to a cluster center. If the distance is less than a pre-defined threshold, the data point will be labeled with that class.

4 Evaluation

4.1 Experimental Design

This paper uses the CIC-IDS2017 dataset [21] for conducting experiments. The dataset spans 5 days of network traffic and comprises 1 class representing normal behavior and 10 classes corresponding to different malicious attack types. The experiments were conducted separately for binary classifiers and multi-class classifiers.

Tables 2 and 3 show the amount of data for binary classification and multi-category classification, respectively.

Table 2. Amount of data flows in binary class

Categories	# of flows
Benign	40,460
Malicious	40,460

Table 3. Amount of data flows in multi-class (four malicious classes with too few data are deleted)

Categories	**# of flows**
Benign	6743
Bruteforce	6383
DoS	14,000
DDoS	15,695
Botnet	1227
Webattack	1955
Port_Scan	1200

4.2 Supervised Learning

Firstly, the data is preprocessed and divided into an 80% training set and a 20% test set. Secondly, after conducting extensive experiments, we configured the input of the 1D CNN model to consist of the first 100 bytes from the first 3 packets of each flow. With this input configuration, the 1D CNN model achieves an accuracy of over 99.7% for both binary and multi-category classification.

4.3 Evaluate Clustering Methods

In our IDS architecture, we first completed the training of the CNN model. Subsequently, we utilized the model to extract class-specific features from the data and applied K-means clustering on these extracted features.

For both binary and multi-class data, we conducted clustering experiments. The clustering results for binary data are presented in Fig. 3, where we observed that the majority of the data points were correctly labeled in both the training and testing sets.

In the case of the multi-class data clustering experiment, shown in Fig. 4, we did observe a few mislabeled data points in the K-means clustering. However, overall, the results were still satisfactory for our task.

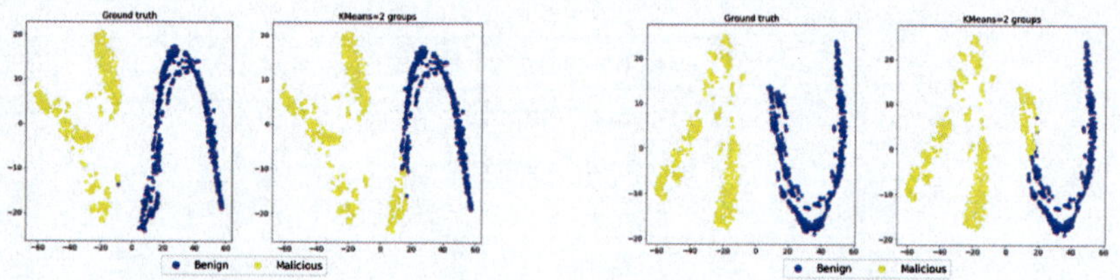

(a) Clustering results for the training set (b) Clustering results for the testing set

Fig. 3. Results of K-means clustering for binary categories

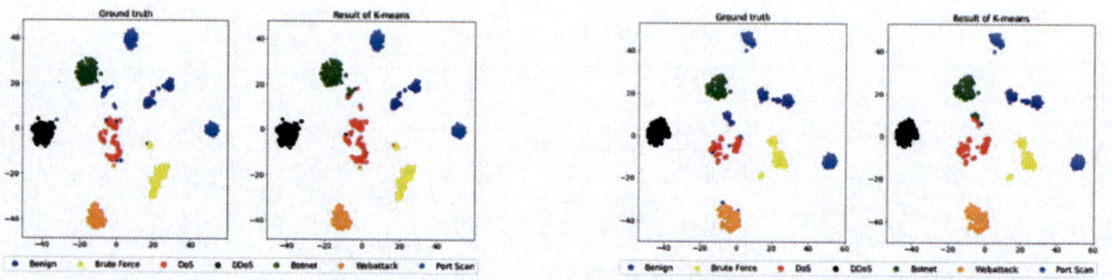

(a) Clustering results for the training set (b) Clustering results for the testing set

Fig. 4. Results of K-means clustering for multi-class

4.4 Performance Comparison: Baseline Versus Proposed Method

In this experiment, we used 1% of the data from the training set as pre-training data to train both binary and multi-class pre-trained models. To simulate the real-world scenario of Self-training, the model requires periodic retraining. Specifically, re-training is conducted when the number of newly labeled data reaches a pre-defined threshold. For the baseline experiment, we solely utilized the "Predict then label" method for data labeling.

The results of the binary classification model training are presented in Fig. 5. The accuracy of each round, except for Loop 0, indicates the accuracy after retraining with the inclusion of 10,000 new labeled data samples in each round. The baseline model attains a stable accuracy, while our proposed method achieves an even higher model accuracy after the final round of retraining. This improvement is attributed to the hybrid labeling method, which achieves a labeling success rate of 99.41% in the final round, surpassing the baseline's rate of 98.56%.

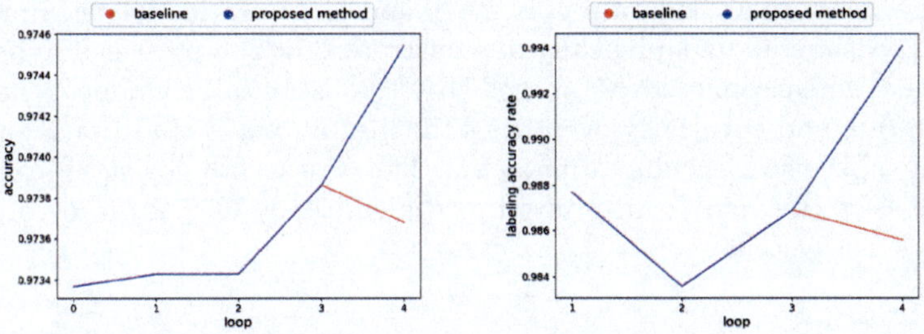

(a) Model accuracy on testing data (b) Labeling accuracy rate

Fig. 5. Performance evaluation of binary classifiers

The multi-class classification results are illustrated in Fig. 6. We observe a steady improvement in the model's accuracy with each round of retraining, and eventually, the accuracy curve stabilizes. Notably, our proposed method outperforms the baseline model in terms of accuracy in the final round. This improvement can be attributed to the success rate of labeling, which reaches 98.38% in the last round, surpassing the baseline's rate of 97.44%.

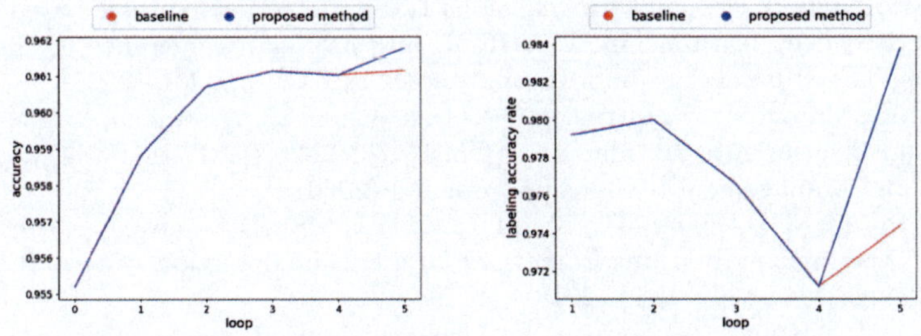

(a) Model accuracy on testing data (b) Labeling accuracy rate

Fig. 6. Performance evaluation of multi-class classifiers

In conclusion, our proposed mixed-labeling method proves to be valuable in data selection, enhancing confidence in the chosen data and enabling the network model to learn more accurate information. As a result, it improves the model's intrusion detection capability.

5 Conclusion

This paper focuses on conducting research and investigation related to ML-based NIDS, with the aim of addressing the challenges encountered by such systems. The study introduces a novel hybrid labeling technique designed to label previously unlabeled data effectively and facilitate model retraining.

The experiments unequivocally demonstrate that the proposed hybrid labeling method, when applied to ML-based NIDS, substantially enhances the model's detection rate and effectively overcomes the limitations of traditional supervised and semi-supervised learning approaches. This research holds significant practical value as it can significantly improve the efficiency and accuracy of cybersecurity defense systems.

References

1. Zhu, X., Goldberg, A.: Introduction to Semi-supervised Learning. Springer Nature (2022)
2. Chen, T., Chen, Y., Lv, M., He, G., Zhu, T., Wang, T., Weng, Z.: A payload based malicious http traffic detection method using transfer semi-supervised learning. Appl. Sci. **11**, 7188 (2021)
3. Khonde, S., Ulagamuthalvi, V.: Ensemble-based semi-supervised learning approach for a distributed intrusion detection system. J. Cyber Secur. Technol. **3**, 163–188 (2019)
4. Zhang, Y., Niu, J., He, G., Zhu, L., Guo, D.: Network intrusion detection based on active semi-supervised learning. In: 2021 51st Annual IEEE/IFIP International Conference on Dependable Systems and Networks Workshops (DSN-W), pp. 129–135 (2021)
5. Lu, W., Ding, Y.: A network malicious traffic detection method based on semi-supervised deep learning. In: 2021 IEEE International Conference on Signal Processing, Communications and Computing (ICSPCC), pp. 1–6 (2021)
6. Hara, K., Shiomoto, K.: Intrusion detection system using semi-supervised learning with adversarial auto-encoder. In: NOMS 2020–2020 IEEE/IFIP Network Operations and Management Symposium, pp. 1–8 (2020)
7. Yao, H., Fu, D., Zhang, P., Li, M., Liu, Y.: MSML: a novel multilevel semi-supervised machine learning framework for intrusion detection system. IEEE Internet Things J. **6**, 1949–1959 (2018)
8. Hyun, M., Jeong, J., Kwak, N.: Class-imbalanced semi-supervised learning. ArXiv:2002.06815 (2020)
9. Kumari, V., Varma, P.: A semi-supervised intrusion detection system using active learning SVM and fuzzy c-means clustering. In: 2017 International Conference on I-SMAC (IoT in Social, Mobile, Analytics and Cloud) (I-SMAC), pp. 481–485 (2017)

10. Lee, J., Kim, J., Choi, M.: SSAE-DeepCNN model for network intrusion detection. In: 2021 22nd Asia-Pacific Network Operations and Management Symposium (APNOMS), pp. 78–83 (2021)
11. Liu, C., Gu, Z., Wang, J.: A hybrid intrusion detection system based on scalable K-means+ random forest and deep learning. IEEE Access **9**, 75729–75740 (2021)
12. Verkerken, M., D'hooge, L., Wauters, T., Volckaert, B., De Turck, F.: Unsupervised machine learning techniques for network intrusion detection on modern data. In: 2020 4th Cyber Security in Networking Conference (CSNet), pp. 1–8 (2020)
13. Cai, S., Han, D., Li, D.: A feedback semi-supervised learning with meta-gradient for intrusion detection. IEEE Syst. J. **17**, 1158–1169 (2022)
14. Viegas, E., Santin, A., Cogo, V., Abreu, V.: A reliable semi-supervised intrusion detection model: one year of network traffic anomalies. In: ICC 2020–2020 IEEE International Conference on Communications (ICC), pp. 1–6 (2020)
15. Din, S., Shao, J., Kumar, J., Ali, W., Liu, J., Ye, Y.: Online reliable semi-supervised learning on evolving data streams. Inf. Sci. **525**, 153–171 (2020)
16. Noorbehbahani, F., Fanian, A., Mousavi, R., Hasannejad, H.: An incremental intrusion detection system using a new semi-supervised stream classification method. Int. J. Commun Syst **30**, e3002 (2017)
17. Das, T., Gangopadhyay, S., Zhou, J.: SSIDS: semi-supervised intrusion detection system by extending the logical analysis of data. ArXiv:2007.10608 (2020)
18. Seo, W., Pak, W.: Real-time network intrusion prevention system based on hybrid machine learning. IEEE Access **9**, 46386–46397 (2021)
19. Hwang, R., Peng, M., Huang, C., Lin, P., Nguyen, V.: An unsupervised deep learning model for early network traffic anomaly detection. IEEE Access **8**, 30387–30399 (2020)
20. Mohammadpour, L., Ling, T., Liew, C., Aryanfar, A.: A survey of CNN-based network intrusion detection. Appl. Sci. **12**, 8162 (2022)
21. Sharafaldin, I., Lashkari, A., Ghorbani, A.: Toward generating a new intrusion detection dataset and intrusion traffic characterization. In: Proceedings of the 4th International Conference on Information Systems Security and Privacy (ICISSP 2018), vol. 1, pp. 108–116 (2018)

Monitor Respiration Rate and Sleep Position Using Multi-task Learning

Hoang-Dieu Vu[1,2,3](✉), Dinh-Hieu Le[4], Khanh-Ly Can[1,2], Hieu Dao-To[1,2,3], Dinh-Dat Pham[1], Sy-Hiep Nguyen[7], Ha-My Nguyen[5], and Duc-Nghia Tran[6]

[1] Phenikaa University, Yen Nghia, Hanoi 12116, Vietnam
{dieu.vuhoang,ly.cankhanh,hieu.daoto}@phenikaa-uni.edu.vn, 20010736@st.phenikaa-uni.edu.vn

[2] PHENIKAA Research and Technology Institute (PRATI), A&A Green Phoenix Group JSC, No. 167 Hoang Ngan, Trung Hoa, Cau Giay, Hanoi 11313, Vietnam

[3] Graduate University of Sciences and Technology, VAST, Hanoi, Vietnam

[4] Hanoi University of Sciences and Technology, Hanoi, Vietnam

[5] Banking Academy of Vietnam, Hanoi, Vietnam

[6] Institute of Information Technology, VAST, Hanoi, Vietnam
nghiatd@ioit.ac.vn

[7] University of Information and Communication Technology, Thai Nguyen, Vietnam
nshiep@ictu.edu.vn

Abstract. The literature strongly supports the significance of gauging breathing rate and analyzing sleep position data as determinants influencing sleep quality. This has given rise to a range of technological solutions which stands out ongoing investigative pursuits directed at wearables have indicated a preference for machine learning models. Yet, the latent potential of more advanced and precise deep learning models remains largely untapped. However, wearables often grapple with limited computational prowess, the adoption of multi-task machine learning becomes imperative. Multi-task learning stands out by enabling shared knowledge utilization across interconnected tasks, optimizing resource allocation. In the context of tasks like breathing rate assessment and sleep posture classification, which share data sources, multi-task models offer an efficient approach for edge devices. Recognizing the potential for improvement, we put forth a system that estimates breathing rate and classifies sleep positions, amalgamating accelerometers, and an LSTM deep learning model in an innovative manner. In this experimentation phase, data was sourced from 10 adolescents who exhibited dynamic breathing patterns across varying sleep postures (left side, right side, supine, prone). Our trial yielded an error rate of ± 2.43 MAE in respiratory rate estimation and a 97.3% accuracy rate in sleep position classification. This methodology holds promise for automated breathing rate estimation and its potential utilization as a home health monitoring device.

Keywords: Multi-task · Sleep posture · Respiratory rate · Accelerometers · Deep learning · Healthcare

P. T. Nghia et al. (Eds.): ICTA 2023, LNNS 847, pp. 86–93, 2023.
https://doi.org/10.1007/978-3-031-49529-8_10

1 Introduction

Recently there has been a great deal of interest, especially among sleep researchers, in sleep positions and postures. Sleep position has been identified as one of the main factors contributing to nightmares [1], sleep apnea [2] and snoring [3]. Other studies have shown that sleeping in the wrong position can cause infants to die prematurely [4]. As a result, identifying sleep positions has become an increasingly important scientific goal. In addition, the development of various techniques for measuring respiratory activity, particularly in mobile settings, has been driven by the need for early detection of potentially harmful conditions such as sleep apnea [5, 6], sudden infant death syndrome [7] and chronic obstructive pulmonary disease [8].

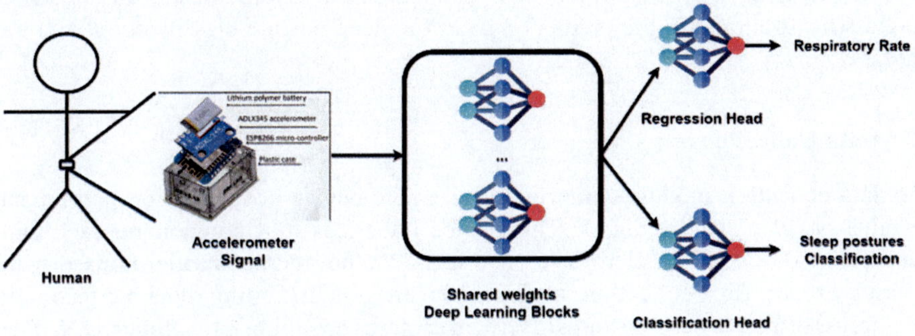

Fig. 1. System's pipeline

In sleep analysis, recent research has focused on sleep posture monitoring, attracting attention from researchers and clinicians. Two approaches stand out: non-contact methods using cameras, Utral-Wireband radar technology (UWB) [9], and wireless devices [10, 11], facing illumination challenges and obstructions, and contact-based techniques with wearables, reducing interference. These contact-based methods [12] employ features like Ballistocardiogram and respiratory signals for posture identification and in-bed pressure sensors, wrist-sensor devices for classification. Acceleration sensors enhance posture and respiratory rate estimation [13]. In respiratory rate measurement, clinical techniques (spirometry, capnometry) coexist with signal-based methods (ECG, PPG). Non-contact solutions offer convenience but are environment-sensitive, while wearables provide cost-effective options for rate estimation.

Nowadays, deep learning models are emerging as strong contenders for downstream tasks due to their automatic feature extraction and enhanced predictive accuracy [14]. While certain deep learning models have been used for posture classification and respiration rate estimation with non-contact devices [15], the potential of deep learning remains untapped in wearable device-based systems and multi-task approaches. Recognizing the existing research gap, our aim is to bridge it with an innovative strategy. Our approach involves utilizing data from an accelerometer sensor-based device for two distinct tasks: sleep posture classification and respiratory rate estimation. To address this multitasking objective, we employ shared LSTM weights. Our contributions encompass:

- Incorporating acceleration sensor, LSTM model, and multi-tasking techniques to estimate respiratory rates and classify sleep postures.
- Applying data augmentation techniques from the audio domain to enrich the dataset.

2 Methodology

2.1 Multi-task Learning

Multi-task learning is widely used across various fields like language processing, healthcare, and more [13]. In healthcare, it's applied to predict health conditions and more, especially with wearable devices [14]. Our focus is on simplifying multi-task learning for respiratory rate estimation and sleep posture classification. We start with training the more challenging task—respiratory rate estimation. Afterward, we freeze the shared-weight structure, modify the prediction head for sleep posture classification, and train that task.

2.2 Data Collection

The data collection module comprises a wearable device positioned on participants' abdomen, as depicted in Fig. 1. This device integrates four components: a Lithium Battery, an ADLK345 accelerometer, and an ESP8266 microcontroller housed within a plastic casing. Data collection involved 10 teenagers exhibiting dynamic respiration rates across different sleep postures. The data captures breathing rates along the X, Y, and Z axes. Each teenager contributed one minute of data for each sleep posture, resulting in a total collection of 160 data files, each containing 3000 samples.

2.3 Data Augmentation

In our experiment, we applied a resampling method for data. For each sub-matrix of 10 samples, we calculated the average values of these samples which allows us to significantly reduce the size of the matrix which stores the information on acceleration data and at the same time makes the signal smoother. Therefore, the length of data decreased from 3000 to 300 samples. In the field of healthcare research, data collection work is very hard. Besides encouraging teenagers to collect data, we also must spend much time guiding and measuring data from them. So, the dataset has 160 files of data, it is quite a small dataset. Therefore, we used some augmentation techniques from the audio domain to enrich our data which are Clipping, Gain, Gain Transition, Reverse, and Shift. Data were generated by augmentation methods which may lose some vital information of data. To reduce this limitation, we applied a find peaks algorithm to validate the augmentative data with the original data. The data augmentations are shown in Fig. 2.

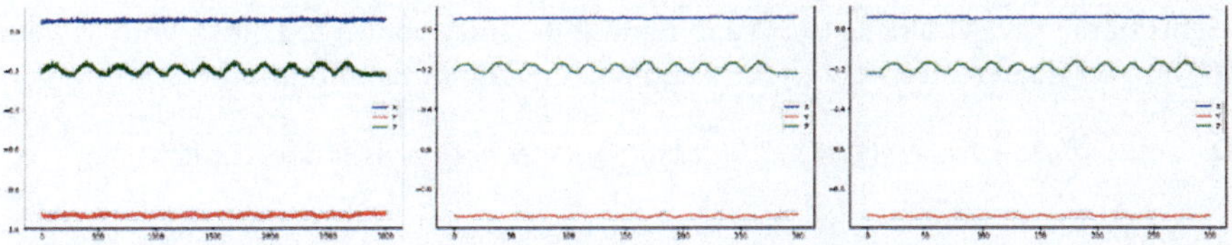

Fig. 2. Raw data, data after resolution reducing, and data after using shift augmentation.

2.4 Modeling

The main architecture employed to tackle this problem is the Long Short Term Memory (LSTM) model – a prevalent architecture for processing sequential data. Within it, the shared module for both sleep posture classification and respiratory rate estimation consists of LSTM blocks and a Dropout layer; the module dedicated solely to respiratory rate estimation features one fully connected (FC) layer; and the module exclusive to sleep posture classification comprises 3 FC layers. The activation functions for the FC layers all utilize the ReLU function, except for the final FC layer of the sleep posture classification module, which employs the Softmax function. The above structure is depicted in Fig. 3.

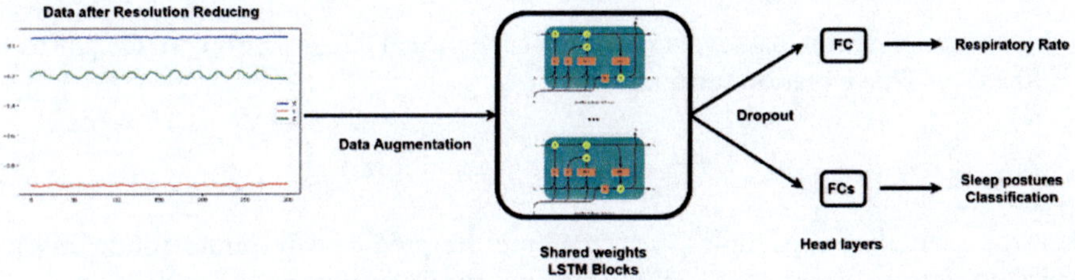

Fig. 3. System's pipeline

We sample x from $X = [x_1, x_2, x_3, \ldots x_n]$, and $X \in R^{N \times D \times S}$, where N is the length of data, D is the number of samples, and S is the number of features that correspond to three axes X, Y, and Z. The initial task involves estimating the respiratory rate. Firstly, the input x is passed through LSTM layers, followed by a dropout layer, before being fed into the final FC layer (called Head layers) to generate prediction results.

$$h_t = LSTM\_blocks(x) \text{Where h is hidden units.}$$

$$o_t = FC\big(Dropout(h_t)\big)$$

$$\widehat{y_t} = ReLU(o_t) = max(0, o_t)$$

where $\widehat{y_t}$ is the output of the FC layer combined with the ReLU activation function. Upon completing the training for the respiratory rate estimation task, we freeze the weights of the LSTM blocks (θ_{reg}) and replace the fully connected block with new ones (*FC_blocks*) dedicated solely to the sleep posture classification task.

$$h_{t\prime} = Freeze(LSTM\_blocks_{\theta_{reg}})(x) \quad \text{where h}\prime \text{ is hidden units.}$$

$$o\prime_t = FC\_blocks\big(Dropout(h\prime_t)\big)$$

$$\widehat{z_t} = Softmax(o\prime_t)$$

RMSProp Optimizer [15] was in used and the model's weights (θ) were updated by the following equation: $\theta_t = \theta_{\{t-1\}} - \alpha \nabla L$.

2.5 Loss Function

In the regression task, we applied the Mean Squared Error (MSE_{loss}) as the loss function in our training process. This equation formulates it:

$$MSE_{loss} = \sum_{i=1}^{N} \left(y_i - \widehat{y_i}\right)^2$$

In the classification task, we applied the Categorical Cross Entropy (CCE_{loss}) as the loss function. This equation formulates it:

$$CCE_{loss} = -\sum_{i=1}^{N} z_i \cdot \log\left(\widehat{z_i}\right)$$

where N is number of samples, $\widehat{y_i}$ and y_i are the predicted respiratory rate and true respiratory rate, $\widehat{z_i}$ and z_i are the predicted and true sleep posture class.

3 Experiments

3.1 Dataset

Data were collected from 10 teenagers with dynamic respiration rates during various sleep postures (lateral left, lateral right, supine, prone, and sit). Each teenager measures data in one minute for each sleep posture. Overall, we amassed a dataset of 160 files, each containing 3000 samples. Subsequently, the dataset was partitioned into three subsets in a 6:2:2 ratio, comprising the training set, validation set, and test set. We applied some augmentation techniques from the audio domain to the dataset to significantly increase data therefore, we had 864 files in the training set, 288 files in the validation set and test set instead of 96 files, and 32 files in the initial training set, validation set and test set respectively.

3.2 Experimental Setup

The shared LSTM blocks is based on 2 LSTM layers with 256 hidden units for each. The dropout we used having the fraction of the input units to drop being 0.25. The FC layer consists of only 1 layer and 1 unit implemented to generate the respiration rates, whereas for the sleep posture classification task, 3 FC layers are utilized with respective units of 256, 128, and the number of classification classes. We used the RMSProp algorithm as optimizer, with a learning rate of 0.001 for 600 epochs and 300 epochs of estimating task and classification task respectively. The Save Best Only model checkpoint was in use with the validation loss monitor to prevent overfitting. During the training phase, we set up 512 samples for a batch size and deployed all of the experiments using Google Colab Tesla T4 GPU. For comparison, we changed the LSTM weights sharing blocks with other deep learning layers:

- **The CNN 1D configuration** consists of 2 convolutional layers, each with 256 filters and a kernel size of 3, and a max pooling layer with a pool size of 2.
- **The Dense setup** involves transforming a 3D input into a 2D format through flattening, followed by two FC (dense) layers with 256 units each, and a dropout layer with a dropout rate of 0.25 placed in between before fed into the Head layers.

4 Results

Table 1 compares the performance of three setups: LSTM, CNN 1D, and Dense weights sharing model on the test set. The LSTM model exhibited the lowest MAE and MSE values, indicating accurate fit to sequential data, along with a balanced F1 score. The CNN 1D model displayed slightly higher MAE, MSE, and Params values, and lower F1 than LSTM. The Dense model indicated the highest MAE and MSE values but achieved a perfect F1 score of 1.0, with lower complexity. In summary, the table underscores LSTM having better practical results in RR estimating and sleep postures classification task.

Table 1. Evaluation multi-task models on Test Set

Model	# MAE	# MSE	#F1	#Params
LSTM	**2.43**	**12.25**	0.973	891,142
CNN 1D	3.64	27.27	0.934	5,036,038
Dense	4.40	31.56	**1.0**	**396,038**

The prediction results on the test set for the respiratory rate estimation task are presented in Fig. 4. Figure 5 depicts the training process of the three models for the classification task, where the Dense model demonstrates strong classification characteristics, easily achieving optimal results, while the LSTM model requires more epochs to converge.

Table 2 illustrates that LSTM multi-task and LSTM single-task share the same MAE because for the multi-task approach, we first train on the RR estimation task, then

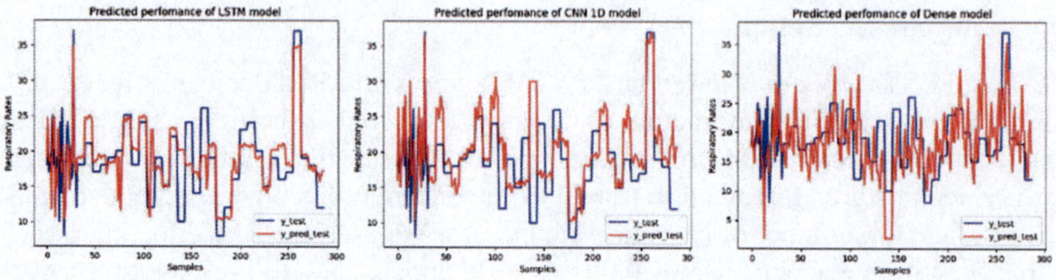

Fig. 4. Prediction performance of each deep learning weights sharing model on Test Set

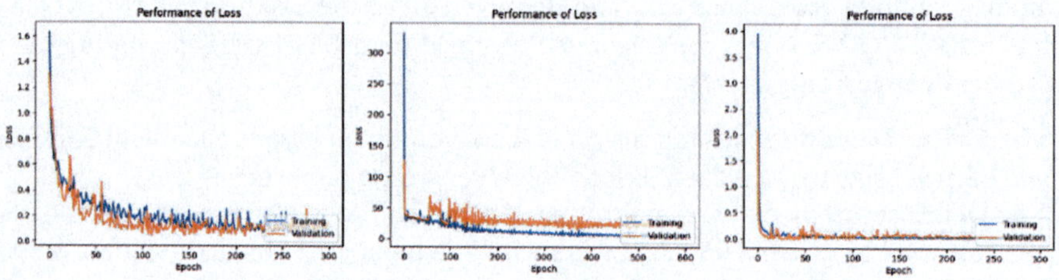

Fig. 5. Training performance of Loss of each deep learning model

Table 2. Compare Multi-task and Single-task, and fine-tune all weights model's performance

Model	# MAE	# Accuracy	# Total Params
LSTM multi-task	**2.43**	0.976	**891,142**
LSTM all-weights	6.77	1.0	891,142
LSTM single-task	2.43	**1.0**	1,682,694

freeze the weights of the shared LSTM network block while simultaneously employing a separate FC block for sleep posture classification. The accuracy of the single-task model reaches 100%, albeit with nearly double the parameter load. In Table 2, we also present the outcomes of fine-tuning the entire model's weights (LSTM all-weights) on the sleep posture classification task, followed by reassessing the model's performance on the respiratory rate estimation task. The experiment demonstrates a notable decrease in performance for the respiratory rate estimation task. With a focus on edge devices, LSTM multi-task demonstrates more suitable performance.

5 Conclusion

This study has proposed a multi-task processing system that combines accelerometer data with an LSTM model, yielding practical results for both respiratory rate estimation (MAE of 2.43) and sleep posture classification (97.3% accuracy). Additionally, we have successfully adapted data augmentation methods from the speech processing domain to this field. In the upcoming phases, we plan to optimize the outcomes of the respiratory rate estimation task to provide more accurate prediction values.

References

1. Agargun, M. Y., Boysan, M., Hanoglu, L.: Sleeping position, dream emotions, and subjective sleep quality. Sleep Hypn. **6**, 8–13 (2004)
2. Akahoshi, R., Yanamoto, S., Sakamoto, Y., Naruse, T., Hayashida, S., Soutome, S., .., Umeda, M.: Effects of sleep position on the treatment response of oral appliance for obstructive sleep apnea. CRANIO® **40**(3), 262–267 (2022)
3. Benoist, L.B., Morong, S., van Maanen, J.P., Hilgevoord, A.A., de Vries, N.: Influence of sleep position on snoring. Positional Therapy in Obstructive Sleep Apnea, pp. 143–150 (2015)
4. Shipstone, R., Thompson, J.M., Young, J., Byard, R.W.: The use of post-mortem lividity to determine sleep position in sudden unexpected deaths in infancy. Acta Paediatr. **109**(6), 1162–1165 (2020)
5. Younes, M.: Role of respiratory control mechanisms in the pathogenesis of obstructive sleep disorders. J. Appl. Physiol. **105**(5), 1389–1405 (2008)
6. Hung, P.D.: Estimating respiration rate using an accelerometer sensor. In: Proceedings of the 8th International Conference on Computational Systems-Biology and Bioinformatics, pp. 11–14 (2017)
7. Rantonen, T., Jalonen, J., Grönlund, J., Antila, K., Southall, D., Välimäki, I.: Increased amplitude modulation of continuous respiration precedes sudden infant death syndrome: detection by spectral estimation of respirogram. Early Human Dev. **53**(1), 53–63 (1998)
8. Hasselgren, M., Arne, M., Lindahl, A., Janson, S., Lundbäck, B.: Estimated prevalences of respiratory symptoms, asthma and chronic obstructive pulmonary disease related to detection rate in primary health care. Scand. J. Prim. Health Care **19**(1), 54–57 (2001)
9. Piriyajitakonkij, M., Warin, P., Lakhan, P., Leelaarporn, P., Kumchaiseemak, N., Suwajanakorn, S., ... & Wilaiprasitporn, T.: SleepPoseNet: Multi-view learning for sleep postural transition recognition using UWB. IEEE J. Biomed. Health Inform. 25(4), pp. 1305-1314 (2020).
10. Roshini, A., Kiran, K.V.D.: An enhanced posture prediction-Bayesian network algorithm for sleep posture recognition in Wireless Body Area networks. Int. J. Telemed. Appl. (2022)
11. Yue, S., Yang, Y., Wang, H., Rahul, H., Katabi, D.: BodyCompass: Monitoring sleep posture with wireless signals. In: Proceedings of the ACM on Interactive, Mobile, Wearable and Ubiquitous Technologies, vol. 4, no. 2, pp. 1–25 (2020)
12. Xing, Z., Gao, W., & Chuai, G, "Research on sleeping position recognition algorithm based on human body vibration signal," In *2022 IEEE 2nd International Conference on Power, Electronics and Computer Applications (ICPECA)*, pp. 403–406 (2022)
13. Vu, H.D., Tran, D.N., Can, K.L., Dao, T.H., Pham, D.D., Tran, D.T: Enhancing sleep postures classification by incorporating acceleration sensor and LSTM model. In: 2023 IEEE Statistical Signal Processing Workshop (SSP), pp. 661–665 (2023)
14. Yang, Y., Wu, Z., Yang, Y., Lian, S., Guo, F., Wang, Z.: A survey of information extraction based on deep learning. App. Sci. **2**(19), 9691 (2022)
15. Brieva, J., Ponce, H., Moya-Albor, E.: Non-contact breathing rate estimation using machine learning with an optimized architecture. Mathematics **11**(3), 645 (2023)
16. Thung, K.H., Wee, C.Y.: A brief review on multi-task learning. Multimed. Tools Appl. **77**, 29705–29725 (2018)
17. Jaques, N., Taylor, S., Nosakhare, E., Sano, A., Picard, R.: Multi-task learning for predicting health, stress, and happiness. In: NIPS Workshop on Machine Learning for Healthcare (2016)
18. Tieleman, T., Hinton, G.: Rmsprop, Divide the gradient by a running average of its recent magnitude. coursera: neural networks for machine learning. COURSERA Neural Netw. Mach. Learn. **17** (2012)

Predicting Drug-Disease Associations Based on Integrated Similarities and Weighted Bi-level Network

Van Tinh Nguyen(✉), Huy Vu Duc, Yen Vu Minh, and Lan Nguyen Thi Huong

Faculty of Information Technology, Hanoi University of Industry, Hanoi, Vietnam
nguyenvantinh_cntt@haui.edu.vn, {huyvd,yenvm,lannth}@fit-haui.edu.vn

Abstract. It is a time-consuming, laborious, and costly process to develop new drugs. Consequently, developing computational methods for drug repositioning by predicting drug-disease associations has become urgent and has attracted a lot of attention from researchers. In this study, a new approach for predicting drug-disease associations was presented. The proposed method integrates multiple similarities to leverage the influence of multiple similarities from different perspectives. It solves the sparsity data in the set of associations between drugs and diseases by using a WKNKN algorithm. And it predicts drug-disease associations using a weighted bi-level network that each drug-disease pair is attached two information properties and information of association types in line with the strategy of information distribution from two different directions. Our proposed method shows noticeable performance, reaching the greatest AUC value of 0.9844 and AUPR value of 0.7672. Compared to other baseline methods on the same Cdataset, our proposed method shows outperformance in the two AUC and AUPR measurements. Hence, it is considerable as a valuable tool for drug-disease association prediction.

Keywords: Drug-disease association prediction · Multiple similarities · Weighted bi-level network · Weighted K nearest known neighbors

1 Introduction

Drugs are known as chemicals that affect human diseases' diagnosis, treatment, curation and prevention [1–3]. They may play essential roles in the disease's etiology. The events of drugs exerting effects on diseases indicate that a drug is related with a disease. It is crucial to identify associations between drugs and diseases for drug discovery and drug repositioning [1]. Identifying drug-disease associations by wet experiments needs a long period of time and a huge cost. It can take approximately 15 years and $800 to $1000 millions to bring a new single drug to market [4]. Therefore, it is needed to build up computational methods to identify drug-disease associations for saving time and money. Recently, various computational methods for predicting associations between drugs and diseases have been proposed and submitted by many researchers. They can roughly be branched into network-based, text mining-based, machine learning-based, and also multi-model and multi-source integration-based methods as follows.

P. T. Nghia et al. (Eds.): ICTA 2023, LNNS 847, pp. 94–106, 2023.
https://doi.org/10.1007/978-3-031-49529-8_11

Firstly, network-based methods have widely applied for predicting relationships between drugs and diseases which relied on the constructing biological networks from heterogeneous data sources idea. For examples, Zhang et al. [1] proposed a method called NTSIM to predict unobserved associations between drugs and diseases by using a label propagation algorithm on a bipartite network. Zhu et al. [5] predicted drug-disease associations with a transductive regression model with graph regularization by constructing a three layer heterogeneous networks from multiple data sources including drugs, diseases and targets information. The network-based methods for predicting drug-disease associations have advantages of good intepretability [6]. Howerver, they often show a convinced preference for drugs which have more known associations between drugs and diseases. Therefore, they can not effectively predict novel associations for drugs that have less known associations or a isolated drug [7].

Secondly, text mining-based methods which identified associations between drugs and diseases from literature and analyzed biological elements shared by diseases and drugs. The methods in this category normally gather identified associations between drugs and diseases from literature. For example, Yu et al. [4] studied drug-disease associations by using common protein complexes associated to drugs and diseases. Wang et al. [8] used drug-gene-disease relations to infer drug-disease associations. This kind of methods can not predict novel associations. Besides that, it is the fact that drugs and diseases rarely share common elements, so these methods' capability is undermined [2].

Thirdly, in recent years, machine learning-based methods have been deployed to forecast associations between drugs and diseases based on their capability of handling complex data. For instance, a SCMFDD method predicted drug-disease associations using a similarity constrained matrix factorization model in which drug-disease associations were projected into two low-rank spaces [3]. Wang et al. [9] introduced a neural network based method named as HNDR to predict drug-disease associations by integrating neighborhood information in heterogeneous network. Meng et al. [10] proposed a method called DRWBNCF to reveal potential drug-disease associations using neighborhood information along with neighborhood interaction based neural collaborative filtering approach. As a part of machine learning-based methods, recently deep learning-based methods have also been widely used to infer drug-disease relationships. For example, Yu et al. [11] predicted drug-disease through a layer attention graph convolutional network. As mentioned before, machine learning-based methods can handle complicated data. However, they have to use different features that are difficult to be available for all diseases and drugs.

And lastly, the combination of multi-model and multi-type of information have been carried out to leverage the advantages of methods in each category. For examples, a model called SNF-NN was proposed by Jarada et al. [12] to predict drug-disease associations by fusing multiple similarities and a highly tuned new neural network model. Combining a sparse auto encoder model and rotation forest in a new model, Jiang et al. [13] introduced a new method to reveal associations between drugs and diseases.

As aforementioned, each category of computational approaches for inferring drug-disease associations contains their own strengths and weaknesses. However, most computational methods have to deal with the sparse data problem because of the fact that the known associations' quantity is very limitted in comparing with the quantity of possible associations among them. To solve this issue, in recent years, some techniques such as collaborative filtering (CF) and Weighted K-nearest known neighbors (WKNKN) have been employed in different works in different fields to reduce the impact of data sparsity [10, 14–17].

In this paper, a new approach is proposed for inferring associations between drugs and diseases. It was inspired by the success a weighted bi-level network in the BLNIMDA method proposed by Shang et al. [18] in predicting associations between micro RNAs and diseases, as well as the efficiency of WKNKN algorithm in solving data sparsity as mentioned before. The proposed method attained a remarkable performance in predicting drug-disease associations with the best averaged AUC is 0.9844 and the best averaged AUPR is 0.7672 under 5-fold cross experiments on CDataset. It outperforms other baseline approaches in AUC and AUPR values. Addionally, some case studies were implemented to support the reliability of prediction performance. Our proposed method could be a considerable tool for drug-disease asscociation prediction.

2 Materials and Method

2.1 Materials

In this paper, we used the benchmarked Fdataset and Cdataset to evaluate prediction performance. The Fdataset already used in many studies including the works of Liu et al., Wang et al. and Meng et al. [6, 9, 10]. The Cdataset were also used in different studies including the studies of Liu et al. [6] and Meng et al. [10]. The Fdataset contains 1933 known associations between 593 drugs and 313 diseases. Drugs were gathered from DrugBank [19] database. Diseases were collected from OMIM [20] database. The Cdataset includes 2532 known associations between 663 drugs obtained from DrugBank database and 409 diseases from OMIM database. The known drug-disese association set is represented as an adjacency matrix $A^{DrD} \in R^{n_r x n_d}$ where each element $A^{DrD}(i,j) \in \{0, 1\}$, n_r means the quantity of drugs whereas n_d indicates the number of diseases. $A^{DrD}(i,j) = 1$ if drug dr_i has a verified association with disease d_j, otherwise $A^{DrD}(i,j) = 0$.

Besides the known drug-disease associations, these datasets also contain drug-drug similarity as well as disease-disease similarity information. The drug-drug similarity was computed relying on the chemical structure of SMILES format [21]. It is presented as Chemical fingerprints' Tanimoto index of drug pair via Chemical Development Kit [22]. The drug-drug similarity is denoted as a matrix $DrSim(dr_i, dr_j)$ with values in range [0, 1].

The disease-disease similarity is based on disease-disease similarity dataset from MimMiner [23]. It computed disease-disease similarity relied on the MeSH terms in the diseases' medical descriptions. The disease-disease similarity is represented by a matrix $DiSim(d_i, d_j)$ with values in range [0, 1].

2.2 Method

The proposed method, in this study, can be summarized as in Fig. 1. Generally, it contains the stages as described below.

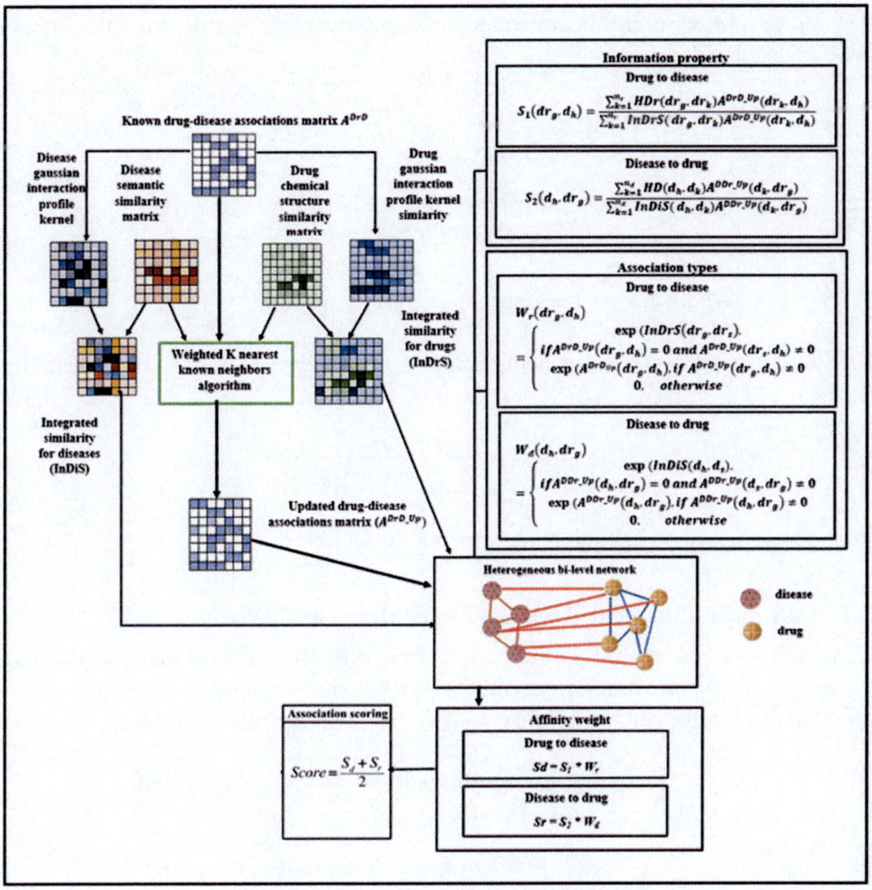

Fig. 1. Flowchart of proposed method.

2.2.1 Calculating Gaussian Interaction Profile Kernel Similarity for Drugs and Diseases

For each drug and disease, different similarity's measurement may differently affect on the contribution to the learning ability of the prediction model. Therefore, in this work, we also considered Gaussian interaction profile (GIP) kernal similarity for both drugs and diseases besides of chemical structure similarity for drugs and disease semantic similarity for diseases. Its objective is to leverage the nodes' topological relation networks. Assuming that the ith row of the A^{DrD} matrix represents the vector related to drug dr_i in A^{DrD}, denoted as $A^{DrD}(dr_i)$. Similarly, the jth column of A^{DrD} matrix represents the vector related to disease d_j in A^{DrD}, denoted as $A^{DrD}(d_j)$. According to [24], we computed GIP kernel similarity for drugs as well as for diseases as below.

For drugs dr_i and dr_j, GIP kernel similarity is as:

$$GIP_{drSim}(dr_i, dr_j) = \exp\left(-\gamma_{drSim} \|A^{DrD}(dr_i) - A^{DrD}(dr_j)\|^2\right) \tag{1}$$

where γ_{drSim} represents an adjustment parameter of the kernel bandwith. This parameter can be updated as:

$$\gamma_{drSim} = \frac{\gamma_{dr'}}{\frac{1}{n_r}\sum_{i=1}^{n_r}\|A^{DrD}(dr_i)\|^2} \tag{2}$$

here γ'_{dr} is assigned to 1 in line with previous study [24].

For diseases d_i and d_j, GIP kernel similarity is as:

$$GIP_{diSim}(d_i, d_j) = \exp\left(-\gamma_{diSim}\|A^{DrD}(d_i) - A^{DrD}(d_j)\|^2\right) \tag{3}$$

where γ_{diSim} represents an adjustment parameter of the kernel bandwith. It can be updated as:

$$\gamma_{diSim} = \frac{\gamma_{d'}}{\frac{1}{n_d}\sum_{i=1}^{n_d}\|A^{DrD}(d_i)\|^2} \tag{4}$$

here γ'_d is assigned to 1 in line with previous study [24].

2.2.2 Calculating Integrated Similarity for Drugs and Diseases

To leverage influence of multiple similarities from different points of view, we integrated drug chemical structure similarity with drug GIP kernel similarity. Simultaneously, we integrated disease semantic similarity with disease GIP kernel similarity as:

$$InDrS(dr_i, dr_j) = \begin{cases} DrSim(dr_i, dr_j)\ if\ DrSim(dr_i, dr_j) > 0; \\ GIP_{drSim}(dr_i, dr_j),\ otherwise; \end{cases} \tag{5}$$

$$InDiS(d_i, d_j) = \begin{cases} DiSim(d_i, d_j)\ if\ DiSim(d_i, d_j) > 0; \\ GIP_{diSim}(d_i, d_j),\ otherwise; \end{cases} \tag{6}$$

2.2.3 Performing a WKNKN Algorithm to Attain Updated Drug-Disease Associations Matrix

To eliminate sparsity data's impact on drug-disease association set, we employed a WKNKN algorithm which previously used in other works including the works of Ezzat et al. [16] and Nguyen et al. [17]. This process is done relying on the information of known neighbors with presumption that the unverified drug-disease pairs which have values of 0 in A^{DrD} matrix could be potentially accurate associations. And the associations among neighboring nodes may reveal latent associations among them to imply richer semantics. In more details, WKNKN algorithm replaces the element $A^{DrD}(i, j) = 0$ with a continuous probability value in range [0, 1] as below.

First, for each drug dr_i, we determined the drug chemical structure similarity of K drugs which are nearest to the drug dr_i to compute the interaction likelihood profile for drug dr_i. Second, for each disease d_j, we took the disease semantic similarity of K diseases which are nearest to disease d_j to measure the interaction likelihood profile for disease d_j. Next, if $A^{DrD}(i, j) = 0$, it is replaced by averaging the two interaction profile values. And lastly, a threshold was used to label the values into two classes as 0 and 1 to have the updated drug-disease associations $A^{DrD_{Up}}$ matrix.

The below Algorithm 1 summarized above actions in pseudocode. In this algorithm, r means a decay term, $r \leq 1$, θ stand for a threshold and was set to 0.2. The *KNN()* turns back, in descending order, the K-nearest known neighbors of their similarities to drug dr_i or disease d_j.

Algorithm 1: WKNKN algorithm to obtain updated drug-disease associations matrix.

```
Inputs: Matrices A^{DrD} ∈ R^{n_r x n_d}, DrSim ∈ R^{n_r x n_r} and
    DiSim ∈ R^{n_d x n_d}, neighborhood size K, decay term r,
    threshold θ.
Outputs: updated drug-disease associations matrix
    A^{DrD_Up}.
begin
    ADr = ADi = 0 # initialize two temporary mactrices ;
    # for drugs
    for l = 1 to n do
        drNN=KNN(l, DrSim, K);
        for i = 1 to K do
            w_i = r^{i-1} DrSim(l, drNN_i);
        end
        Z_l = Σ_{i=1}^{K} DrSim(l, drNN_i);
        ADr(l) = (1/Z_l) Σ_{i=1}^{K} w_i A^{DrD}(drNN_i);
    end
    # for diseases
    for d = 1 to m do
        dNN=KNN(d, DiSim, K);
        for i = 1 to K do
            w_i = r^{i-1} DiSim(d, dNN_i);
        end
        Z_d = Σ_{i=1}^{K} DiSim(d, dNN_i);
        ADi(d) = (1/Z_d) Σ_{i=1}^{K} w_i A^{DrD}(dNN_i);
    end
    ADrD = (ADr+ADi)/2;
    A^{DrD_Up} = max(A^{DrD}, ADrD);
    # classify A^{DrD_Up} by threshold θ.
    for i = 1 to n do
        for j = 1 to m do
            if A^{DrD_Up} ≥ θ then
                A^{DrD_Up}[i, j] = 1;
            else
                A^{DrD_Up}[i, j] = 0;
            end
        end
    end
    return A^{DrD_Up};
end
```

2.2.4 Generation of Two Side Information Properties Through Constructing of Bidirectional Information and Assigning All Drug-Disease Associations into Three Association Types

On the basis of the hepothesis that similar drugs are more likely to be related with similar diseases [25] and the integration of multiple integrated similarities and updated drug-disease association set, we generated each drug-disease association pair's two side information properties through bidirectional information construction as follows:

For drugs to diseases direction, the information property for each drug-disease association is defined as:

$$S_1(dr_g, d_h) = \frac{\sum_{k=1}^{n_r} HDr(dr_g, dr_k)A^{DrD\_Up}(dr_k, d_h)}{\sum_{k=1}^{n_r} InDrS(dr_g, dr_k)A^{DrD\_Up}(dr_k, d_h)} \tag{7}$$

where

$$HDr(dr_g, dr_k) = \begin{cases} InDrS(dr_g, dr_k)\ if\ InDrS(dr_g, dr_k) \geq T, T\ is\ a\ threshold \\ 0,\ otherwise \end{cases} \tag{8}$$

considering that the weak similarity nodes also influence the prediction performance, we set a threshold parameter T to eliminate the nodes which have weak similarity. Bases on this threshold parameter T, drug-disease associations are divided into three association types: strong, potential and no associations. These types of associations in drugs to disease direction are shown in the below formular:

$$W_r(dr_g, d_h) = \begin{cases} exp(InDrS(dr_g, dr_s),\ if\ A^{DrD\_Up}(dr_g, d_h) = 0\ and\ A^{DrD\_Up}(dr_s, d_h) \neq 0 \\ exp(A^{DrD\_Up}(dr_g, d_h),\ if\ A^{DrD\_Up}(dr_g, d_h) \neq 0 \\ 0,\ otherwise \end{cases} \tag{9}$$

where dr_s is the drug with greatest similarity to dr_g.

Similarly, for diseases to drugs direction, the information property for each drug-disease association is defined as:

$$S_2(d_h, dr_g) = \frac{\sum_{k=1}^{n_d} HD(d_h, d_k)A^{DDr\_Up}(d_k, dr_g)}{\sum_{k=1}^{n_d} InDiS(d_h, d_k)A^{DDr\_Up}(d_k, dr_g)} \tag{10}$$

where A^{DDr_Up} is transposition of the A^{DrD_Up} matrix, and

$$HD(d_h, d_k) = \begin{cases} InDiS(d_h, d_k),\ if\ InDiS(d_h, d_k) \geq T, T\ is\ a\ threshold \\ 0,\ otherwise \end{cases} \tag{11}$$

In addition, the types of associations from in diseases to drugs direction are shown as:

$$W_d(d_h, dr_g) = \begin{cases} exp(InDiS(d_h, d_s),\ if\ A^{DDr\_Up}(d_h, dr_g) = 0\ and\ A^{DDr\_Up}(d_s, dr_g) \neq 0 \\ exp(A^{DDr\_Up}(d_h, dr_g),\ if\ A^{DDr\_Up}(d_h, dr_g) \neq 0 \\ 0,\ otherwise \end{cases} \tag{12}$$

2.2.5 Two Affinity Weights' Estimation for Each Drug-Disease Association Through Bidirectional Information Construction and its Association Type, and Then Averaged as the Final Drug-Disease Association Score

For drugs to diseases direction, the affinity weight of each drug-disease association is defined through a bidirectional information construction strategy and its association types according to formula:

$$S_d = S_1 * W_r \tag{13}$$

Similarly, for diseases to drugs direction, the affinity weight of each drug-disease association is defined as:

$$S_r = S_2 * W_d \tag{14}$$

And the final prediction score of each drug-disease association is defined as:

$$Score = \frac{S_d + S_r}{2} \tag{15}$$

3 Experimental Results

3.1 Performance Evaluation

To assess the method's performance prediction, we already empolyed 5-fold cross validation experiments, and measured AUC as well as AUPR values of proposed method on two Fdataset and Cdataset as previously mentioned. We already computed TP (True possitive), FP (False positive), TN (True negative) and FN (False negative) values of predicted results in each running time of experiments where TP and TN reflect the positive samples and negative samples that were accurately predicted, respectively, whereas TN and FN indicate the positive and negative samples that were incorrectly inferred. Then we calculated TPR (True positive rate), FPR (False positive rate), Precision and Recall values as in the followings:

$$TPR = \frac{TP}{TP + FN} \tag{16}$$

$$FPR = \frac{FP}{FP + TN} \tag{17}$$

$$Precision = \frac{TP}{TP + FP} \tag{18}$$

$$Recall = \frac{TP}{TP + FN} \tag{19}$$

then we drew the curves of ROC [26] and Precision-Recall (P-R) [27] and measured the AUC and AUPR values of proposed method in each running time.

We again and again altered the values of K nearest known neighbors, decay factor r and threshold T parameters. The proposed method achieves best performance on Cdataset when K is equal to 5, r is equal to 0.7 and T is equal to 0.05. Figure 2. Shows the prediction performance of the proposed method in AUC and AUPR under 5-fold cross validation on Cdataset when K range from 3 to 7, r is set to 0.7 and T is set to 0.05.

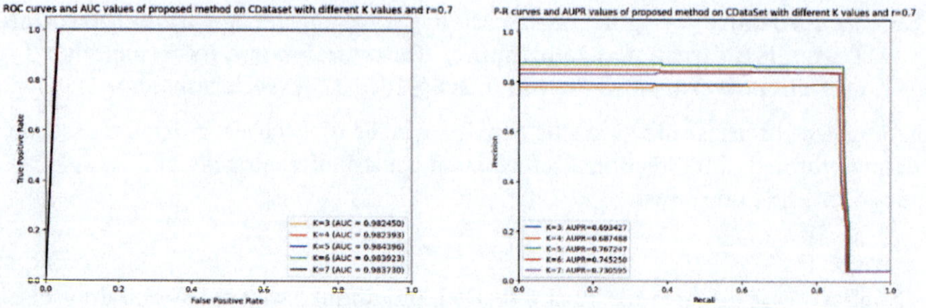

Fig.2. ROC curves, AUC values, P-R curves and AUPR values of proposed method on Cdataset with different K values, r is 0.7 and *T* is 0.05.

3.2 Ablation Study

In this study, to demonstrate the impact of WKNKN algorithm on solving sparsity data problem and integration of multiple similarities, we already performed some ablation study cases. The Fig. 3 illustrates the ROC and P-R curves along with AUC and AUPR values of proposed method and these ablation study cases. As can be seen, the proposed method achieves highest values in both AUC and AUPR metrics in all different cases.

3.3 Comparison with other Methods

To indicate the outperformance of our method in comparing to other relevant methods, we used the results provided by authors in line with their corresponded papers [6, 10, 28] which used the same Cdataset with our method.

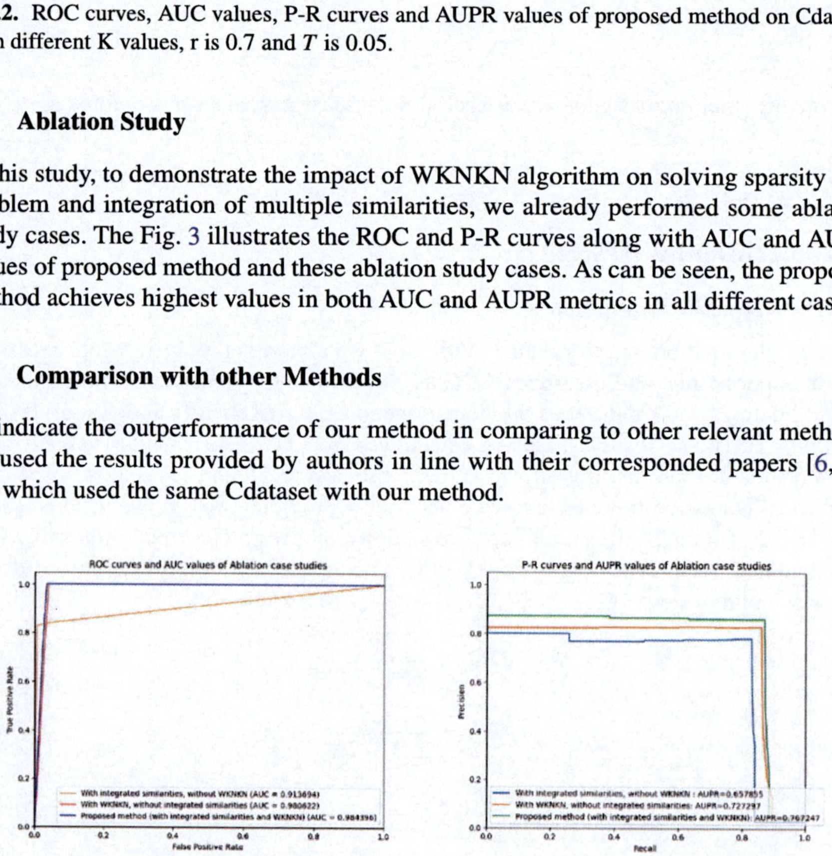

Fig. 3. ROC curves and AUC values, P-R curves and AUPR values of proposed method and some Ablation study cases on Cdataset.

As demonstrated in Table 1, our method is superior to all other related methods in both AUC and AUPR measurements. Therefore, it could be a valuable tool to predict drug-disease associations.

Table 1. Drug-disease association prediction performance in comparison with other methods on the same Cdataset

Method	AUC	AUPR
GLGMPNN [6]	0.957	0.599
DRWBNCF [10]	0.9413	0.5663
WNMFDDA [28]	0.953	–
Our method	**0.9844**	**0.7672**

3.4 Case Studies

To increase the reliablility of our method performance in predicting associations between drugs and diseases, we already selected two popular diseases **including Obesity (D601665)** and **Breast Cancer (D114480)** as case studies.

3.4.1 Obesity Case Study

As can be known, Obesity is regarded as a major public health issue and the fifth leading cause of death worldwide [29]. Identifying asociations between Obesity and drugs can provide clinical indications for prevention and treatment. Therefore, we choose Obesity as one case study to support the prediction performance of our proposed method. As can be seen in Table 2, in top 10 predicted Obesity-related drugs inferred by our method, there are 8 associations have been confirmed in Comparative Toxicogenomics Database (CTD).

Table 2. Top 10 predicted obesity-related drugs inferred by our method.

Rank	DrugID	Drugname	Evidence
1	DB00865	Benzphetamine	CTD database
2	DB00937	Diethylpropion	CTD database
3	DB00945	Aspirin	CTD database
4	DB01104	Sertraline	CTD database
5	DB00983	Formoterol	Not confirmed
6	DB01110	Miconazole	CTD database
7	DB00181	Baclofen	CTD database
8	DB00938	Salmeterol	CTD database
9	DB01012	Cinacalcet	Not confirmed
10	DB00182	Amphetamine	CTD database

3.4.2 Breast Cancer Case Study

Breast Cancer is one of the most cancer determined in women worldwide [30]. In 2020, there are more than 2 million new identified cases of breast cancer. Identifying and approving drugs for breast cancer is costly and time consuming. It is the reason that we choose Breast cancer as a case study to support our method's prediction performance. Table 3 shows top 10 predicted Breast Cancer-associated drugs inferred by our method. One again, 8 out of top 10 predicted associations have been confirmed in CTD database.

Table 3. Top 10 predicted Breast Cancer-related drugs inferred by our method.

Rank	DrugID	Drugname	Evidence
1	DB00104	Octreotide	CTD database
2	DB00158	Folic acid	CTD database
3	DB00253	Medrysone	Not confirmed
4	DB00255	Diethylstilbestrol	CTD database
5	DB00257	Clotrimazole	CTD database
6	DB00282	Pamidronic acid	CTD database
7	DB00351	Megestrol acetate	CTD database
8	DB00367	Levonorgestrel	CTD database
9	DB00374	Treprostinil	Not confirmed
10	DB00396	Progesterone	CTD database

4 Conclusion and Discussions

Identifying drug-disease associations is crucial for comprehensively understanding mechanisms and functions of drugs. However, it is laborious, time-consuming and costly to develop a new drug for a disease. Therefore, it is needed to develop computational methods for drug repositioning. In this paper, we proposed a new computational method to predict associations between drugs and diseases. It offers some contributions as follows. Firstly, integrating multiple sources of information and solving the sparsity data using a WKNKN algorithm can help to improve prediction performance. Secondly, by considering the information property as well as dividing associations into different types, our method can take into account the associations following both of directions and removing weak associations. Finally, by averaging the contribution score in both directions, the prediction results become more impressive. However, in the future, it is needed to incorporate additional information from different sources including information of side effects and other types of similarity to improve prediction performance and increase the reliability of prediction results.

References

1. Zhang, W. et al.: Predicting drug-disease associations based on the known association bipartite network. IEEE Int. Conf. Bioinforma. Biomed. (2017)
2. Zhang, W., Yue, X., Huang, F., Liu, R., Chen, Y., Ruan, C.: Predicting drug-disease associations and their therapeutic function based on the drug-disease association bipartite network. Methods **145**(2018), 51–59 (2018)
3. Zhang, W., et al.: Predicting drug-disease associations by using similarity constrained matrix factorization. BMC Bioinform. **19**(1), 1–12 (2018)
4. Yu, L., Huang, J., Ma, Z., Zhang, J., Zou, Y., Gao, L.: Inferring drug-disease associations based on known protein complexes BMC Med. Genomics **8**(Suppl 2), S2 (2015)
5. Zhu, Q., Luo, J., Ding, P., Xiao, Q.: GRTR: Drug-disease association prediction based on graph regularized transductive regression on heterogeneous network. In: Lecture Notes in Computer Science, pp. 13–25 (2018)
6. Liu, B. M. et al.: A new framework for drug-disease association prediction combing light-gated message passing neural network and gated fusion mechanism. Brief. Bioinform. **23**(6) (2022)
7. Zhou, R., Lu, Z., Luo, H., Xiang, J., Zeng, M., Li, M.: NEDD: a network embedding based method for predicting drug-disease associations. BMC Bioinform. **21**(Suppl 13), 1–12 (2020)
8. Wang, L., Wang, Y., Hu, Q., Li, S.: Systematic analysis of new drug indications by drug-gene-disease coherent subnetworks. CPT Pharmacometrics Syst. Pharmacol. **3**(11) (2014)
9. Wang, Y., Deng, G., Zeng, N., Song, X., Zhuang, Y.: Drug-disease association prediction based on neighborhood information aggregation in neural networks. IEEE Access **7**, 50581–50587 (2019)
10. Meng, Y., Lu, C., Jin, M., Xu, J., Zeng, X., Yang, J.: A weighted bilinear neural collaborative filtering approach for drug repositioning. Brief. Bioinform. **23**(2) (2022)
11. Yu, Z., Huang, F., Zhao, X., Xiao, W., Zhang, W.: Predicting drug-disease associations through layer attention graph convolutional network. Brief. Bioinform.Bioinform. **22**(4), 1–11 (2021)
12. Jarada, T. N., Rokne, J. G., Alhajj, R.: SNF - NN : computational method to predict drug - disease interactions using similarity network fusion and neural networks. BMC Bioinform, 1–20 (2021)
13. Jiang, H.J., Huang, Y.A., You, Z.H.: SAEROF: an ensemble approach for large-scale drug-disease association prediction by incorporating rotation forest and sparse autoencoder deep neural network. Sci. Rep. **10**(1), 1–11 (2020)
14. Yu, J., Xuan, Z., Feng, X., Zou, Q., Wang, L.: A novel collaborative filtering model for LncRNA-disease association prediction based on the Naïve Bayesian classifier. BMC Bioinform. **20**(1), 1–13 (2019)
15. Nguyen, V. T., Tran, D. II.: An improved computational method for prediction of lncRNA-disease associations based on collaborative filtering and resource allocation. In: 2021 13th Int. Conf. Knowl. Syst. Eng., pp. 1–6 (2021)
16. Ezzat, A., Zhao, P., Wu, M., Li, X.L., Kwoh, C.K.: Drug-target interaction prediction with graph regularized matrix factorization. IEEE/ACM Trans. Comput. Biol. Bioinforma. **14**(3), 646–656 (2017)
17. Nguyen, V.T., Le, T.T.K., Than, K., Tran, D.H.: Predicting miRNA–disease associations using improved random walk with restart and integrating multiple similarities. Sci. Rep. **11**(1), 1–16 (2021)
18. Shang, J., Yang, Y., Li, F., Guan, B., Liu, J.X., Sun, Y.: BLNIMDA: identifying miRNA-disease associations based on weighted bi-level network. BMC Genomics **23**(1), 1–12 (2022)
19. Wishart, D. S. et al.: DrugBank: a comprehensive resource for in silico drug discovery and exploration. Nucl Acids Res 34(Database), 668–672 (2006)

20. Hamosh, A., Scott, A. F., Amberger, J. S., Bocchini, C. A., McKusick, V. A.: Online mendelian inheritance in man (OMIM), a knowledgebase of human genes and genetic disorders. Nucl Acids Res 33(DATABASE), 514–517 (2005)
21. Weininger, D.: SMILES, a chemical language and information system: 1: introduction to methodology and encoding rules. J. Chem. Inf. Comput. Sci.Comput. Sci. **28**(1), 31–36 (1988)
22. Steinbeck, C., Han, Y., Kuhn, S., Horlacher, O., Luttmann, E., Willighagen, E.: The chemistry development kit (CDK): an open-source Java library for chemo- and bioinformatics. J. Chem. Inf. Comput. Sci.Comput. Sci. **43**(2), 493–500 (2003)
23. van Driel, M.A., Bruggeman, J., Vriend, G., Brunner, H.G., Leunissen, J.A.M.: A text-mining analysis of the human phenome. Eur. J. Hum. Genet. **14**(5), 535–542 (2006)
24. Hu, P., et al.: Learning from low-rank multimodal representations for predicting disease-drug associations. BMC Med. Inform. Decis. Mak.Decis. Mak. **21**, 1–12 (2021)
25. Luo, H., et al.: Drug repositioning based on comprehensive similarity measures and Bi-Random walk algorithm. Bioinformatics **32**(17), 2664–2671 (2016)
26. Karimollah, H.-T.: Receiver operating characteristic (ROC) curve analysis for medical diagnostic test evaluation. Casp. J Intern Med 4(2), 627–635 (2013)
27. Saito, T., Rehmsmeier, M.: The precision-recall plot is more informative than the ROC plot when evaluating binary classifiers on imbalanced datasets. PLoS ONE **10**(3), e0118432 (2015)
28. Wang, M.N., Xie, X.J., You, Z.H., Ding, D.W., Wong, L.: A weighted non - negative matrix factorization approach to predict potential associations between drug and disease. J. Transl. Med. **1**, 1–12 (2022)
29. Safaei, M., Sundararajan, E. A., Driss, M., Boulila, W., Shapi'i, A.: A systematic literature review on obesity: Understanding the causes & consequences of obesity and reviewing various machine learning approaches used to predict obesity. Comput Biol Med **136**(August), 104754 (2021)
30. Łukasiewicz, S., Czeczelewski, M., Forma, A., Baj, J., Sitarz, R., Stanisławek, A.: Breast cancer—epidemiology, risk factors, classification, prognostic markers, and current treatment strategies—an updated review. Cancers (Basel) **13**(4287), 1–30 (2021)

A Real-time Crowd Density Level Detection System

Dinh-Lam Pham[1,2], Thanh-Hai Nguyen[2], Kyong-Sook Kim[1], and Kwanghoon Pio Kim[1,2(✉)]

[1] Contents Convergence Software Research Institute, Kyonggi University, 154-42 Gwangkyosan-Ro, Youngtong-Gu, Gyeonggi-Do 16227, Republic of Korea
{phamdinhlam,khmjmc,kwang}@kyonggi.ac.kr

[2] Data and Process Engineering Research Laboratory, 154-42 Gwangkyosan-Ro, Youngtong-Gu, Gyeonggi-Do 16227, Republic of Korea
nguyenthanhai@kyonggi.ac.kr

Abstract. Crowd density level evaluates the number of people gathering in a particular area or scene, typically based on videos generated from real-time closed-circuit television (CCTV) cameras. The goal of estimating the levels of a crowd is to provide accurate and reliable information about the crowd size and density, which can be helpful in various situations. Furthermore, detecting and analyzing crowds in real-time is a crucial task for many applications such as monitoring large crowds in public areas, detecting any suspicious activities or potential threats; managing crowds in public places to prevent overcrowding and ensure the safety of individuals; monitoring the flow of people in an event to optimize the use of security officers and facilities. In this paper, we present a deep learning-based architecture and system for detecting levels of crowd density from CCTV cameras. In particular, the system can automatically generate alert information when identifying high-density camera areas, enabling security staff to take corrective measures to manage the crowd and prevent overcrowding before it happens. The proposed system provides a reliable approach for real-time crowd detection, which can be integrated into existing surveillance systems.

Keywords: Crowd density level detection system · Deep learning-based crowd detection system · Public safety · Surveillance management · CCTV Camera

1 Introduction

Crowd management is controlling and directing large groups of people in a safe and organized manner, especially during events such as festivals, demonstrations, and mass gatherings. It involves various methods to ensure that crowds can safely move and gather without creating a hazard or risk to public safety as well as the safety of people in the crowd [1, 2]. Crowd detection and management are important in event planning, public gatherings, and emergencies. It requires the cooperation of many stakeholders, including event organizers, security officers, emergency responders, and other relevant agencies.

P. T. Nghia et al. (Eds.): ICTA 2023, LNNS 847, pp. 107–113, 2023.
https://doi.org/10.1007/978-3-031-49529-8_12

Additionally, managing crowds and large numbers of CCTV cameras can be challenging due to several characteristics [3, 4]. One of the first limitations is resource constraints. Managing crowds with multiple cameras can be resource-intensive and requires personnel, equipment, and technology investments. Resource constraints can make it difficult to effectively manage crowds and monitor all CCTV cameras in real-time. For example, some organizations may not have enough staff to monitor all camera screens simultaneously. Another problem is information overload. CCTV Cameras can generate tens of Gigabytes of data in an hour, which makes it difficult to process and analyze efficiently. This can lead to information overload and complicate identifying potential risks and hazards. For example, if there are too many cameras or too much data to analyze, important information may be missed, making it difficult to respond effectively to potential threats. Besides, camera surveillance brings privacy concerns. Using cameras for crowd surveillance can raise privacy issues, especially if the cameras record sensitive information or personal information that could be used for identity theft or other misconduct purposes.

There are several approaches for detecting the density of a crowd, i.e., Using computer vision and neural network to detect the number of people in a given area through videos captured by CCTV cameras [5, 6]; Using signals of WIFI to detect the number of mobile devices in a specific location, which can then be used to estimate the number of people in that area [7]; or using Bluetooth-based methods to detect the existence of people in a given area by catching the signal of their Bluetooth devices [8]. Among these methods, the computer vision approach is popular because of taking advantage of the available CCTV cameras in public areas and does not require installation on people's devices.

In this paper, we introduce a deep learning-based architecture and system for detecting different crowd levels from CCTV cameras. In particular, we propose an architecture with the P2PNet method [9] to estimate the number of people in a crowd from a camera frame and then present all the information in real-time. We divided the crowd density in each camera into 5 (five) different levels and used different thresholds to define these levels. A notification will be sent when the number of people detected in the camera exceeds a certain limit, indicating that the crowd density has been reached at this level. In the rest of this paper, we organized the content as follows. In the next section, we discuss some related and recent studies on using deep learning approaches in crowd density detection. Then, we present the architecture and components to create a deep learning-based real-time crowd density detection system. Section 4 describes the system and presents the experimental results, while Sect. 5 discusses some of the results obtained in this study and its limitations.

2 Related works

In this section, we discuss some recent research related to the use of deep learning and neural networks in crowd detection and crowd density assessment. The results obtained from these studies and some limitations.

In [9], Qingyu Song and his colleagues introduced a purely point-based framework, an approach using the VGG16 neural networks (P2PNet) [10] combining with the Hungarian algorithm to detect and count number of people in a crowd. In this study, the authors proposed a novel approach to crowd counting and localization using a purely point-based framework. They introduced a new metric, called normalized average precision (nAP), to provide a more complete and accurate performance evaluation. In the method, they represented each person in a crowd as a single point, simplifying the counting and localization problem. The method used VGG-16 to learn a mapping from the input image to the set of points representing the people in the crowd. Furthermore, they also proposed a post-processing step that uses clustering to group nearby points into individual people. The framework directly predicts a set of point proposals to represent heads in an image, being consistent with the human annotation results. This study outperforms previous state-of-the-art methods on both counting and localization tasks.

In [11], Zhi-Qi Cheng and his colleagues introduced their study about rethinking the spatial invariance of convolutional networks for object counting. This research assumes that strict spatial invariance is necessary for object counting and proposes a new approach that relaxes this assumption. The authors argue that strict pixel-level spatial invariance can lead to overfitting noise in density map generation, resulting in inaccurate object counting. To address this problem, the authors propose a new method that uses locally connected Gaussian kernels to estimate the spatial position in density maps. This approach allows the feature extraction process to stimulate the density map generation process and overcome annotation noise. The authors also introduce a low-rank approximation with translation invariance to implement massive Gaussian convolution efficiently. The proposed approach is evaluated on 4 (four) mainstream object counting networks (MCNN, CSRNet, SANet, and ResNet-50) and 7 (seven) popular benchmarks. The results show that their proposed method outperforms other methods and achieves promising learning of the spatial position of objects.

In [12], Jiwei Chen and his colleagues proposed a method called Crowd Attention Convolutional Neural Network (CAT-CNN), which can adaptively estimate the importance of a human head at each pixel location by encoding a confidence map. With this approach, the position of human heads in the estimated density map gets more attention to encoding the final density map, avoiding misjudgments. In particular, counting the total number of people in a crowd of each image is classified in a designed classification assignment to obtain a highly refined density map. In the CAT-CNN model, the authors introduced three stages of the model. The feature extraction module adapts to different scales and crowd count groups in the first stage. In the second stage, two middle modules encode the confidence and estimated density maps. The third or final stage uses the confidence map to refine the estimated density map into the final density map. This study also presented experiments on three datasets, showing that the proposed method outperforms state-of-the-art methods.

In our study presented in this paper, we introduce an architecture to take advantage of these state-of-the-art methods to create a real-time crowd density level detection to send alarm information to the security manager automatically. In particular, we select the P2PNet method [10] for detecting the number of people in the frame transferred from the real-time frame processing module and then classify the level of the crowd based on thresholds of different levels. In the next section, we present the proposed architecture and its components.

3 Architecture and Components

Figure 1 illustrates the proposed architecture for crowd density level classification based on the data obtained from the frames and the number of people in the crowd detected using the P2PNet. As we can see on the figure, the architecture contains the following components.

- **CCTV camera management**: This component manages the CCTV cameras that are used to record video of the crowd. These cameras may be attracted to the street, shopping mall, festival location, etc. The CCTV camera management will transfer video captured from their devices to the real-time frame processing component for processing.
- **Real-time frame processing**: The real-time frame processing component manages input frames in a queue and uses VidGear [13] and FFmpeg [14] libraries to analyze the video. This component will process the frame and then transfer it to the P2PNet processing component.
- **P2PNet processing**: The P2PNet processing component uses a modified VGG16 neural network to predict the number of points (persons) from the frame and then passes this information to the Density level processing component.
- **Density level processing**: This component classifies the density of the crowd into five levels using threshold values. These threshold values are set based on the number of people present in a particular area from the view of a camera. Since different areas will have different concentration limits, this component allows the security officer to set a limited number of people in each different camera view. The crowd density is divided into 5 levels. E.g., If we set the limited number of people in a crowd for the angle view from Camera A as 90 people, we get the limit of persons with different levels as follows.
- Level 1: $\leq 20\%$ (≤ 18 persons)

  - Level 2: 20–40% (19–36 persons)
  - Level 3: 40–60% (37–54 persons)
  - Level 4: 60–80% (55–72 persons)
  - Level 5: >80% (>72 persons).

Additionally, this component includes an alarm function that can be triggered when the density level reaches the certain threshold. The alarm function notifies the authorities or event organizers of the situation so that they can take appropriate action to manage the crowd. The threshold for triggering the alarm can be set based on the specific event needs or location being monitored. This feature can help ensure the crowd's safety and prevent potential incidents.

Using this architecture, we can manage all the existing cameras in an area, and furthermore, the security officer can use the system from his computer and mobile phone anywhere. The user right to access the system related to the issue of user permissions, allowing IP addresses as well as blurring the video frames to protect privacy are necessary, but it is not within the scope of this study, so we do not discuss the matter in this paper.

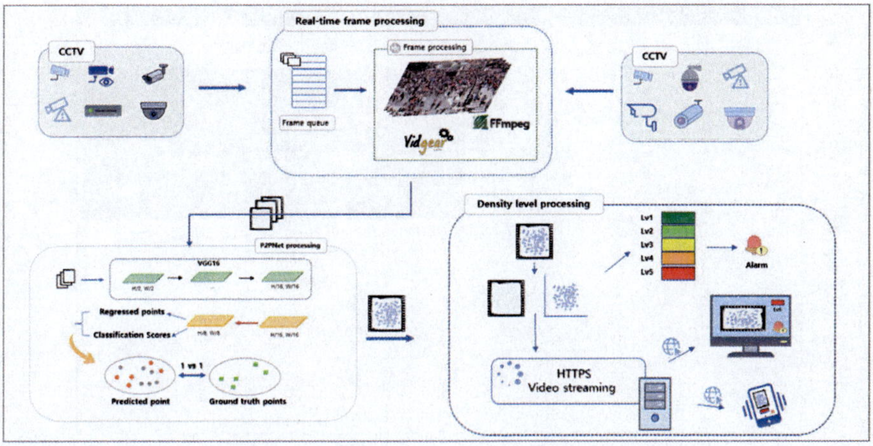

Fig. 1. Architecture of the proposed method for real-time crowd density level detection system.

4 Implementation and Experimental Results

Our approach is to create a web-based real-time crowd density level detection system, so we choose Python Django [15] as the framework to implement the modules described in the architecture. After that, we evaluate the proposed system with some real-life cases. In the first case study, we set the limit people for this camera's viewing angle to a maximum of 40 people. As shown in Fig. 2, the number of people detected is 19 at 14.98 frames per second, and the detected crowd density level is Level 3.

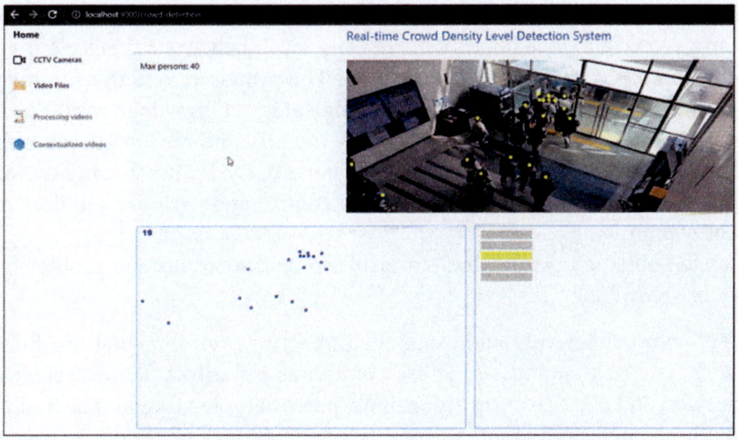

Fig. 2. Experiment results on real-time CCTV camera.

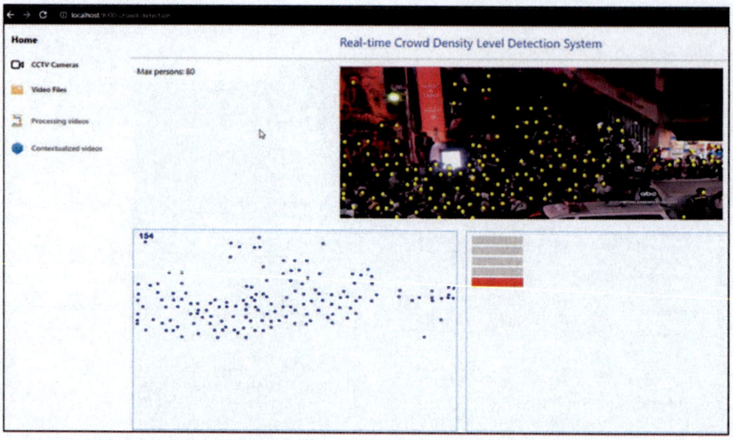

Fig. 3. Experimental result on the video of Itaewon crowd crush on 29th October, 2022, Korea.

In the second trial, without loss of generality, we read input from a video file instead of reading data from the CCTV camera. This video file captures the crowd crush in Itaewon, Korea, on October 29th, 2022 [16]. As shown in Fig. 3, we set the maximum number of people in this camera view to 80 people, the number of people detected in this frame is 154, so the detected crowd density level is Level 5.

5 Conclusion and Future Work

This paper discusses the importance of real-time crowd density detection and classification for different applications. In particular, we proposed a system to evaluate the number of people in a crowd and then classify the density level, hence notifying security officers about each camera's current crowd-level state. The proposed system can integrate into real-life CCTV management centers, enhancing safety in crowded areas. Furthermore, the system can help security personnel quickly identify and respond to potential safety risks, helping to prevent potentially serious incidents early. By monitoring crowd density in real-time, the system helps security officers control and optimize the flow of people and prevent overcrowding.

Besides the obtained results, there are still some limitations and problems that this study has not yet solved:

We have not comprehensively addressed the impact of environmental conditions, such as weather or low-light situations. These conditions can affect the performance of the crowd detection and classification algorithms, potentially leading to inaccurate results. This research has not studied the scalability of the system to handle larger and more complex situations. With the increasing of more CCTV cameras, the system performance may experience bottlenecks with low processing times.

In future studies, we will overcome the above limitations and put the system into practice.

Acknowledgement. This research was supported by Basic Science Research Program through the National Research Foundation of Korea (NRF) funded by the Ministry of Education (grant number 2020R1A6A1A03040583).

References

1. Saleh, S.A.M., Suandi, S.A., Ibrahim, H.: Recent survey on crowd density estimation and counting for visual surveillance. Eng. Appl. Artif. Intell. **41**, 103–114 (2015). https://doi.org/10.1016/J.ENGAPPAI.2015.01.007
2. Sindagi, V.A., Patel, V.M.: A survey of recent advances in CNN-based single image crowd counting and density estimation. Pattern Recognit. Lett. **107**, 3–16 (2018). https://doi.org/10.1016/J.PATREC.2017.07.007
3. Porikli, F., et al.: Video surveillance: Past, present, and now the future. IEEE Signal Process. Mag. **30**(3), 190–198 (2013). https://doi.org/10.1109/MSP.2013.2241312
4. Elharrouss, O., Almaadeed, N., Al-Maadeed, S.: A review of video surveillance systems. J. Vis. Commun. Image Represent. **77**, 103116 (2021). https://doi.org/10.1016/J.JVCIR.2021.103116
5. Tzelepi, M., Tefas, A.: Human crowd detection for drone flight safety using convolutional neural networks.In: 25th Eur. Signal Process. Conf. EUSIPCO 2017, vol 2017-Janua, pp. 743–747 (2017). https://doi.org/10.23919/EUSIPCO.2017.8081306
6. Marsden, M., McGuinness, K., Little, S., O'Connor, N. E.: ResnetCrowd: A residual deep learning architecture for crowd counting, violent behaviour detection and crowd density level classification. In: 2017 14th IEEE Int. Conf. Adv. Video Signal Based Surveillance, AVSS 2017 (2017). https://doi.org/10.1109/AVSS.2017.8078482
7. Yoshida, T. Y. T.-I. I. Institute: Estimating the number of people using existing WiFi access point based on support vector regression, *search.proquest.com*, Accessed: Mar. 15, 2023. https://search.proquest.com/openview/a2ef087865b62f09021d32c569903093/1?pq-origsite=gscholar&cbl=936334
8. Schauer, L., Werner, M., Marcus, P.: Estimating crowd densities and pedestrian flows using Wi-Fi and Bluetooth. In: MobiQuitous 2014 - 11th Int. Conf. Mob. Ubiquitous Syst. Comput. Netw. Serv., pp. 171–177 (2014). https://doi.org/10.4108/ICST.MOBIQUITOUS.2014.257870
9. Song, Q. et al.: rethinking counting and localization in crowds: a purely point-based framework. pp. 3365–3374 (2021)
10. Simonyan, K., Zisserman, A.: Very deep convolutional networks for large-scale image recognition. In: 3rd Int. Conf. Learn. Represent. ICLR 2015 - Conf. Track Proc. (2014). https://doi.org/10.48550/arxiv.1409.1556
11. Cheng, Z.-Q., Dai, Q., Li, H., Song, J., Wu, X., Hauptmann, A.G.: Rethinking spatial invariance of convolutional networks for object counting. pp. 19638–19648, 2022, Accessed: Mar. 15, 2023. https://github.com/zhiqic/Rethinking-Counting
12. Chen, J., Su, W., Wang, Z.: Crowd counting with crowd attention convolutional neural network. Neurocomputing **382**, 210–220 (2020). https://doi.org/10.1016/J.NEUCOM.2019.11.064
13. Overview - VidGear. https://abhitronix.github.io/vidgear/latest/ (accessed Aug. 17, 2023)
14. FFmpeg, Open-source multimedia framework. https://ffmpeg.org/ (accessed Aug. 15, 2023)
15. Python Django web framework. https://www.djangoproject.com/ (accessed Aug. 19, 2023)
16. Itaewon crowd crush - BBC News. https://www.bbc.com/news/world-asia-63443044 (accessed Aug. 24, 2023)

An Architecture for More Fine-Grained Hidden Representation in Named Entity Recognition for Biomedical Texts

Bui Duc Tho[1,2], Son-Ba Giang[1], Minh-Tien Nguyen[1(✉)], and Tri-Thanh Nguyen[2]

[1] Hung Yen University of Technology and Education, Hung Yen, Vietnam
buiductho@utehy.edu.vn , tiennm@utehy.edu.vn

[2] Vietnam National University, Hanoi, University of Engineering and Technology, 144, Xuan Thuy, Cau Giay, Hanoi, Vietnam
ntthanh@vnu.edu.vn

Abstract. This paper introduces a model for Biomedical Named Entity Recognition (BioNER). Different from existing models that mainly rely on pre-trained models, i.e., PubMedBERT, the proposed model is empowered by using PubMedBERT as the main backbone for mapping input sequences to contextual vectors. To learn more fine-grained hidden representation and effectively adapt to the recognition downstream task, the model stacks BiLSTM and CRFs on top of PubMedBERT. Given an input sentence, the model first maps the sentence into contextual vectors by PubMedBERT. The vectors are next fed into a BiLSTM layer for learning a more fine-grained hidden representation that serves as the input for sequence labeling by using CRFs. We confirm the efficiency of the model on benchmark corpora. Experimental results on 29 diverse datasets indicate that the proposed model obtains promising results compared to good as well as state-of-the-art baselines. The ablation study also shows the behavior of the model in several aspects.

Keywords: BioNER · Sequence labeling · Pre-trained language models

1 Introduction

Named Entity Recognition (NER) is a fundamental task in Natural Language Processing (NLP), involving the identification and categorization of named entities in unstructured texts [1,17,24]. In actual scenarios, NER is not a standalone solution and is usually a part of AI pipelines, i.e., its outputs can serve several applications such as text understanding [3,34], information retrieval [10], question answering [22], and knowledge base construction [32].

The first and second authors contribute equally with no order distinction.

P. T. Nghia et al. (Eds.): ICTA 2023, LNNS 847, pp. 114–125, 2023.
https://doi.org/10.1007/978-3-031-49529-8_13

With the growing amount of biomedical data, there is a need to develop accurate and efficient NER models to extract relevant information and facilitate downstream tasks. Bio-Named Entity Recognition (BioNER) [23] focuses on identifying and classifying named entities in biomedical and biological texts. These named entities include genes, proteins, chemicals, diseases, anatomical terms, etc. BioNER plays a critical role in various applications, including biomedical information retrieval, drug discovery, clinical decision support systems [27,29], and biomedical text mining [28]. The task of Bio-NER is particularly challenging due to the complex terminology, abbreviations, and diverse naming conventions used in the biomedical domain. Recent machine learning models, such as Conditional Random Fields (CRFs) [16], Bidirectional Long Short-Term Memory (BiLSTM) [25] networks, or pre-trained language models (PLMs) [6], have demonstrated good performance in identifying biomedical entities. Transformers-based models, such as ClinicalBERT [13] or PubMedBERT [9] have proven to be highly effective in capturing contextual information and improving BioNER accuracy. Among PLMs for BioNER, PubMedBERT has shown competitive performance for the biomedical domain [9], including BioNER. However, most of prior studies directly utilize PLMs for BioNER. We argue that the performance of PLMs can still be improved with additional layers, i.e., BiLSTM and CRFs.

This paper introduces a model for BioNER. The model is empowered by using PubMedBERT as the backbone. It allows the model to effectively represent input sequences in the form of contextual vectors. More precisely, input sequences are first fed into PubMedBERT for the purpose of obtaining contextual vectors. The vectors are then put into a BiLSTM layer for learning more fine-grained hidden features. A sequence of labels is finally predicted by using CRFs to ensure the global optimization of sequence labeling. This paper makes three main contributions.

- It introduces a model that composes of PubMedBERT, BiLSTM, and CRFs for BioNER. The model relies on the power of PubMedBERT, which is pre-trained on a huge amount of biomedical text. The combination of a PLM and additional layers allows the model to effectively learn hidden features for sequence labeling. The model is simple but effective for recognizing bio named entities. The source code of this paper is also publicly accessible.[1]
- It demonstrates the efficiency of the proposed model by comparing it to strong baselines on benchmark corpora. Experimental results on 29 datasets show that additional layers are beneficial for BioNER.
- It conducts an ablation study to show the behavior of the model in several aspects. This observation facilitates the next studies of BioNER.

[1] Code and data: https://github.com/Jester6136/pubmedBERT-BiLSTM-CRF.

2 Related Work

Named Entity Recognition NER can be addressed by rule-based [20] and supervised learning. The rule-based approach is easy to implement, however, the creation and maintenance of rules is a non-trivial task that requires the involvement of domain experts. Whereas, the supervised approach uses a labeled dataset for training the model. For this approach, the NER task can be addressed by sequence labeling [6,31], span extraction [8,19], or text generation formulation [2,7,11]. The sequence labeling formulation formulates the NER task as a sequence tagging problem, in which NER models assign a tag (an entity type) to each token of an input sentence. In contrast, the span extraction formulation considers an entity as a span and adapts machine reading comprehension techniques to extract these spans. Recently, NER has been converted to a text generation problem in which NER models learn to generate labels for corresponding entities [2,7,11]. The supervised learning methods learn hidden representations by using a wide range of neural networks such as BiLSTM, CRFs [31], or Transformers in the form of PLMs [2,7,11].

Bio Named Entity Recognition BioNER adapts the success of NER in common domains to biomedical texts with promising results [15,31]. Compared to the common methods of NER, BioNER mainly formulates the task as a sequence labeling problem with features learned from neural networks [15]. Other recent studies also show the contribution of additional context for BioNER, with promising results [31]. PLMs have also been adapted to BioNER with good performance [9,13,14,18]. We follow the direction of using supervised learning methods for BioNER. However, instead of using only PLMs, we design a model that combines BiLSTM for learning more fine-grained features and CRF for global optimization of sequence labeling.

3 Bio Named Entity Recognition

3.1 Problem Formulation

BioNER aims to identify named entities, such as genes, proteins, diseases, and drugs, within biomedical documents/texts. Given a sentence S containing biomedical named entities $S = \{w_1, w_2, \ldots, w_n\}$ with n tokens, the task is to assign a label $y \in Y$ to each token in S, where Y is a predefined set of labels.

3.2 The Proposed Model

Figure 1 shows the architecture of the proposed model. Given an input sentence (i.e., a sequence of tokens), the model first maps the sequence to contextual vectors by PubMedBERT. The vectors are then fed into a BiLSTM layer to learn more fine-grained features for sequence labeling with CRFs. In the context of sequence labeling, CRFs help to improve the coherence and consistency of the predicted entity labels by considering the global structure of the sequence.

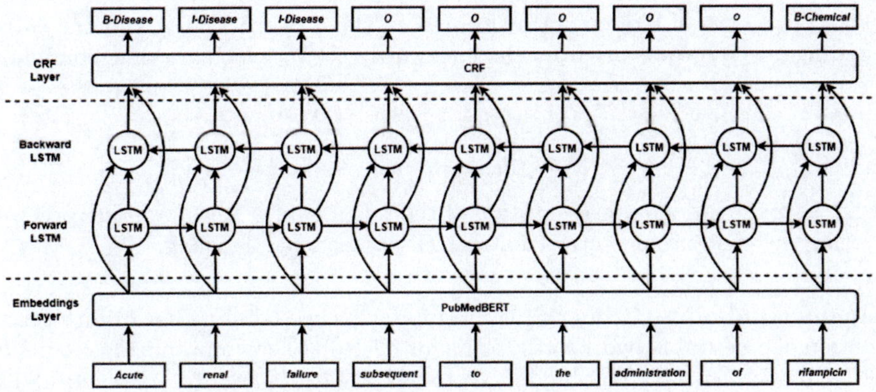

Fig. 1. The PubMedBERT-BiLSTM-CRF model.

Input Representation The input representation is to convert an input sequence into contextual vectors for later layers. It uses PubMedBERT [9] pre-trained with a massive amount of biomedical data by using the Transformer [30]. PubMedBERT [9] was pre-trained with text from the PubMed database. By employing pre-training techniques, the model acquires word vectors that correspond to individual characters in the labeled corpus. These word vectors, commonly referred to as word embeddings, encode semantic information and serve as foundational input features for subsequent processing stages.

PubMedBERT was used to map an input sequence into contextual vectors. Given a sequence with n tokens $S = \{w_1, w_2, \ldots, w_n\}$, the first step is to concatenate all the tokens in to a single sequence. This sequence is fed into PubMedBERT to obtain contextual vectors as follows.

$$\boldsymbol{V} = \{\boldsymbol{v}_1, \boldsymbol{v}_2, \ldots, \boldsymbol{v}_n\} = PubMedBERT(S)$$

where $\boldsymbol{v}_i$ is the contextual vector of the i-th token.

Learning Hidden Representation Following PubMedBERT, the model stacks a BiLSTM [25] layer on the top of PubMedBERT for two reasons. First, adding BiLSTM usually helps to learn more fine-grained hidden features that are beneficial for the learning process [15,21,31]. By considering both the forward and backward directions of the input sequence, the BiLSTM layer captures comprehensive contextual information from neighboring words. Second, BiLSTM facilitates the adaptation of PubMedBERT for the BioNER downstream task.

BiLSTM [25] is a variant of the traditional LSTM architecture [12]. This bidirectional nature enables BiLSTM to capture contextual information from the past and future time steps simultaneously, making it highly effective in understanding and modeling the dependencies in sequential data. The input to the BiLSTM is the word embedding sequence $\boldsymbol{v}_i$ of each token, and it produces output sequences of hidden states from both the forward LSTM and the corresponding output sequences from the backward LSTM, which are then combined.

$$\overrightarrow{\boldsymbol{h}_i} = LSTM(\overrightarrow{\boldsymbol{h}}_{i-1}, \boldsymbol{v}_i); \qquad \overleftarrow{\boldsymbol{h}_i} = LSTM(\overleftarrow{\boldsymbol{h}}_{i+1}, \boldsymbol{v}_i)$$

$$\boldsymbol{h}_i = [\overrightarrow{\boldsymbol{h}_i}; \overleftarrow{\boldsymbol{h}_i}]$$

where $\boldsymbol{h}_i$ is the hidden representation of the token i-th. A set of n hidden vectors $\boldsymbol{h}$ forms the contextual vector (matrix) $\boldsymbol{H}$ for sequence labeling.

Sequence Labeling CRFs [16] is used for sequence labeling because the combination of a neural network with CRFs for NER has shown promising results on the NER task [15,31]. The hidden vector $\boldsymbol{H} = \{\boldsymbol{h}_1, \boldsymbol{h}_2, \ldots, \boldsymbol{h}_n\}$ from BiLSTM was fed into CRFs to compute the conditional probability $p_\theta(\bar{y}|\boldsymbol{H})$.

$$\psi(y', y, \boldsymbol{h}_i) = \exp(\boldsymbol{W}_y^T \boldsymbol{h}_i + \boldsymbol{b}_{y',y})$$

$$p_\theta(\bar{y}|S) = \frac{\prod_{i=1}^{n} \psi(y_{i-1}, y_i, \boldsymbol{h}_i)}{\sum_{\bar{y} \in \mathcal{Y}(S)} \prod_{i=1}^{n} \psi(y'_{i-1}, y'_i, \boldsymbol{h}_i)}$$

where ψ is the potential mapping function and θ is the parameter of the model. $\mathcal{Y}(S)$ represents all possible label sequences of the given input S. $\boldsymbol{W} \in \mathbb{R}^{t \times d}$ and $\boldsymbol{b} \in \mathbb{R}^{t \times t}$ are parameters computing emission and transition scores. d is the hidden size of $\boldsymbol{h}_i$ and t is the size of the label set.

Training and Inference For training, the negative log-likelihood loss for a single training example was used.

$$\mathcal{L}(\theta) = -\log p_\theta(\bar{y}|S)$$

The overall loss for the entire training dataset, which consists of m training examples, is the average of the individual losses:

$$\mathcal{L}_{\text{total}}(\theta) = \frac{1}{m} \sum_{i=1}^{m} \mathcal{L}(\theta)$$

The inference uses the Viterbi algorithm to find the optimal path in the prediction by using dynamic programming.

$$y^* = \arg\max_{\bar{y}}(p_\theta(\bar{y}|\mathbf{H}))$$

4 Settings and Evaluation Metrics

Datasets Evaluation uses MTL-Bioinformatics-2016 [4] that includes 29 diverse biomedical datasets (detailed information is shown in Table 1). These datasets are in various domains which have been widely used in prior studies [5,15,31]. Table 1 summarizes the information of 29 datasets.

Table 1. 29 datasets used in experiments.

Dataset	Entities	Train	Dev	Test
AnatEM	Anatomy	5861	2118	3830
BC5CDR	Chemical, disease	4560	4581	4797
BC5CDR-chem	Chemical	4560	4581	4797
BC5CDR-disease	Disease	4560	4581	4797
BioNLP09	Protein	7462	1448	2446
BioNLP11EPI	Protein	5698	1955	4122
BioNLP11ID	4 NEs[4]	2496	721	1961
BioNLP11ID-chem	Chemical	2496	721	1961
BioNLP11ID-ggp	Gene/protein	2496	721	1961
BioNLP11ID-species	Organism	2496	721	1961
BioNLP13CG	16 NEs[5]	3033	1003	1906
BioNLP13CG-cc	Cellular	3033	1003	1906
BioNLP13CG-cell	Cell	3033	1003	1906
BioNLP13CG-chem	Chemical	3033	1003	1906
BioNLP13CG-ggp	Gene/protein	3033	1003	1906
BioNLP13CG-species	Organism	3033	1003	1906
BioNLP13GE	Protein	2449	2737	3391
BioNLP13PC	4 NEs[6]	2499	857	1695
BioNLP13PC-cc	Cellular	2499	857	1695
BioNLP13PC-chem	Chemical	2499	857	1695
BioNLP13PC-ggp	Gene/protein	2499	857	1695
CRAFT	6 NEs[7]	10,875	3730	7425
CRAFT-cc	Gene ontology	10,875	3730	7425
CRAFT-cell	Cell line	10,875	3730	7425
CRAFT-chem	Chemical	10,875	3730	7425
CRAFT-ggp	Gene/protein	10,875	3730	7425
CRAFT-species	Taxonomy	10,875	3730	7425
Ex-PTM	Gene/protein	1377	437	1839
Linnaeus	Species	11,935	4078	7142

[4] Protein, organism, chemical, regulon-operon

[5] Gene/protein, cell, chemical, cancer, organ, organism, tissue, amino acid, cellular component, organism substance, pathological formation, anatomical system, immaterial anatomical, organism subdivision, multi-tissue structure, developing anatomical structure

[6] Gene/protein, complex, chemical, cellular/component

[7] Gene/protein, taxonomy, chemical, cell line, sequence ontology, gene ontology

Evaluation Metrics The evaluation uses the micro F-score as the main metric for comparing NER models.

Implementation The training and inference used an NVIDIA T4 GPU boasting 16 GB of GDDR6 memory, operating on PCIe 3.0. The model was trained in 20 epochs with the batch size of 64. The BiLSTM layer consisted of 256 units, and the learning rate was set to 1e−5.

5 Results and Discussion

This section reports the F-score comparison of the proposed model with good baselines and shows the ablation study to understand the behavior of the model.

5.1 F-Score Comparison

The Contribution of Additional Layers The F-scores in the left hand side of Table 2 shows the contribution of additional layers to the model.

As observed, PubMedBERT+BiLSTM+CRF demonstrates better performance compared to PubMedBERT. Specifically, for the CRAFT-ggp dataset, the PubMedBERT+BiLSTM+CRF model outperforms the PubMedBERT model by 7.42 of the F1 score. For the BioNLP09, BioNLP11, BioNLP13, and CRAFT datasets, the PubMedBERT-BiLSTM-CRF model consistently exhibits better performance compared to PubMedBERT, with an average F1 score improvement of 1.32. These results show that the addition of the BiLSTM layer allows the model to capture more complex contextual dependencies and sequential patterns in the input data. This enhances the model's ability to understand the relationships between tokens, leading to improve entity recognition. Furthermore, the CRF layer helps to model the transition probabilities between different named entity labels, enforcing global consistency in the predicted entity sequences. This is particularly beneficial for BioNER, where named entities often exhibit specific patterns, dependencies, and ambiguity. The PubMedBERT+CRFs model also achieves competitive performance, without the BiLSTM layer. This suggests that the combination of PubMedBERT and CRFs is capable of achieving good results on the datasets. However, achieving optimal results requires extensive tuning of the learning rate over multiple iterations.

Comparison with SOTA Methods We challenge the proposed model to state-of-the-art (SOTA) BioNER methods. This experiment is to observe the gap between the model and the SOTA methods. To do that, we tried our best to collect the SOTA BioNER methods for comparison [15,26,33]. We can only gather six available results out of 29 datasets, since the six SOTA results were trained and validated specifically on their respective datasets, while the other datasets either lack training and performance evaluation or have undergone modifications with the inclusion of external context techniques, making it inappropriate

Table 2. The F-scores on 29 datasets. **Bold** text is the best. The left shows the contribution of additional layers. The right compares BioBERT and PubMedBERT.

Datasets	PubMedBERT	PubMedBERT CRF	PubMedBERT BiLSTM CRF	BioBERT	PubMedBERT
AnatEM	90.02	89.99	**90.06**	88.84	**90.06**
BC5CDR	89.53	89.92	**89.98**	89.21	**89.98**
BC5CDR-chem	93.00	93.17	**93.60**	92.86	**93.60**
BC5CDR-disease	85.28	**86.28**	86.07	84.05	**86.07**
BioNLP09	89.67	**90.57**	90.25	89.30	**90.25**
BioNLP11EPI	87.52	88.52	**88.63**	85.89	**88.63**
BioNLP11ID	84.19	84.14	**85.84**	**86.09**	85.84
BioNLP11ID-chem	71.73	70.14	**74.49**	68.90	**74.49**
BioNLP11ID-ggp	88.66	90.36	**91.28**	91.26	**91.28**
BioNLP11ID-species	78.90	80.80	**81.38**	79.76	**81.38**
BioNLP13CG	87.16	87.51	**87.69**	86.50	**87.69**
BioNLP13CG-cc	81.79	**83.85**	81.11	**81.33**	81.11
BioNLP13CG-cell	87.94	**89.93**	89.37	**89.60**	89.37
BioNLP13CG-chem	82.12	**86.31**	84.70	83.25	**84.70**
BioNLP13CG-ggp	91.99	**92.74**	92.28	91.34	**92.28**
BioNLP13CG-species	86.37	88.21	**88.30**	86.85	**88.30**
BioNLP13GE	82.61	**84.42**	82.53	76.11	**82.53**
BioNLP13PC	91.25	91.61	**91.78**	90.80	**91.78**
BioNLP13PC-cc	88.89	**89.29**	88.38	87.83	**88.38**
BioNLP13PC-chem	88.37	90.68	**91.92**	85.48	**91.92**
BioNLP13PC-ggp	92.88	93.37	**93.39**	91.62	**93.39**
CRAFT	85.39	87.59	**88.08**	85.30	**88.08**
CRAFT-cc	70.43	**73.43**	73.39	72.82	**73.39**
CRAFT-cell	90.14	**90.56**	90.49	88.97	**90.49**
CRAFT-chem	85.71	**85.74**	83.77	**85.05**	83.77
CRAFT-ggp	79.70	83.57	**87.12**	81.62	**87.12**
CRAFT-species	98.89	98.87	**98.93**	97.95	**98.93**
Ex-PTM	82.80	**84.08**	83.60	78.92	**83.60**
Linnaeus	**88.75**	88.57	88.14	86.98	**88.14**
Average	86.26	87.39	**87.47**	85.65	**87.47**

Table 3. Comparison of performance with SOTA models.

Datasets	PubMedBERT-BiLSTM-CRF	SOTA
AnatEM	90.06	**91.65** [15]
BC5CDR	89.98	**91.90** [33]
BC5CDR-chem	93.60	**94.88** [15]
BC5CDR-disease	86.07	**88.50** [26]
BioNLP13CG	**87.90**	87.83 [15]
Linnaeus	**88.14**	86.26 [15]

direct comparisons. To keep the best settings, results were directly extracted from [15,26,33] and report in Table 3.

The F-scores in Table 3 show that the proposed model is better than other models on the BioNLP13CG and Linnaeus datasets. The SOTA models on AnatEM, BC5CDR-chem, BioNLP13CG, and Linnaeus employed Spark NLP, an open-source natural language processing library built on top of Apache Spark. For BC5CDR, the best model optimizes bi-encoder for NER via contrastive learning [33]. For BC5CDR-disease, the SOTA model utilizes biomegatron [26], another pre-trained model with sophisticated fine-tuning techniques. However, our model follows the SOTA methods with small margins. On the other hand, for BC5CDR-chem and Linnaeus, the SOTA methods utilize the BiLSTM+CNN+Char [15] model, which combines bidirectional LSTM and convolutional neural network (CNN) layers with character-level embeddings. The BiLSTM+CNN+Char architecture is capable of capturing both local and global contextual information, making it effective for tasks that require understanding the relationships between tokens in a sequence. However, the proposed model is still better than the BiLSTM+CNN+Char model on the BioNLP13CG and Linnaeus datasets, further confirming our idea state in Sect. 3. We also aware that these SOTA methods should be tested for the remaining datasets to create benchmark results and references for next studies. We leave it as a minor future task that slightly affects the comparison of the proposed model.

5.2 Performance with Different PLMs

The performance of the model using different PLMs was observed. To do that, we replaced PubMedBERT by BioBERT (kept the same architecture of using BiLSTM and CRFs) and trained the model on all 29 datasets. The results in the right hand side of Table 2 show that the utilization of PubMedBERT leads to better results. The primary reason for this improvement is the domain-specific pre-training of PubMedBERT. Unlike prior biomedical-related BERT models, PubMedBERT is specifically pre-trained on a large corpus of biomedical texts, particularly from PubMed articles [9]. This enables PubMedBERT to capture and understand the specific language and context of biomedical literature more effectively compared to BioBERT, which is pre-trained on more general texts. The vocabulary of PubMedBERT is also more domain specific than that of BioBERT. As a result, PubMedBERT can provide more accurate results, particularly in the field of BioNER and medical-related tasks.

5.3 The Performance with Epochs

We observed the performance of the proposed model in different epochs. Due to space limitations, we only tested with three datasets: BC5CDR, BioNLP11ID, and BioNLP13CGspecies. Figure 2 shows that the model demonstrates fast convergence, reaching its best performance after approximately 13 epochs. The graph clearly illustrates a steady increase in performance during the initial epochs, followed by a plateau where the model's performance stabilizes. The rapid convergence of the model is an encouraging observation, as it indicates that the model is able to quickly learn and capture the underlying patterns and dependencies in the data. This fast convergence not only saves computational resources but also suggests that the model effectively utilizes the available training data to optimize its parameters and achieve the best possible performance.

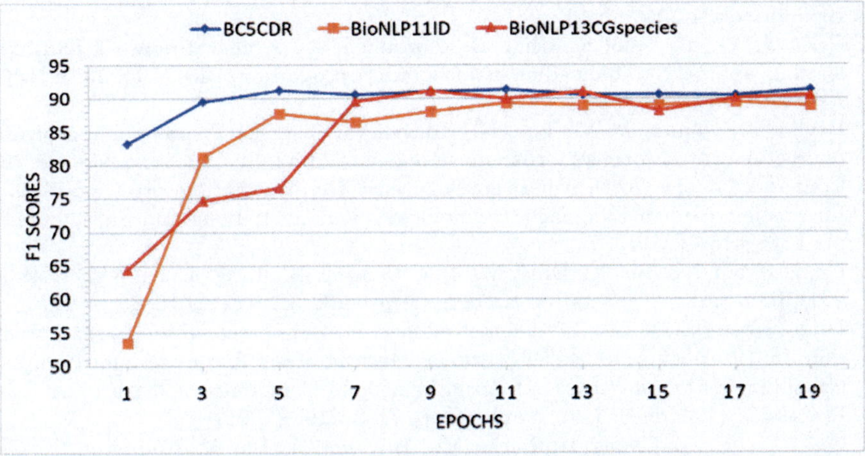

Fig. 2. The performance across different epochs.

6 Conclusion

This paper investigated named entity recognition for biomedical data. We introduce an architecture that stacks BiLSTM and CRFs on the top of PubMedBERT for effectively learning the hidden representation of input sequences. Experimental results on 29 benchmark biomedical datasets show that adding additional layers is beneficial for improving the performance of PubMedBERT. The results serve as important reference points for future research and provide a benchmark for assessing the effectiveness of methodologies in the field of Bioinformatics.

Future work will investigate other architectures to improve its performance. In addition, the model may combine multiple information channels from different sources to enrich the representation of input sequences. Finally, SOTA methods should be tested on all 29 datasets to create benchmark results for next studies.

References

1. Angeli, G., Premkumar, M.J., Manning, C.D.: Leveraging linguistic structure for open domain information extraction. In: Proceedings of the 53rd Annual Meeting of the Association for Computational Linguistics and the 7th International Joint Conference on Natural Language Processing (Volume 1: Long Papers), pp. 344–354 (2015)

2. Brown, T., Mann, B., Ryder, N., Subbiah, M., Kaplan, J.D., Dhariwal, P., Neelakantan, A., Shyam, P., Sastry, G., Askell, A., et al.: Language models are few-shot learners. Adv. Neural. Inf. Process. Syst. **33**, 1877–1901 (2020)
3. Cheng, P., Erk, K.: Attending to entities for better text understanding. In: Proceedings of the AAAI Conference on Artificial Intelligence, vol. 34, no. 05, pp. 7554–7561 (2020)
4. Crichton, G.: MTL-bioinformatics-2016. https://github.com/cambridgeltl/MTL-Bioinformatics-2016 (2016)
5. Crichton, G., Pyysalo, S., Chiu, B., Korhonen, A.: A neural network multi-task learning approach to biomedical named entity recognition. BMC Bioinform. **18**(1), 1–14 (2017)
6. Devlin, J., Chang, M.W., Lee, K., Toutanova, K.: BERT: pre-training of deep bidirectional transformers for language understanding. In: Proceedings of the 2019 Conference of the North American Chapter of the Association for Computational Linguistics: Human Language Technologies (Volume 1: Long and Short Papers), pp. 4171–4186 (2019)
7. Du, Z., Qian, Y., Liu, X., Ding, M., Qiu, J., Yang, Z., Tang, J.: All NLP tasks are generation tasks: a general pretraining framework. arXiv:2103.10360 (2021)
8. Fu, J., Huang, X.J., Liu, P.: Spanner: named entity re-/recognition as span prediction. In: Proceedings of the 59th Annual Meeting of the Association for Computational Linguistics and the 11th International Joint Conference on Natural Language Processing (Volume 1: Long Papers), pp. 7183–7195 (2021)
9. Gu, Y., Tinn, R., Cheng, H., Lucas, M., Usuyama, N., Liu, X., Naumann, T., Gao, J., Poon, H.: Domain-specific language model pretraining for biomedical natural language processing (2020)
10. Guo, J., Xu, G., Cheng, X., Li, H.: Named entity recognition in query. In: Proceedings of the 32nd International ACM SIGIR Conference on Research and Development in Information Retrieval, pp. 267–274 (2009)
11. He, Y., Tang, B.: SetGNER: general named entity recognition as entity set generation. In: Proceedings of the 2022 Conference on Empirical Methods in Natural Language Processing, pp. 3074–3085 (2022)
12. Hochreiter, S., Schmidhuber, J.: Long short-term memory. Neural Comput. **9**(8), 1735–1780 (1997)
13. Huang, K., Altosaar, J., Ranganath, R.: ClinicalBERT: modeling clinical notes and predicting hospital readmission (2020)
14. Raj Kanakarajan, K., Kundumani, B., Sankarasubbu, M.: Bioelectra: pretrained biomedical text encoder using discriminators. In: Proceedings of the 20th Workshop on Biomedical Language Processing, pp. 143–154 (2021)
15. Kocaman, V., Talby, D.: Accurate clinical and biomedical named entity recognition at scale. Softw. Impacts **13**, 100373 (2022)
16. Lafferty, J., McCallum, A., Pereira, F.C.: Conditional random fields: probabilistic models for segmenting and labeling sequence data (2001)
17. Lample, G., Ballesteros, M., Subramanian, S., Kawakami, K., Dyer, C.: Neural architectures for named entity recognition. In: Proceedings of the 2016 Conference of the North American Chapter of the Association for Computational Linguistics: Human Language Technologies, pp. 260–270 (2016)
18. Lee, J., Yoon, W., Kim, S., Kim, D., Kim, S., So, C.H., Kang, J.: BioBERT: a pre-trained biomedical language representation model for biomedical text mining. Bioinformatics **36**(4), 1234–1240 (2020)

19. Li, F., Lin, Z., Zhang, M., Ji, D.: A span-based model for joint overlapped and discontinuous named entity recognition. In: Proceedings of the 59th Annual Meeting of the Association for Computational Linguistics and the 11th International Joint Conference on Natural Language Processing, pp. 4814–4828 (2021)
20. Li, T., Hu, L., Li, H., Sun, C., Li, S., Chi, L.: TripleRank: an unsupervised keyphrase extraction algorithm. Knowl.-Based Syst. **219**, 106846 (2021)
21. Li, X., Feng, J., Meng, Y., Han, Q., Wu, F., Li, J.: A unified MRC framework for named entity recognition. In: Proceedings of the 58th Annual Meeting of the Association for Computational Linguistics, pp. 5849–5859 (2020)
22. Mollá, D., Zaanen, M.V., Smith, D.: Named entity recognition for question answering. In: Proceedings of the Australasian Language Technology Workshop 2006, pp. 51–58 (2006)
23. Ning, J., Li, J., Yang, Z., Sun, Y., Wang, L., Zhang, Y., Lin, H., Wang, J.: BioNER-CFEM: biomedical named entity recognition based on character feature enhancement with multimodal method. In: 2022 IEEE International Conference on Bioinformatics and Biomedicine (BIBM), pp. 787–790 (2022). https://doi.org/10.1109/BIBM55620.2022.9995331
24. Puccetti, G., Chiarello, F., Fantoni, G.: A simple and fast method for named entity context extraction from patents. Expert Syst. Appl. **184**(2021), 115570 (2021)
25. Schuster, M., Paliwal, K.: Bidirectional recurrent neural networks. IEEE Trans. Signal Process. **45**(11), 2673–2681 (1997). https://doi.org/10.1109/78.650093
26. Shin, H.C., Zhang, Y., Bakhturina, E., Puri, R., Patwary, M., Shoeybi, M., Mani, R.: BioMegatron: larger biomedical domain language model (2020)
27. Tzitzivacos, D.: International classification of diseases 10th edition (ICD-10). CME: Your SA Journal of CPD **25**(1), 8–10 (2007)
28. Uzuner, Ö., Luo, Y., Szolovits, P.: Evaluating the state-of-the-art in automatic de-identification. J. Am. Med. Inform. Assoc. **14**(5), 550–563 (2007)
29. Uzuner, Ö., South, B.R., Shen, S., DuVall, S.L.: 2010 i2b2/VA challenge on concepts, assertions, and relations in clinical text. J. Am. Med. Inform. Assoc. **18**(5), 552–556 (2011)
30. Vaswani, A., Shazeer, N., Parmar, N., Uszkoreit, J., Jones, L., Gomez, A.N., Kaiser, L., Polosukhin, I.: Attention is all you need. In: Advances in Neural Information Processing Systems, pp. 6000–6010 (2017)
31. Wang, X., Jiang, Y., Bach, N., Wang, T., Huang, Z., Huang, F., Tu, K.: Improving named entity recognition by external context retrieving and cooperative learning. In: Proceedings of the 59th Annual Meeting of the Association for Computational Linguistics and the 11th International Joint Conference on Natural Language Processing (Volume 1: Long Papers), pp. 1800–1812 (2021)
32. Xu, J., Kim, S., Song, M., Jeong, M., Kim, D., Kang, J., Rousseau, J.F., Li, X., Xu, W., Torvik, V.I., Bu, Y., Chen, C., Akef Ebeid, I., Li, D., Ding, Y.: Building a PubMed knowledge graph. Sci. Data **7**(1), 205 (2020)
33. Zhang, S., Cheng, H., Gao, J., Poon, H.: Optimizing bi-encoder for named entity recognition via contrastive learning. arXiv:2208.14565 (2022)
34. Zhang, Z., Han, X., Liu, Z., Jiang, X., Sun, M., Liu, Q.: Ernie: enhanced language representation with informative entities. In: Proceedings of the 57th Annual Meeting of the Association for Computational Linguistics, pp. 1441–1451 (2019)

Application of Deep Learning Methods for Forest Fire Intelligent Image Processing

Nguyen The Long[1(✉)], Nguyen Thu Huong[1,2,3], Shmeleva Anna. G.[1], and Pham Thi Lien[3]

[1] Department of Informatics, Institute of Cybersecurity and Digital Technologies, MIREA – Russian Technological University, 119454 Moscow, Russia
thelongit88@gmail.com, Shmeleva_a@mirea.ru

[2] Laboratory of Artificial Intelligence and Machine Learning, Institute of Information Technology and Data Science, Irkutsk National Research Technical University, 664074 Irkutsk, Russia

[3] University of Information and Communication Technology, Thai Nguyen University, 70000 Thai Nguyen, Vietnam
{nthuongcntt,ptlien}@ictu.edu.vn

Abstract. There are many challenges for a Deep Learning application project. Many problems need to be solved in an optimal way to improve the performance of the system. We can focus on a specific step or on the whole process. The aim of the work is to show a new way of using convolutional neural networks is proposed to image intelligent process. Due to the influence of climate change, natural disasters are becoming more complicated and have serious consequences. In particular, the phenomenon of forest fires is affected by factors such as humidity, temperature and vegetation characteristics, etc., so the detection and prediction of forest fires face many challenges. In this study, we focus on using image processing techniques and visual algorithms to preprocess data and filter noise, in order to create a complete database for training and testing model. Natural elements are also used in combination with object features (forest fire images) to build feature vectors. This study will allow to develop the technology and open up new opportunities for its subsequent application.

Keywords: Deep Learning · Forest Fire · Image Processing · VGG-16 · Convolutional Neural Network · Artificial Intelligent

1 Introduction

In this day and when technology is rapidly advancing and automation is reaching every process of human life, it is not surprising that the computer vision once described by science fiction writers has evolved and reached a point where it is no longer the future, but a very real present. The rapid development of image processing theories has led to the emergence of image processing technologies. In a nutshell, machine vision is a technology that allows equipment to recognize the manufacturing process, analyze data

P. T. Nghia et al. (Eds.): ICTA 2023, LNNS 847, pp. 126–133, 2023.
https://doi.org/10.1007/978-3-031-49529-8_14

and make an informed decision. All in a fraction of a second [1, 2] Machine learning-based approaches are used in fire detection applications. This study [3] proposes the use of a color model to detect flames. The authors have demonstrated the importance and strengths of the approach compared to other methods. A fire pixel can be determined by the color pixels which lie in a predetermined range of intensity values. Usually, there are pixels that resemble the fire color pixels like orange, sunset, sunshine, etc. There is a need to discriminate these pixels into fire and non-fire pixels. The color models will help to choose among them to propose a new method. As AUV technology advances, scientists are moving towards using AUVs to monitor forest fires [4]. Because the cost and efficiency of this technology is very effective and suitable for the objective factors of nature, as well as topography, it is being applied and becoming more popular. Fractal codes are proposed [5] for use in smoke detection systems. By determining the smoke zone properties, they proceed to define the boundary between them. This work is similar to the process of classifying objects in an image. The main purpose is to find out the characteristics of the smoke to raise the alarm level, and predict the direction of the fire. The Kruskal set of rules is used to assemble the minimum spanning tree and alter the criterion in accordance. The degree of variability in neighboring superb pixels applies to the development method, as a consequence obeying certain global residences despite the fact that grasping choices are made. Though, it's far tough for this technique to control the favored range of supper pixels [6]. The traditional generation of woodland fireplace detection by means of temperature sensing is prone to environmental factors, which include spatial height and width, air flow fee, dust, humidity, and so forth., and fake alarm charge is high. Image-kind fire detection has the characteristics of non-contact detection and has grown to be a new research subject of fireplace detection technology. The existing photograph-based totally fireplace detection is more often than not primarily based on algorithms which include the assist vector gadget (SVM) adopts the precept of structural threat minimization and takes into consideration the education mistakes and generalization potential [7], which can clear up the sensible issues of small samples, nonlinearity, high dimensionality, and local minimum factors, it turns into a research hotspot in the subject of device studying [8]. Some Deep Learning technology have been implemented to hit upon woodland fires [9, 10], however neural net-work set of rules has gaining knowledge of risks, slow convergence, and clean to fall into the nearby minimal point [11, 12], sensitivity and stability of reduced fire detection [13]. Consequently, we propose a top-of-the-line technique to overcome this disadvantage, that's to enhance photograph segmentation within the picture pre-processing step to improve efficiency and accuracy, in addition to the time to stumble on wooded area hearth. This work [14] defined the usage of a neural function extractor for generating a feature set to be used in a neural network based totally character and numeral recognition gadget. The effects obtained are then in comparison with the ones in advance obtained from a similar application and a PCA-based totally function extraction method. Six device mastering algorithms to categories exclusive Pedestrians events based at the Inertial Dimension Unit and global Positioning gadget alerts are applied in [15].

2 Methodology

A convolutional neural network is any such special form of feed-forward neural networks, and feed-forward method that the variable neutrals on this network are divided into corporations known as layers. And whilst the sort of layered neural network is carried out to the facts, then the activation of the layers – the fee of those variables – is calculated sequentially: first the activation cost of the primary layer, then the activation price of the second one layer, and so on till the final layer. The activation results of the closing layer function the output endpoint of the effects of the convolutional neural community.

The neural style transfer process consists of several steps, of which the following can be distinguished as the main ones:

- preparation of a convolutional neural network;
- preparation of initial images and creation of noisy;
- creating an output image.

2.1 Preparing the Convolutional Neural Network

Preparing a convolutional neural network means training or loading this very network, namely its entire model with the weights of each neuron. From a mathematical point of view, the neural network itself, or rather its model (see Fig. 1. Structure of VGG-16), is a collection of matrices of each layer of this neural network. That is, there is a matrix of convolution layers, a matrix of two hidden fully connected layers, and a final N-spatial vector.

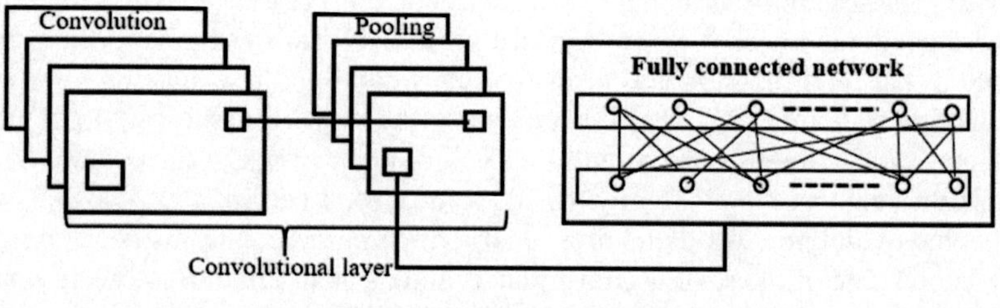

Fig. 1. Structure of VGG-16

The backpropagation method can be divided into 4 separate blocks:

- Direct distribution;
- Loss function;
- Backpropagation;
- Weight update.

For the duration of forward propagation, a training image is taken – as we remember, this is a matrix with size 32 × 32 × 3. In the first training example, due to the fact that all weights or clear out values have been initialized randomly, the output cost will be something like [1], i.e., such a value, which will not give preference to any particular number. A network with such weights cannot discover base layer properties and cannot reasonably decide the magnificence of an image. This results in a loss function. What we're using is training information. Such facts have both an image and a label. Shall we say the first training picture is the quantity three. The label for the picture is [0, 0, 0, 1, 0, 0, 0, 0, 0, 0]. The loss feature can be expressed in different methods (Eq. 1), but the RMSE (Root Mean Square Error – E_{total}) is frequently used.

$$E_{total=} = \frac{1}{2}(target - output)^2 \tag{1}$$

where *target* is reality, *output* is a prediction.

Backpropagate through the network based on determines which weights had the most impact on the loss, and locate methods to track them to reduce the loss. All the filter allows out weights and update them so that they trade inside the course of the error reduction gradient, according to the method, according to the formula (Eq. 2)

$$W = W_i - \eta \frac{dE_{total}}{dW} \tag{2}$$

where W is the weight of the neural, W_i - initial weight, η is the learning step size.

In fact, we minimize the number of losses that result from the processing of a convolutional neural network. On the calculus side, we minimize the loss (error) by adjusting the independent variables of the weights.

2.2 Preparation of Initial Images and Creation of Noisy

After the neural network has been prepared by loading or training it, it is necessary to load images into the system. In our case, content and style. As mentioned in the previous chapter, which are matrices of values that are further normalized, which will help speed up the generation of the output image.

After that we create a noisy image. A noisy image is a combined image from two source images by overlaying the style image on the content image with a certain blending percentage. This is done mainly to speed up obtaining good results. Of course, it would be possible to use a purely noisy image, but as practice shows, we will still come to the same result, but spending an order of magnitude more time on it. In what follows, we will use this particular noisy image, called the combined image.

2.3 Creating an Output Image

To find this middle, it is necessary to calculate the difference between the activation layers of the content image (or style image) and the combined one. This is done using functions called image loss functions or image cost functions.

We take conv4_2, one of the sophisticated VGG-16 convolutional layers, as the content image activation layer. This is due to the desire to store objects from the content image. We do this thanks to the squared difference function between the content image and the combined image, shown in the formula (Eq. 3), which helps us to transfer the objects captured by the chosen activation layer from the content image ($L_{content}$) to the combined image.

$$L_{content}p, x, l = \frac{1}{2}i, j\left(F_{ij}^{l} - P_{ij}^{l}\right) \tag{3}$$

where F_{ij}^{l} are the activation values of the combined image layer, P_{ij}^{l} are the activation values of the content image layer.

This is quite enough to calculate the loss of the image of the style (L_{style}), , but in practice, to get the best result, not one activation layer is used in the calculations, but several. An example is shown in formula (Eq. 4).

$$L_{style}a, x = {}_{l=0}{}^{L}W_{l}E_{l} \tag{4}$$

The final function for calculating the image difference will be the total cost function. Total loss measures how the original image differs from the contents of the style image. To have control over how we want our combined image to look like content or style (L_{total}), , two variables are introduced: α and β. Calculated by formula (Eq. 5).

$$L_{total}p, a, x = \alpha L_{content}p, x + \beta L_{style}a, x \tag{5}$$

Finally, having in hand a great way to measure the similarity of the combined image with the original. We need to update it so that the next time the resulting image has a smaller difference, in other words, a smaller loss function value. It turns out that step by step we will update the combined image so that it becomes more and more similar to both the content image and the style images.

We update it by subtracting the derivation (deviation) value from the loss function. The derivation value itself simply tells us which way to change the image. This process is called cost minimization. As a result, we have nothing more than stochastic gradient descent, which tends to zero. To do this, we will use Adam (Adaptive Moment Estimation) – a method that calculates adaptive learning rates for each parameter.

3 Result

The key parameters of the neural style transfer algorithm are: the speed and quality of output image generation. The speed is understood as the number of iterations until an acceptable result is achieved, since the speed of execution of these will depend on the power of the computer's CPU. But in order to have a general idea of the amount of time spent on generating one image, we will fix the execution time of each iteration. In turn, quality is an undefined feature, on the one hand it depends on the opinion of the person (expert) who evaluates the final result, but on the other hand, it was said earlier that the algorithm with each iteration goes to an equal ratio of the content image and the image of the style according to the parameters that are for it are given, that is, the image quality for the algorithm is measured in the difference coefficient, therefore, in order not to start from abstract values, we will take this parameter to evaluate the quality. To decide on the result of the developed algorithm, it is necessary to conduct a series of tests that will allow you to check the performance of the application in real conditions.

During the training process, the models were optimized using the accuracy and loss metrics. The accuracy expression essentially describes which proportions of the two groups (fire or no fire) were correctly classified on average. The loss value is derived from the objective function of the neural network and represents the error to be minimized. After about 50 epochs, the values for the validation sets were stable. However, to truly ensure training success, 100 epochs were chosen for the entire training. No overfitting was detected that would have been noticeable by a decrease in accuracy. To illustrate this research, the following figure shows the training and validation accuracy and loss (see Fig. 2. The training and validation accuracy and loss of proposed method) plotted at the genus level for an example model of the original dataset.

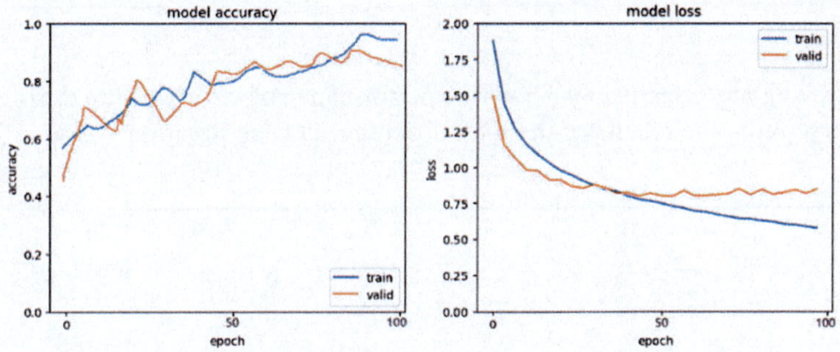

Fig. 2. The training and validation accuracy and loss of proposed method

To evaluate the performance of the algorithm according to the required criteria. The test was conducted with different pairs of content and style images (see Fig. 3. Original of forest image and Result of Intelligent image process). During each test, the execution time (Time spent), the total difference (Total Loss) was recorded and the generated image was saved at every 50th iteration.

Fig. 3. Original of forest image and Result of Intelligent image process

The first test was carried out with the images shown in Table 1. Results of the test. A content image and a style image with the same resolution size were selected, namely, the image data has dimensions of 800 × 600 pixels.

During the third test, 5000 iterations were completed. And generated 101 images. The execution time of one iteration is similarly 12 s. This is also clearly seen in the test results presented in Fig. 4 (see Fig. 4. Graph of the total difference by test iterations).

Table 1. Results of the test.

No iteration	The total difference (total loss)	The execution times (time spent)
1	4,20E+13	12
51	1,18E+12	13
101	4,92E+11	12
151	2,94E+11	12
201	2,08E+11	12
251	1,60E+11	11
301	1,30E+11	11

Before drawing any conclusions about the performance of the algorithm during testing, it is worth plotting the results of the total difference at each iteration.

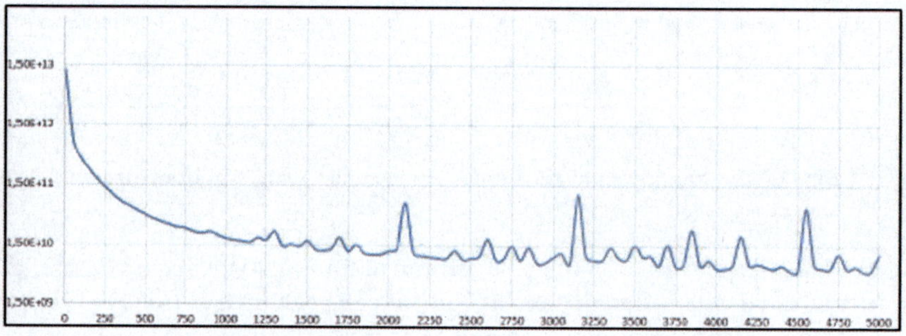

Fig. 4. Graph of the total difference by test iterations

4 Conclusion

The performed work is devoted to the study of convolutional neural networks and the development of a new way to use them. The purpose of the work determined its main direction - the development of an algorithm for neural style transfer using a convolutional neural network to create images with a pair of necessary content and style images as input. For this, the ways of using convolutional neural networks in modern realities were considered and analyzed. During the analysis, an area was identified where convolutional networks are of little use, and a new task and ways of its implementation were formulated.

This system allows you to create an image, having one source image and a second style image as input, taking the features of the first and the drawing style of the second. The system is quite demanding on the computing power of the computer, as it involves the execution of numerous mathematical operations, and to a lesser extent on the computer's RAM to load the convolutional neural network architecture into it. As a result of the work, the ways of developing this work were demonstrated, such as: reducing the time of image generation and improving the quality of the created image.

References

1. Kulikov, A.A.: The model is a reprint of an object in the image. Russ. Technol. J. **8**(3), 7–13 (2020)
2. Kulikov, A.: A: Application of biometric systems in face identification technologies. Russ. Technol. J. **9**(1), 7–14 (2021)
3. Alves, J.S.C.: Automatic forest fire detection based on a machine learning and image analysis pipeline. New Knowl. Inf. Syst. Technol, 240–251 (2019)
4. Arridge, S., H. A.: Networks for nonlinear diffusion problems in imaging. IEEE Transactions on Image Processing, 471 – 482 (2020)
5. Bangare, S., Dubal, A.E.: Reviewing OTSU method for image thresholding. Int. J. Appl. Eng. Res. **10**, 21777–21783 (2015)
6. Bhanumathi, V., Sangeetha, R.: CNN based training and classification of MRI brain images. In: processing 5th International Conference on Advanced Computing Communication Systems (ICACCS), pp. 129–133 (2019)
7. Browne, M.G.S.: Convolutional neural networks for image processing: An application in robot vision. Adv. Artif. Intell., 641–652 (2020)
8. Chen, Q., Xu, J., K.V.: Fast image processing with fully-convolutional networks. In: International Conference on Computer Vision (ICCV), pp. 235–250 (2017)
9. Nguyen, T.H., Nguyen, T.L.: Machine learning algorithms application to road defects classification. Intell. Decis. Technol. **12**(4), 1–8 (2018)
10. Wang, C.: Research on support vector machines. Comput. Eng. Appl. **37**, 58–61 (2001)
11. Fujiwara, N., Terada, K.: Extraction of a smoke region using fractal coding. In: Australasian Joint Conference on Artificial Intelligence, pp. 659–662 (2020)
12. Huang, X., Shan, J.: Lung nodule detection in CT using 3d convolutional neural networks. In: IEEE 14th International Symposium on Biomedical Imaging, pp. 379–383 (2017)
13. Hui, T.H.A.: Multi-type flame detection combined with faster R-CNN. J. Image Graph. **24**, 73–83 (2019)
14. Zhao, J., Zhong, Z.: SVM based forest fire detection using static and dynamic features. Comput. Sci. Inf. Syst. **8**, 821–841 (2011)
15. Zhao, W., Zhang, H.: A semantic segmentation algorithm using FCN with combination of BSLIC. Appl. Sci. **8**, 501–525 (2018)

Choosing Data Points to Label for Semi-supervised Learning Based on Neighborhood

Dinh-Minh Vu[1], Thanh-Son Nguyen[2(✉)], Trung-Nghia Phung[3], and Trinh-Hoang Vu[4]

[1] Hanoi University of Industry, No. 298 Cau Dien St., Bac Tu Liem Dist., Hanoi, Vietnam
minhvd@haui.edu.vn

[2] Academy of Finance, No. 58, Le Van Hien St., Duc Thang Wrd., Bac Tu Liem Dist., Hanoi, Vietnam
Sonnt@hvtc.edu.vn

[3] University of Information and Communication Technology, Z115 Street, Quyet Thang Commune, Thai Nguyen City, Vietnam
ptnghia@ictu.edu.vn

[4] VNU University of Science, No. 334, Nguyen Trai St., Thanh Xuan Dist., Hanoi, Vietnam

Abstract. Semi-supervised fuzzy clustering is an extension of fuzzy clustering, it uses additional information to guide and monitor the clustering process. The additional information can be seed form (labeled data), constraint form (must-link/cannot-link), or predefined membership form. In fact, with based clustering, the clustering result depends on the selected data point, so with different selected data points will produce different clustering results. In some cases, the performance of clustering can be reduced if the data points are not selected properly. This is a really problem with semi-supervised clustering algorithms. In addition, the selection of the right data points helps to reduce the number of queries from experts. For this purpose, we propose a new learning algorithm (called WHFMN) for the seed's collection activity, this activity identifies candidates for additional information using fuzzy min max neural network. Four of the most interesting characteristics of WHFMN are that (1) it automatically calculates data points to add additional information, (2) it automatically adds new clusters without retraining, (3) it automatically finds the noise objects according to the density of the natural clusters and (4) it can find natural cluster structure even if density and shape are different between clusters. Experimental results prove that the proposed algorithm can improve the efficiency of the label selection method more than the previously published methods.

Keywords: Hyperboxe · Fuzzy min max · Seed form · Data point · Neighborhood

P. T. Nghia et al. (Eds.): ICTA 2023, LNNS 847, pp. 134–141, 2023.
https://doi.org/10.1007/978-3-031-49529-8_15

1 Introduction

Clustering is one of the techniques in data mining, it belongs to the class of unsuper-vised learning methods. The training processes to separate groups according to the similarity of data is called clustering. Fuzzy semi-supervised clustering is an extension of fuzzy clustering that has also received much attention from scientists who are interested in research [3, 6, 8–10, 12], and is applied in many fields of recognition, information processing, healthcare, etc. [1, 2].

Semi-supervised fuzzy clustering uses additional information to guide and monitor the clustering process, thereby increasing the quality of the cluster. In fact, semi-supervised clustering can improve clustering results with a small amount of additional information.

Based-on seed form method, the clustering result depends on the selected data point (choosing seed), and the number of selected data points. So, with different selected data points will produce different clustering results. Therefore, with different selected data points, and different number of selected data points will produce different clustering results. Sometimes the performance of clustering can be degraded if the selected data point is not good. Newly proposed learning algorithms aim to increase the clustering quality and reduce the ratio of labeled data samples compared to unlabeled samples in the training dataset.

In [12], the research team developed an active learning algorithm for the seed collection task, identifying data points to receive user labels using the k-nearest neighbor algorithm and min-max called SSGC (Semi-Supervised Graph based Clustering). SSGC is capable of detecting noise based on k-nearest neighbor algorithm. In [7], the authors proposed an algorithm that combines K-Means and min-max for seed selection called SKMMM (semi-supervised clustering based on K-Means and Min-Max). SKMMM collects seeds using an iterative and minimal strategy-based approach. However, both SSGC [12] and SKMMM [7] still need a lot of queries from experts.

In [5], the authors mentioned the method of randomly selecting seeds using FMNN (Fuzzy Min-max Neural Network) [9] neural network called GFMNN (General FMNN). This model shows the advantages of FMNN in semi-supervised clustering. However, this method is heavily influenced by the order of the data points and requires more seeds per training to increase the quality of the cluster. Also, based on the advantages of FMNN, in 2018, the research team proposed a new model that significantly improves the quality of GFMNN called SCFMN (Semi-supervised Clustering in FMNN) [11]. SCFMN model helps to reduce the number of queries from experts and still gives good results. However, SCFMN still requires more time for training the network (it increases algorithm complexity). To improve SCFMN, in 2022, Minh, V. D, et al. proposed the MSCFMN (Modified SCFMN) model [14]. MSCFMN has been proven with its advantages over SCFMN. In addition, MSCFMN also significantly improved its quality and good applicability in the healthcare field. However, in both GFMNN [5], SCFMN [11] and MSCFMN [14], the research teams hadn't evaluated the boundary data, which causes ambiguity between data classes/clusters. In other words, these models didn't capable of detecting noise.

From the above studies, it is shown that FMNN can help reduce queries from experts and increase the quality of clusters. For this purpose, we deal with the problem of choosing good labels for semi-supervised clustering using the SS-FMM (Semi-Supervised clustering in FMNN) [13]. Our new algorithm performs the task of selecting neighborhoods and data points to label (performs queries from experts), and then propagates information to neighboring samples based-on fuzzy values using SS-FMM. This method reduces ex-pert queries and increases the number of labeled data points, thereby increasing the clustering performance. Especially the ability to select labeled data points at the edge of the cluster.

In summary, the contributions of this paper are as follows: (i) chooses good seeds for the center of each cluster; (ii) self-determining noise (boundary regions) according to the density of natural clusters and selecting good seeds for these regions; (iii) it is possible to find natural cluster structure even if the densities between clusters are very different; and (iv) it automatically adds new clusters without retraining.

2 The structure of SS-FMM

2.1 Fuzzy min max hyperbox

A fuzzy min-max hyperbox (HX) in an n-dimensional data space is defined through two limit points min (V) and max (W) denoted as $B(V, W)$. HX B_j is a set of four ingredients, be defined by (1) [13].

$$B_j = \{A_h, V_j, W_j, b_j(A_h, V_j, W_j)\}, \left(j = \overline{1, k}, B_j \in B, k = |B|\right) \tag{1}$$

where: V_j, W_j are the min and max points of the HX B_j; $b_{(A_h, V_j, W_j)}$ is the membership function or dependency of sample A_h to B. $b_{(A_h, V_j, W_j)}$ is defined according to (4) [13]; $A_h = (a_{h1}, ..., a_{hn}) \in R^n$, $(h = \overline{1, m})$ is the hth sample of training set D $(m = |D|)$.

The dimension on the ith dimension of the B_j (denoted as θ_i), and the size of B_j (denoted as θ) are calculated by (2) [13]:

$$\theta = \frac{1}{n} \sum_{i=1}^{n} \left(\max(w_{ji}, v_{ji}) - \min(w_{ji}, v_{ji})\right) \tag{2}$$

2.2 Function belonging to fuzzy hyperbox

The membership function $b_{(A_h, B_j)}$ determines the degree of belonging of A_h to HX B_j. A_h $(A_h \in D)$ is contained in HX B_j if $b_{(A_h, V_j, W_j)} = 1$ (A_h has full membership in B_j). $b_{(A_h, V_j, W_j)}$ is defined according to (3) [13].

$$b_{(A_h, V_j, W_j)} = \frac{1}{n} \sum_{i=1}^{n} \left[1 - f\left(a_{hi} - w_{ji}, \gamma\right) - f\left(v_{ji} - a_{hi}, \gamma\right)\right] \tag{3}$$

where γ is the sensitivity used to correct the attenuation when the sample comes out of HX [13].

2.3 The architecture of the fuzzy min max neural network

Figure 1(a) show the FMNN architecture representation [13]. The F_A input class consists of n nodes, each of which is an attribute of the input data. The second layer F_B consists of q nodes, each node is a superbox, each node q is added during training.

The F_C output layer consists of k nodes, each node is a HX. Each jth neuron in the F_B layer (Fig. 1(b)) is connected to the inputs through a set of double weights, respectively two vectors V_j and W_j. The connection between the jth node in F_B and the ith node in F_A through two weights v_{ji} and w_{ji}.

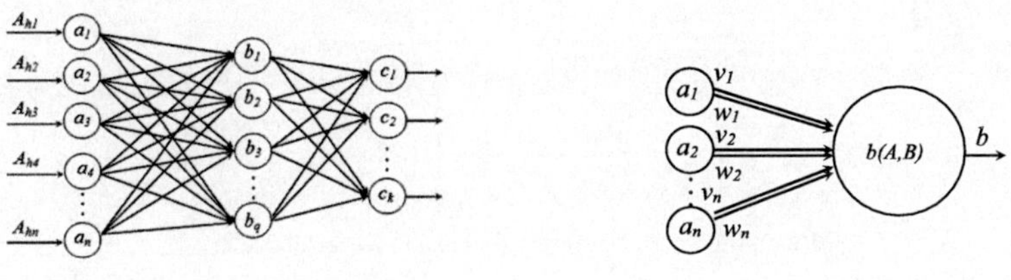

Fig. 1. Architecture of SS-FMM

3 The proposed method

In this section, we present the WHFMN (Weighted of Hyperbox in FMNN) model. WHFMN chooses seeds based on the HX partition of SS-FMM. The difference of WHFMN compared to SS-FMM is the ability to detect interference at the boundary regions. For the HXs in the boundary region, WHFMN assigns it a weight of 1 (and increases if overlap continues), the center of this HX is the seed selected for expert opinion. For the non-noise HXs, WHFMN assigns it a weight of 0. In other words, the HXs in the boundary region have a higher weight than the remaining regions.

The learning algorithm of WHFMN builds a large seed set and with a low number of user queries. WHFMN performs SS-FMM training to generating HXs and weighted them. The steps of the WHFMN learning algorithm are shown in Table 1.

The time complexity (denoted T) of WHFMN algorithm is calculated by (4):

$$T = (T_1 + T_2) \tag{4}$$

where:

- T_1 is the time complexity of stage divides D dataset into c HXs, then Experts provide a label for this HXs (line 1): $T_1 = (M \times N \times K)$.
- T_2 is the time complexity of the phase of selecting and evaluating boundary data points to receive labels from the expert (from lines 2 to 17): $T_2 = (M \times N \times K)$.
- The time complexity of WHFMN algorithm: $T = O(M \times N \times K)$

Table 1. WHFMN learning algorithm.

Input: D, c;
Output: B;
Using SS-FMM [13] divides D into c HXs, Experts provide a label for this HXs;
For $A_h \in D$ *with* $A_h^{label} = null$
    For $B_j \in B$
        If $b_j\left(A_h, B_j\right) = 1$ then
            $A_h^{label} = B_j^{label}$
        Else
            If there are any B_j satisfying $\left(b_{(A_h,B_j)} = max\left\{b_{(A_h,B_j)} : j = \overline{1,k}\right\}\right) and \left(\theta_{(A_h,B_j)} \le \theta_{max}\right)$ then
                Expand B_j; $A_h^{label} = B_j^{label}$;
                If B_j overlaps with any other B_i ($i \ne j$, B_i and B_j belong to two different clusters) then
                    Weighting for B_j; Experts provide a label for this HX;
            Else not any Bj satisfy $\left(b_{(A_h,B_j)} = max\left\{b_{(A_h,B_j)} : j = \overline{1,k}\right\}\right) and \left(\theta_{(A_h,B_j)} \le \theta_{max}\right)$ then
                A new HX is created (called HX_{new}); Experts provide a label for this HX;
                $B = B \cup HX_{new}; A_h^{label} = HX_{new}^{label}$
            Endif;
        Endif;
    Endfor;
Endfor;
Rerurn B;

4 Experimental results

4.1 Data description and Evaluation method

In this section, the performance of WHFMN model is evaluated and compared to other methods such as Random, SSGC [12], SKMMM [7], MSCFMN [14], SS-FMM [13]. To do this, we implement WHFMN on datasets from UCI [16] and CS [15]. The details of each data set are given in Table 2.

Table 2. The data set was used for the experiments and evaluation.

#	Data set	No. Objects	No. Attributes	No. Clusters
1	Iris	150	4	3
2	Soybean	47	35	4
3	Zoo	101	16	7
4	Yeast	1484	8	10

Four indicators are used to evaluate including Rand Index (*RI*) [7], number of expert queries (*NoQ*), and number of seeds (*NoS*).

The *RI* evaluates the fit between the model's partitioning result (*X*) and its actual partitioning (*Y*). Assume, x_i and x_j are data points in data set *D* (with $i \neq j, n = |D|$), y_i and y_j are labels of x_i and x_j in both *X* and *Y*. Let *u* be the number of decisions where $y_i = y_j$ in *X*, and *v* is number of decisions where $y_i \neq y_j$ in *Y*. The *RI* measure is determined by the formula (5):

$$RI_{(X,Y)} = \frac{2(u+v)}{n(n-1)} \tag{7}$$

The *RI* takes 1 of the values in the range [0..1], the closer the value of *RI* is to 1 the better.

Test was conducted on a computer with Intel CoreTM i3 3.40 GHz processor with 8GB RAM on 64bit MATLAB R2014a.

4.2 Evaluation results

Figures 2, 3, and 4 present and compare the experimental results of WHFMN, SKMMM, MSCFMN, SS-FMM, and Random methods. Evaluation indicators include *RI*, *NoQ*, and *NoS*.

Fig. 2. Comparison of *RI* among 5 models on experimental datasets

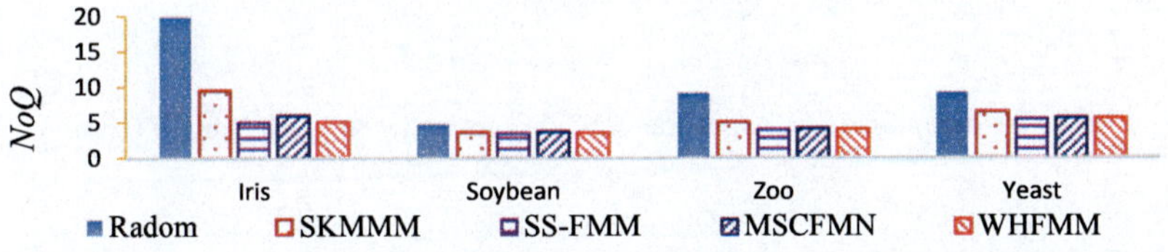

Fig. 3. Comparison of *NoQ* among 5 models on experimental datasets.

Figure 2 is a graphical comparison of the *RI* value of the WHFMN with the SKMMM, MSCFMN, SS-FMM, and Random methods. It can be seen from the figure that the clustering quality of WHFMN is better than that of other methods. WHFMN has an *RI* equivalent to MSCFMN and higher than SKMMM, SS-FMM, and random method. This can be explained by the fact that the WHFMN has 2 stages of getting the user label (choosing seeds). Stage 1 of WHFMN performs cluster center determination and receives labels at each cluster center. This stage is very good for label propagation (breeding) in

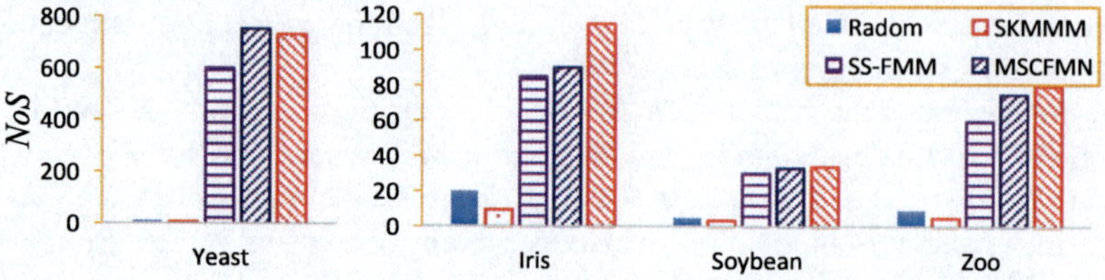

Fig. 4. Comparison of *NoS* among 5 models on experimental datasets

WHFMN algorithm because it is inherited from SS-FMM. Stage 2 performs identification of boundary regions to receive labels from experts (avoiding false partitioning due to noise). The SKMMM, SS-FMM and MSCFMN methods are not capable of determining the cluster boundary region. However, the WHFMN required fewer queries from the expert (*NoQ* in Fig. 3) and still produced a large seed set (*NoS* in Fig. 4). Although the MSCFMN is capable of controlling label propagation across low opacity sample spaces near the edge of the cluster, the boundaries of the cluster are still undetermined. The cluster quality of the random method is unstable and low.

Figure 5 is a graphical comparison of the Rand Index between SSGC and WHFMN on the Iris and Thyroidnew datasets with different number of seeds. The results show that WHFMN performs better than SSGC. This can be explained by the ability of the WHFMN to evaluate and query users at the cluster center to spread information and the ability to determine the cluster boundary to query users.

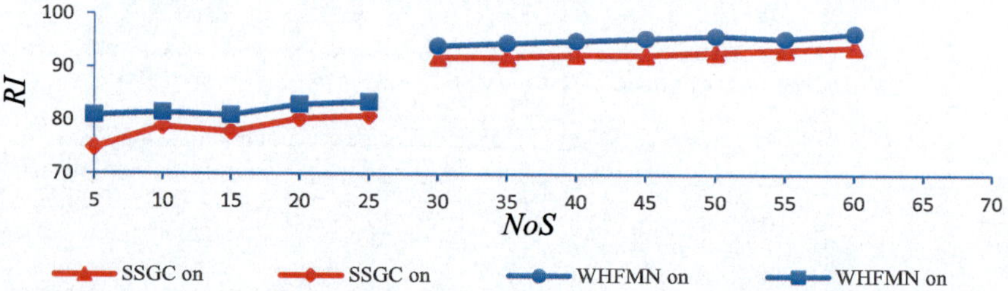

Fig. 5. *RI* comparison on experimental data sets between SSGC and WHFMN.

5 Conclusion

This paper presents a model for selecting data points for expert opinion using FMNN with neighborhood identification. Compared with methods such as random seed selection Random, SKMMM [7], MSCFMN [14], SS-FMM [13], SSGC [12], our method has the advantage of minimizing the number of user queries. The experimental results show that WHFMN has good results when dealing with experimental data. WHFMN has the ability to add new particles when the dataset forms new clusters without retraining. This feature is obtained by inheriting from SS-FMM [13]. However, the adaptive learning processes of WHFMN takes time and experience by "trial and error". This also is a limitation of the min-max fuzzy clustering algorithm, and a further research direction to be considered.

References

1. Allahyar, A., Yazdi, H.S., Harati, A.: Constrained SemiSupervised Growing Self-Organizing Map. Neurocomputing **147**, 456–471 (2015)
2. Gabrys, B., Bargiela, A.: General fuzzy min-max neural network for clustering and classification. IEEE Trans. Neural Net. **11**(3), 769–783 (2000)
3. Wagstaff, K., Cardie, C., Rogers, S., Schrodl, S.: Constrained k-means clustering with background knowledge. In: In Proc. of Intl. Conf. on Machine Learning, pp. 577–584 (2001)
4. Khuat, T.T., Gabrys, B.: Accelerated learning algorithms of general fuzzy min-max neural network using a novel hyperbox selection rule. Inf. Sci. **547**, 887–909 (2021)
5. Khuat, T.T., Gabrys, B.: A comparative study of general fuzzy min-max neural networks for pattern classification problems. Neurocomputing **386**, 110–125 (2020)
6. Lelis, L., Sander, J.: "Semi-supervised density-based clustering," in In proc. IEEE Intl. Conf. on Data Mining, pp. 842–847 (2009)
7. Le, C., Vu, V.V., Yen, N.T.H.: Choosing seeds for semi-supervised graph based clustering. J. Comput. Sci. Cybern. **35**(4), 373–384 (2019)
8. Yan, R., Zhang, J., Yang, J., Hauptmann, A.: A discriminative learning framework with pairwise constraints for video object classification. IEEE Trans. Pattern Anal. Mach. Intell. **28**(4), 578–593 (2004)
9. Simpson, P.K.: Fuzzy min-max neural network-Part II: clustering. IEEE Trans. Fuzzy Syst. **1**(1), 32–45 (1993)
10. Basu, S., Banerjee, A., Mooney, R.: Semi-supervised clustering by seeding. In: Proc. of 19th Intl. Conf. on Machine Learning, pp. 281–304 (2002)
11. Tran, T.N., Vu, D.M., Tran, M.T., Le, B.D.: The combination of fuzzy min-max neural network and semi-supervised learning in solving liver disease diagnosis support problem. Arab. J. Sci. Eng. **44**(4), 2933–2944 (2019)
12. Vu, V.-V.: An efficient semi-supervised graph based clustering. Intell. Data Anal. **22**(2), 297–307 (2018)
13. Vu, D.M., Nguyen, V.H., Le, B.D. (2016). Semi-supervised clustering in fuzzy min-max neural network. In: International Conference on Advances in Information and Communication Technology, pp. 541–550). Springer International Publishing
14. Minh, V.D., Ngan, T.T., Tuan, T. M., Duong, V.T., Cuong, N.T.: (2108–6838) An improvement in integrating clustering method and neural network to extract rules and application in diagnosis support. Iran. J. Fuzzy Syst. (2022)
15. https://cs.joensuu.fi/sipu/datasets
16. https://archive.ics.uci.edu/ml/datasets.html

Classification of Fermentation Levels of Cacao Beans (*Theobroma cacao* L.) Using Sensing Technology and Neural Networks

Rikiya E. Yamamoto[1], James Leo C. Grimaldo[1], Mary Meliza V. Parcon[1], Roselyn A. Maaño[1], Ronaldo C. Maaño[1], Jose B. Tan Jr.[1], Rodrigo C. Belleza Jr.[1], Pedro Jose L. De Castro[1], and Nguyen Duc-Binh[2](✉)

[1] Manuel S. Enverga University Foundation, 4301 Quezon, Philippines
t21-32986@student.mseuf.edu.ph, {jc.grimaldo,mv.parcon, ronaldo.maano,jose.tanjr,rodrigo.belleza, pj.decastro}@mseuf.edu.ph, roselyn.maano@gmail.com

[2] Thai Nguyen University of Information and Communication Technology, Thai Nguyen City, Vietnam
ndbinh@ictu.edu.vn

Abstract. The global chocolate market is proliferating, and this growth has a significant economic impact on cacao-producing regions. Smallholder farmers in these regions rely heavily on the chocolate industry's income, providing them a much-needed boost. The quality assessment of cocoa beans is essential in the cocoa industry as it directly affects the value and price of the final product. The combination of artificial neural networks with the electronic nose device has emerged as a promising method to assess the quality of cocoa beans. The study classifies the fermentation degree of cacao (*Theobroma cacao* L.) beans through sensing technology and deep learning. A moving average is applied to the raw data to eliminate volatility. A comparison of pre-processing techniques to better understand the data and identify critical patterns and relationships within the dataset. ANN, DNN, and TabNet were employed as data modeling techniques. Deep Neural Network (DNN) shows remarkable results to the other two architectures. Visualization techniques using a dashboard display gas data graphs and fermentation degree classifications throughout fermentation. The dashboard serves as monitoring to reduce the waste of unfermented cacao beans cut during cut-testing.

Keywords: Fermentation Levels of Cacao Beans (*Theobroma cacao* L.) · Sensing Technology · Neural Networks

1 Introduction

Cacao was pivotal in pre-Hispanic Maya's economic, social, religious, and political life [1]. The global chocolate market proliferates and has a significant economic impact on smallholder farmers in cacao-producing regions. Annual crop yields in the cacao industry are precarious, and sustainability is at risk due to various challenges. The International

P. T. Nghia et al. (Eds.): ICTA 2023, LNNS 847, pp. 142–157, 2023.
https://doi.org/10.1007/978-3-031-49529-8_16

Cocoa Organization has identified four major strategic issues: sustainable production, maintaining a sustainable industry chain, promoting sustainable cocoa consumption, and strategic management [2]. They aim to address these challenges so that cacao's economy can contribute towards achieving Sustainable Development Goals related to food and job security. The Philippines has significantly contributed to the global cacao industry in recent years. Small-scale farmers in the country have played an integral role by processing their cocoa beans into a locally beloved delicacy known as "tablea" [3]. Chocolate is a popular food that can be eaten as candy or used as a flavoring in beverages and confectionaries. Due to its widespread consumption, the chocolate market has impacted the global economy [4]. According to the Philippines Statistics Authority, the Philippines is regarded as one of the top producers in the world, with the Davao region contributing the most in the country, yielding 2260 metric tons in 2022. Despite this, there is still a significant difference in the country's supply to neighboring West African countries such as Côte d'Ivoire and Ghana, which account for approximately 70% of the world's cacao supply [5]. Cacao processing involves collecting the pods, gathering the beans, and postharvest processing, including fermentation, roasting, winnowing, and conching [6]. The development of flavor and aroma in chocolate is dependent on the processing procedures. These stages are particular to fermentation. Bioactive substances present during fermentation enhance the flavor of cocoa beans by generating biochemical changes in their composition [7]. The growth of yeasts and lactic acid bacteria, which are precursors to gases such as ethanol, carbon dioxide, and other metabolites, is indicated by the release of a nutty and floral aroma [8]. The resulting factors are monitored and managed to provide optimal fermentation conditions. The parameters could be examined as indicators of the degree of fermentation of the cacao beans.

Electronic devices, such as electronic noses, identify the volatile compounds during fermentation. The gas sensors attached to the e-nose can detect various gases, including ethanol and carbon dioxide. Electronic noses benefit from a faster and cheaper way that farmers may utilize. Artificial Intelligence combined with the e-nose automates the process of monitoring and analyzing the mentioned substances, enabling the automation of manual labor to speed up and enhance chocolate production [9]. Integrating Internet of Things (IoT) technology and Artificial Intelligence (AI) revolutionizes the fermentation monitoring process in chocolate production. These technologies enable remote monitoring, predictive analytics, and automation, opening the path for improved chocolate quality, efficiency, and cost-effectiveness. By connecting electronic noses and gas sensors to the network, real-time data collection and analysis can be achieved, enabling farmers to make informed decisions and optimize fermentation conditions. Adopting this technology can move the industry ahead, bringing up novel prospects for precision fermentation control and addressing the expanding needs of chocolatiers worldwide. The study is focused on preprocessing the cacao beans dataset by extracting essential attributes through normalization and feature scaling and identifying the best-performing algorithm based on the training conducted with different machine learning and deep learning algorithms.

2 Materials and Methods

The study's conceptual framework in Fig. 1 is composed of hardware input phase hardware, including components such as ESP32, Raspberry Pi, SD Card Module, and Gas Sensors used to build the electronic nose. The software implements Keras, Python, Django, and TensorFlow on platforms such as Raspberry Pi OS and Visual Studio Code. The process involves assembling materials for an electronic nose, programming codes, and gathering data for analysis and interpretation. Emphasis is on analyzing cacao fermentation degree using Artificial Neural Networks through an electronic nose device. A dashboard-based website with Artificial Intelligence will be deployed in the outcome phase to determine the cacao fermentation degree accurately. The study aims to enhance cacao yield in the Philippines by contributing valuable and critical insights into its agricultural industry.

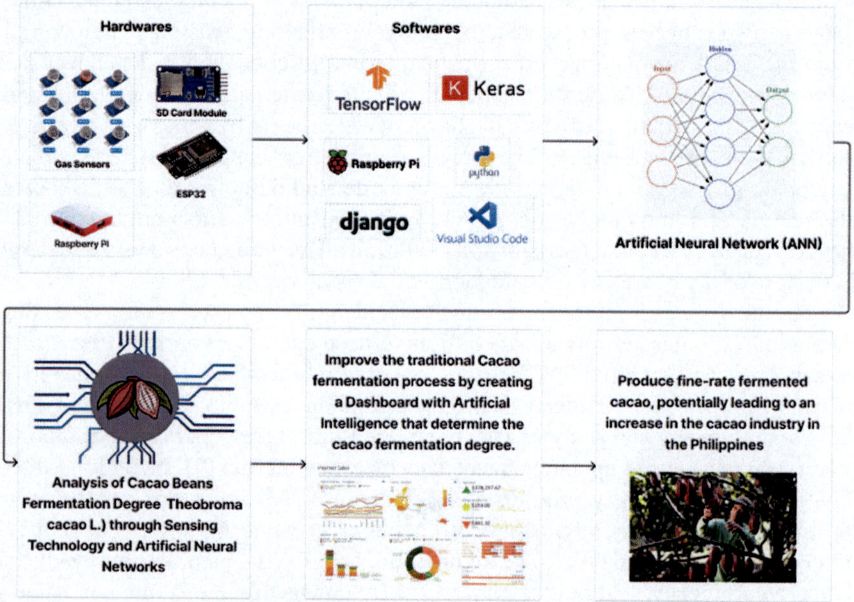

Fig. 1. Conceptual framework of the study

2.1 Data Used in the Study

Four kilograms of cacao beans were procured from a local cacao farm in Lucena City. The beans were distributed among four Styrofoam coolers, each containing one kilogram, mixed once at the beginning of each day, from the third to the sixth day of fermentation. The coolers were replaced with different ones to prevent beans from becoming saturated with their juices released during fermentation.

2.2 ESP32 Microcontroller

The ESP32 Microcontroller board is a single 2.4 GHz chip that combines Wi-Fi and Bluetooth functionalities at the heart of the system. The microcontroller was designed using the TSMC low-power 40 nm technology with an input supply of 5V and is connected to an AC adapter. The microcontroller board can operate at −40 to +105 degrees Celsius. Sensor connectivity is achieved by linking sensors to GPIO pins, which are used to receive data from them. The SD card Module is connected to the GPIO23 pin to transmit data to serve as a backup storage device for data. The ESP32 board interfaces between the hardware and software, enabling real-time monitoring and analysis of gas levels.

2.3 Gas Sensors

Five gas sensors, MQ-2, MQ-3, MQ-7, MQ-9, and MQ-135, are used in this study. The specifications of each gas sensor are presented, and the gases each detect. The output of the MQ gas sensors is analog voltage signals proportional to the concentration of the gases detected. The signals are converted into digital values using the developed program in the ESP32 Microcontroller. Cacao bean fermentation is a process in the production of chocolate that gives its distinct flavor and aroma. The data was gathered by monitoring and measuring gas emission signals during fermentation (Table 1).

Table 1. Gas sensors specifications

Sensor type	Detectable gas
MQ-2	Hydrogen, LPG, Methane, Carbon Monoxide, Alcohol, smoke, Propane
MQ-3	LPG, Methane, Carbon Monoxide, Alcohol, Benzene, Hexane
MQ-7	Carbon Monoxide
MQ-9	Carbon Monoxide, Flammable Gases
MQ-135	Carbon Monoxide, Ammonia, Benzene, Alcohol, smoke

2.4 Cacao Fermentation

Cacao bean fermentation is a process in the production of chocolate that gives its distinct flavor and aroma. The data was gathered by monitoring and measuring gas emission signals during fermentation. The data is collected for seven days, at three-second intervals, with the duration of cocoa bean fermentation. There are four sets of data obtained from the Styrofoam boxes. Each batch of data contains 200,000 records of gas emission readings. The number of readings and days were recorded alongside the digital data and classified as fermented or unfermented. Normalization was used to adapt the data to match a consistent number line, allowing for easier comparison and analysis [10]. Time series, consisting of a succession of data across time, can be smoothened using the moving average, reducing slight fluctuations. It was then analyzed and summarized with Explanatory Data Analysis (EDA) [11].

3 Results and Discussion

3.1 Data Pre-processing

The study applied data-cleaning techniques to enhance the quality of the dataset. Normalization involves transforming the data to fit a uniform number line to facilitate comparison and analysis. A moving average applied to a time series data is smoothed out to remove minor fluctuations. Exploratory Data Analysis (EDA) and data visualization techniques were employed. ANOVA table and R-squared analysis were performed to determine significant variables. Principal Components Analysis (PCA) used scaled data to cluster and visualize the data more effectively. Implementing the data-cleaning techniques ensures the dataset is suitable for statistical analysis and visualization, enabling to obtain more meaningful and insightful results (Fig. 2).

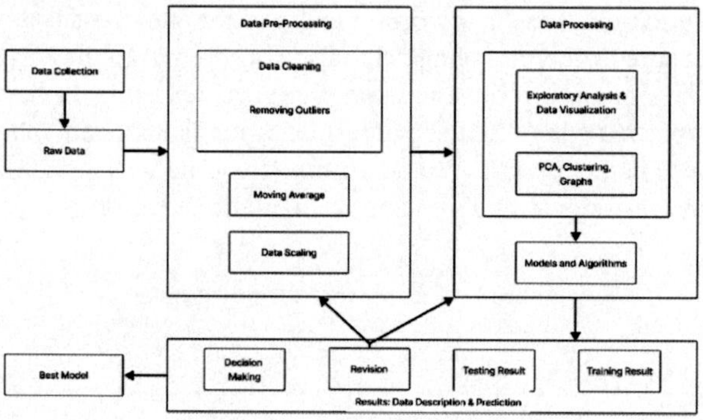

Fig. 2. Data pre-processing flow

3.2 Data Modeling

Artificial neural networks (ANNs) are a type of machine learning model that can adjust their parameters independently through learning algorithms. This means they can adapt and improve their performance over time as they are exposed to new data. This characteristic makes ANNs well-suited for modeling complex, non-linear relationships within statistical data. DNNs are a type of ANN with multiple hidden layers that allow them to model complex non-linear relationships within data. Their hierarchical approach to learning and ability to adapt to new data make them a powerful tool for a wide range of tasks. TabNet is a deep learning architecture designed for tabular data, where each column represents a feature or variable, and each row represents an observation or example. In training the ANN model, various optimizers and activation functions were utilized, specifically, RMSProp, Adam, and SGD as optimizers and ReLU and LeakyReLU as activation functions. The researchers employed a combination of these optimizers to evaluate their impact on accuracy. Table 2 provides insights into the architecture of the ANN model when using RMSProp, Adam, and ReLU. This model consists of three Dense layers with a total of 2,497 parameters.

Table 2. ANN Model Architecture using RMSProp, Adam, and ReLU

Layer (type)	Output shape	Param #
dense (Dense)	(None, 64)	384
dense_1 (Dense)	(None, 32)	2080
dense_2 (Dense)	(None, 1)	33

Table 3 displays the structure of the ANN model when utilizing SGD, Adam, and LeakyReLU. This model also consists of three Dense layers, incorporating LeakyReLU layers, enhancing its performance.

Table 3. ANN Model Architecture using SGD, Adam, and LeakyReLU

Layer (type)	Output shape	Param #
dense (Dense)	(None, 64)	384
leaky_re_lu (LeakyReLU)	(None, 64)	0
dense_1 (Dense)	(None, 32)	2080
leaky_re_lu_1 (LeakyReLU)	(None, 32)	0
dense_2 (Dense)	(None, 1)	33

The same optimizers used in the previous model were applied to the DNN model. The architecture of the DNN model, which incorporates RMSProp, Adam, and ReLU, is illustrated in Table 4. This model comprises five dense layers with dropout layers inserted in between them. Dropout layers have been incorporated to reduce data noise and mitigate the risk of overfitting. There are a total of 50,433 trainable parameters.

Table 4. DNN model architecture using RMSProp, Ada, and ReLU

Layer (type)	Output shape	Param #
dense (Dense)	(None, 128)	768
dense_1 (Dense)	(None, 128)	16512
dropout (Dropout)	(None, 128)	0
dense_2 (Dense)	(None, 128)	16512
dropout_1 (Dropout)	(None, 128)	0
dense_3 (Dense)	(None, 128)	16512
dropout_2 (Dropout)	(None, 128)	0
dense_4 (Dense)	(None, 1)	129

The alternative architecture of the DNN model employed SGD, Adam, and LeakyReLU optimizers, as detailed in Table 5. This configuration maintains the same number of dense and dropout layers as the previous DNN model, with the addition of LeakyReLU layers. Including LeakyReLU layers introduces non-linearity and addresses potential issues of dead neurons in the network, enhancing its learning capacity and overall performance.

Table 5. DNN model architecture using SGD, Adam, and LeakyReLU

Layer (type)	Output shape	Param #
dense (Dense)	(None, 128)	768
leaky_re_lu (LeakyReLU)	(None, 128)	0
dense_1 (Dense)	(None, 128)	16512
leaky_re_lu_1 (LeakyReLU)	(None, 128)	0
dropout (Dropout)	(None, 128)	0
dense_2 (Dense)	(None, 128)	16512
leaky_re_lu_2 (LeakyReLU)	(None, 128)	0
dropout_1 (Dropout)	(None, 128)	0
dense_3 (Dense)	(None, 128)	16512
leaky_re_lu_3 (LeakyReLU)	(None, 128)	0
dropout_2 (Dropout)	(None, 128)	0
dense_4 (Dense)	(None, 1)	129

3.3 TabNet Architecture

TabNet operates through a sequence of steps as shown in Fig. 3, including the feature transformer, attentive transformer, and feature masking [12]. At the outset, the initial layer serves as the starting point for the architecture, where data from sensor readings undergoes processing without going through a prior feature engineering process. It then proceeds through a batch normalization layer to ensure the data is normalized. The feature transformer stage processes the data, utilizing a 4-layer network with 2-shared and 2-dependent layers specific to each step's requirements. Subsequently, it enters the attention transformer step, encompassing a fully connected layer, batch normalization layer, prior scales layer, and sparsemax layer. The outcome of this step is directed toward the attention mask, facilitating the identification of the selected features. The TabNet model in this study has three steps, each including eight decision and attention layers.

3.4 Model Evaluation

Precision, Recall, and F1-Score are pivotal in determining the model performance exceptional in this study where the imbalanced dataset is utilized. Accuracy is also considered

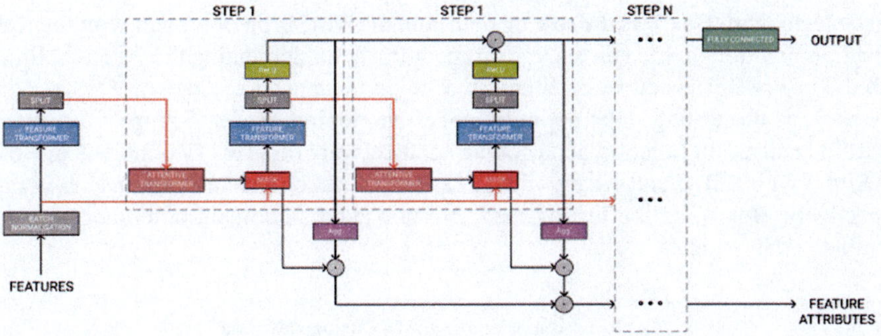

Fig. 3. TabNet Architecture

for such cases, which can be scored as F1-score = 0 or marking the classifier as useless [13]. Precision evaluates the accuracy of positive predictions, indicating the ratio of correctly predicted positive instances to the total predicted positives, while Recall measures the classifier's effectiveness in capturing positive instances across all positive data. The F1-Score, a harmonious mean of precision and recall, offers a balanced assessment. Accuracy quantifies the total correct predictions across all model assertions (Table 6).

Table 6. Precision, recall, F1-score, and accuracy

Model evaluation criteria	Formula
Precision	$\frac{TP}{TP+FP}$
Recall	$\frac{TP}{TP+FN}$
F1-Score	$2 * \frac{Precision*Recall}{Precision+Recall}$
Accuracy	$\frac{TP+TN}{TP+TN+FP+FN}$

3.5 Data Modeling using Artificial Neural Network (ANN)

Table 7 shows results in four boxes using raw data as the testing dataset. The collected data in Box 1 did not yield significant outcomes across all models examined. In Box 2, the model employing the Adam optimizer with the Rectified Linear Unit (ReLU) activation function exhibited favorable performance, achieving an accuracy level of 80%. The accuracy level is comparatively lower when contrasted with the Adam optimizer's two models. Using Box 3 as the testing data, the accuracy rate is 91%. The two models in Box 4 displayed commendable accuracy rates, using the RMSProp optimizer and ReLU activation function, yielding an accuracy of 86%.

Table 8 indicates the results in four boxes using scaled data as testing data. Box 1 testing data yielded significant outcomes for all ANN models except for the one employing the SGD optimizer and Leaky ReLU activation function. The two models performed satisfactorily with Box 1 as the testing data, and the RMSProp optimizer with the ReLU activation function yielded an accuracy rate of 96%. Implementing the Adam optimizer with the Leaky ReLU activation function resulted in an accuracy rate of 99% in Box 1. In Box 3, as the testing data, a single model employing the RMSProp optimizer and the ReLU activation function achieved an accuracy rate of 83%. Two models in Box 3, RMSProp with ReLU and Adam with LeakyReLU, accomplished 91% accuracy rates, respectively. Box 4, as the testing data, did not yield significant outcomes across all examined ANN models.

Table 7. Implementing ANN using raw data

Box	Optimizer	Activation Function	Precision		Recall		F1 Score		Accuracy
			0	1	0	1	0	1	
1	RMSProp	ReLU	0.71	0	1	0	0.83	0	0.71
2			0.29	0.08	0.13	0.18	0.18	0.11	0.15
3			0.99	0.34	0.21	1	0.35	0.5	0.43
4			**0.83**	**1**	**1**	**0.5**	**0.91**	**0.67**	**0.86**
1	SGD	Leaky ReLU	0.71	0	1	0	0.83	0	0.71
2			0.71	0	1	0	0.83	0	0.71
3			0.71	0	1	0	0.83	0	0.71
4			0.71	0	1	0	0.83	0	0.71
1	Adam	ReLU	0.71	0	1	0	0.83	0	0.71
2			**0.84**	**0.68**	**0.9**	**0.56**	**0.86**	**0.62**	**0.80**
3			1	0.75	0.87	1	0.93	0.86	0.91
4			0.77	0.38	0.67	0.49	0.72	0.43	0.62
1	Adam	Leaky ReLU	0.71	0	1	0	0.83	0	0.71
2			0	0.29	0	1	0	0.45	0.29
3			**0.89**	**1**	**1**	**0.69**	**0.94**	**0.82**	**0.91**
4			0.83	0.98	1	0.49	0.91	0.66	0.85

Table 9 reveals the results in four boxes using smoothed data as testing data. Box 1 and Box 2 did not yield significant outcomes. In Box 3, notable performance was observed in two ANN models, particularly the one utilizing the RMSProp optimizer and ReLU activation function, which achieved an accuracy rate of 99%. No ANN models exhibited noteworthy performance when Box 4 was utilized as the testing data.

3.6 Modeling Using Deep Neural Network (DNN)

Table 10 reveals observations and the performance of the Deep Neural Network (DNN) model. Box 1, the Adam optimizer with LeakyReLU activation function achieved an

Table 8. Implementing ANN using scaled data

Box	Optimizer	Activation Function	Precision		Recall		F1 Score		Accuracy
			0	1	0	1	0	1	
1	RMSProp	ReLU	**1**	**0.88**	**0.95**	**1**	**0.97**	**0.94**	**0.96**
2			**0.8**	**1**	**1**	**0.39**	**0.89**	**0.56**	**0.83**
3			**0.89**	**0.99**	**1**	**0.69**	**0.94**	**0.82**	**0.91**
4			0.71	0	1	0	0.83	0	0.71
1	SGD	Leaky ReLU	0.75	1	1	0.18	0.86	0.31	0.77
2			0.71	0	1	0	0.83	0	0.71
3			0.75	1	1	0.16	0.86	0.28	0.76
4			0.71	0	1	0	0.83	0	0.71
1	Adam	ReLU	1	0.84	0.92	1	0.96	0.91	0.95
2			0.77	1	1	0.25	0.87	0.4	0.79
3			0.8	1	1	0.37	0.89	0.54	0.82
4			0.71	0	1	0	0.83	0	0.71
1	Adam	Leaky ReLU	**1**	**0.96**	**0.99**	**1**	**0.99**	**0.98**	**0.99**
2			0.79	1	1	0.34	0.88	0.51	0.81
3			**0.89**	**1**	**1**	**0.69**	**0.94**	**0.82**	**0.91**
4			*0.83*	*0.71*	*0*	*1*	*0*	*0.83*	*0*

accuracy of 100%. Box 2, Box 3, and Box 4 with RMSProp optimizer with ReLU activation function achieved an accuracy of 97% to 97%.

Table 11 reveals an interesting and favorable performance across all tested scenarios, except the DNN model utilizing the SGD optimizer and Leaky ReLU activation function. The models demonstrated commendable accuracy rates, with the lowest being 81% and the highest reaching 97%.

The analysis in Table 12 shows a trend whereby a sizable portion of the models exhibited exceptional performance across all the tested scenarios, except the DNN model utilizing the SGD optimizer and Leaky ReLU activation function. The models achieved accuracies of 94% and above.

3.7 Modeling Using TabNet

In Table 13, using Box 3 as testing data, TabNet achieves a satisfactory performance, achieving an accuracy of 83%.

In Table 14, it can be observed that using scaled data for training, TabNet shows a performance accuracy of 71%.

In Table 15, the TabNet model exhibited satisfactory performance when tested with Box 4 as the testing data, attaining an accuracy of 91% (Table 15).

Table 9. Implementing ANN using smoothened data

Box	Optimizer	Activation Function	Precision		Recall		F1 Score		Accuracy
			0	1	0	1	0	1	
1	RMSProp	ReLU	0.71	0	1.0	0	0.83	0	0.71
2			0.81	0.68	0.9	0.5	0.86	0.57	0.78
3			**1.0**	**0.99**	**1.0**	**1.0**	**1.0**	**0.99**	**0.99**
4			0.74	0.34	0.63	0.46	0.68	0.39	0.58
1	SGD	Leaky ReLU	0.71	0	1.0	0	0.83	0	0.71
2			0.71	0	1.0	0	0.83	0	0.71
3			0.71	0	1.0	0	0.83	0	0.71
4			0.71	0	1.0	0	0.83	0	0.71
1	Adam	ReLU	0.71	0	1.0	0	0.83	0	0.71
2			0.71	0	1.0	0	0.83	0	0.71
3			1.0	0.93	0.97	1.0	0.98	0.96	0.98
4			0.78	0.32	0.32	0.78	0.45	0.45	0.45
1	Adam	Leaky ReLU	0.71	0	1	0	0.83	0	0.71
2			0.68	0.04	0.87	0.01	0.77	0.02	0.62
3			0.78	1	1	0.31	0.88	0.48	0.80
4			0.63	0.08	0.64	0.08	0.64	0.08	0.48

4 Web Application

With TensorFlow's ability to save the best-performing model, the researchers are currently developing a dashboard using the Django framework and Python as the back-end language. This web server dashboard receives data from each fermentation box via ESP32 Wi-Fi. The collected data will then be inputted into the embedded model to classify the fermentation degree of the cacao. This real-time process will enable the classification and monitoring of the cacao fermentation process.

5 Conclusion and future works

This study discovered that the development of a sensing technology system for the traditional cacao fermentation process, along with the incorporation of Artificial Intelligence using data collected through an e-nose, could enhance the efficiency of the process, explicitly benefiting cacao farmers. When analyzing the sensor data from each fermen-

Table 10. Implementing DNN using raw data

Box	Optimizer	Activation Function	Precision		Recall		F1 Score		Accuracy
			0	1	0	1	0	1	
1	RMSProp	ReLU	0.86	1.0	1.0	0.58	0.92	0.73	0.88
2			**0.96**	**1.0**	**1.0**	**0.9**	**0.98**	**0.94**	**0.97**
3			**0.99**	**1.0**	**1.0**	**0.97**	**0.99**	**0.98**	**0.99**
4			**0.99**	**1.0**	**1.0**	**0.97**	**0.99**	**0.98**	**0.99**
1	SGD	Leaky ReLU	0	0	0	0	0	0	0
2			0	0	0	0	0	0	0
3			0	0	0	0	0	0	0
4			0	0	0	0	0	0	0
1	Adam	ReLU	0.82	1.0	1.0	0.43	0.9	0.61	0.84
2			0.71	0	1.0	0	0.83	0	0.71
3			0.71	0	1.0	0	0.83	1.0	0.71
4			0.71	0	1.0	0	0.83	0	0.71
1	Adam	Leaky ReLU	**1.0**	**1.0**	**1.0**	**1.0**	**1.0**	**1.0**	**1.0**
2			0.9	1.0	1.0	0.71	0.95	0.83	0.92
3			1.0	1.0	1.0	1.0	1.0	1.0	1.0
4			1.0	1.0	1.0	1.0	1.0	1.0	1.0

tation box, models that utilized raw data demonstrated a higher classification rate than models that used scaled data. The researcher achieved even better results by applying a moving average to the raw data, eliminating fluctuations in the dataset. Furthermore, after comparing various models such as ANN, DNN, and TabNet, the researchers concluded that Deep Neural Network (DNN) yielded remarkable results to the other two architectures. A dashboard system displays gas data graphs and the classification results of fermentation degrees throughout the fermentation process. This approach will minimize the waste of unfermented cacao beans typically cut during the traditional cut-test method. Moreover, farmers will no longer need to rely on visual assessment to determine the optimal timing for conducting the cut test.

Table 11. Implementing DNN using scaled data

Box	Optimizer	Activation Function	Precision		Recall		F1 Score		Accuracy
			0	1	0	1	0	1	
1	RMSProp	ReLU	0.90	1	1	0.54	0.95	0.70	0.91
2			0.91	1	1	0.59	0.95	0.74	0.92
3			0.84	1	1	0.51	0.91	0.68	0.86
4			0.84	1	1	0.51	0.91	0.68	0.86
1	SGD	Leaky ReLU	1	1	0	0.89	0	0.81	1
2			0.81	1	1	0	0.89	0	0.81
3			0.75	1	1	0.18	0.86	0.30	0.76
4			0.75	1	1	0.18	0.86	0.30	0.76
1	Adam	ReLU	0.97	1	1	0.87	0.98	0.93	0.97
2			0.92	1	1	0.63	0.96	0.77	0.93
3			0.91	1	1	0.75	0.95	0.86	0.93
4			0.91	1	1	0.75	0.95	0.86	0.93
1	Adam	Leaky ReLU	0.95	1	1	0.80	0.98	0.89	0.96
2			0.95	1	1	0.75	0.97	0.86	0.95
3			0.89	1	1	0.68	0.94	0.81	0.91
4			0.89	1	1	0.68	0.94	0.81	0.91

Table 12. Implementing DNN using smoothened data

Box	Optimizer	Activation Function	Precision		Recall		F1 Score		Accuracy
			0	1	0	1	0	1	
1	RMSProp	ReLU	0.92	1	1	0.8	0.96	0.89	0.94
2			0.83	1	0	0.49	0.9	0.66	0.85
3			0.95	1	1	0.87	0.97	0.93	0.96
4			0.95	1	1	0.87	0.97	0.93	0.96
1	SGD	Leaky ReLU	0	0	0	0	0	0	0
2			0	0	0	0	0	0	0
3			0	0	0	0	0	0	0
4			0	0	0	0	0	0	0
1	Adam	ReLU	1	1	0.81	0.96	0.89	0.94	1
2			0.99	1	1	0.98	1	0.99	0.99
3			0.95	1	1	0.88	0.98	0.94	0.97
4			0.95	1	1	0.88	0.98	0.94	0.97
1	Adam	Leaky ReLU	1	1	0.99	1	0.99	1	1
2			0.95	1	1	0.86	0.97	0.93	0.96
3			0.94	1	1	0.85	0.97	0.92	0.96
4			0.94	1	1	0.85	0.97	0.92	0.96

Table 13. Implementing Tabnet using raw data

Box	Precision		Recall		F1 Score		Accuracy
	0	1	0	1	0	1	
1	0.71	0	1	0	0.83	0	0.71
2	0.84	0.42	0.6	0.72	0.7	0.53	0.63
3	**0.99**	**0.63**	**0.77**	**0.99**	**0.87**	**0.77**	**0.83**
4	0.81	0.55	0.84	0.5	0.82	0.53	0.74

Table 14. Implementing Tabnet using scaled data

Box	Precision		Recall		F1 Score		Accuracy
	0	1	0	1	0	1	
1	0.71	0	1	0	0.83	0	0.71
2	0.71	0	1	0	0.83	0	0.71
3	0.71	0	1	0	0.83	0	0.71
4	0.71	0	1	0	0.83	0	0.71

Table 15. Implementing Tabnet using smoothened data

Box	Precision		Recall		F1 Score		Accuracy
	0	1	0	1	0	1	
1	0.71	0	1	0	0.83	0	0.71
2	0.87	0.32	0.19	0.93	0.32	0.48	0.41
3	1	0.72	0.84	1	0.91	0.83	0.88
4	**0.98**	**0.78**	**0.89**	**0.95**	**0.93**	**0.86**	**0.91**

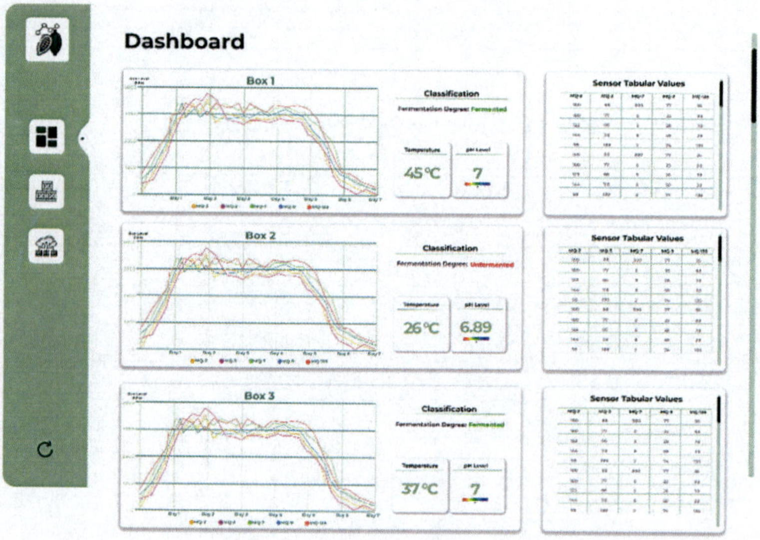

Fig. 4. Dashboard of the real-time monitoring of data.

References

1. Vega, C., Kwik-Uribe, C.: Theobroma cacao—an introduction to the plant, its composition, uses, and health benefits. In Cocoa Butter and Related Compounds, Elsevier, 2012, pp. 35–62. doi: https://doi.org/10.1016/B978-0-9830791-2-5.50005-0
2. Gayi, S.K., Tsowou, K.: Cocoa industry: Integrating small farmers into the global value chain (2016). https://unctad.org/publication/cocoa-industry-integrating-small-farmers-global-value-chain
3. Medenilla, V.: The market potential of value-added cacao products in the Philippines." Accessed: Aug. 30, 2023. https://mb.com.ph/2022/4/12/the-market-potential-of-value-added-cacao-products-in-the-philippines
4. S. Bermudez, V. Voora, C. Larrea, and E. Luna, "Cocoa prices and sustainability sustainable commodities marketplace series market overview," 2022
5. Ghosh, D.: The Top Cocoa Producing Countries in The World. https://www.worldatlas.com/industries/the-top-cocoa-producing-countries-in-the-world.html
6. Rohmah, M., Sari, K., Rahmadi, A.: The taste development of cocoa bean: evidence from the tropical rain forest to the table. In: Galanakis, C.M. (ed.) Trends in sustainable chocolate production, pp. 41–74. Springer International Publishing, Cham ((2022)) https://doi.org/10.1007/978-3-030-90169-1_2
7. Bobiles, S.C., Elegado, F.B., Millena, C.G., Merca, F.E.: Small-scale cacao (Theobroma cacao L.) fermentation process utilizing cacao pod husk. Food Res. **6**(4), 236–245 (2022). https://doi.org/10.26656/fr.2017.6(4).502
8. Muñoz, M.S., Cortina, J.R., Vaillant, F.E., Parra, S.E.: An overview of the physical and biochemical transformation of cocoa seeds to beans and chocolate: flavor formation. Crit. Rev. Food Sci. Nutr. **60**(10), 1593–1613 (2020). https://doi.org/10.1080/10408398.2019.1581726
9. Maaño, R.C., Chavez, E.P., Maaño, R.A.: Towards the development of a smart photovoltaic-powered temperature-controlled poultry egg incubator. Int. J. Simul.: Syst., Sci. Technol. **19**(3), 19.1–19.5 (2018). https://doi.org/10.5013/IJSSST.a.19.03.19
10. Bhandari, A.: Feature engineering: scaling, normalization, and standardization (Updated 2023). Accessed: Aug. 30, 2023. https://www.analyticsvidhya.com/blog/2020/04/feature-scaling-machine-learning-normalization-standardization/
11. Iglesias Moreno, A.: Moving averages with Python. Accessed: Aug. 30, 2023. https://towardsdatascience.com/moving-averages-in-python-16170e20f6c
12. Arik, S.O., Pfister, T.: TabNet: attentive interpretable tabular learning (2019). http://arxiv.org/abs/1908.07442
13. Kanstrén, T.: A look at precision, recall, and F1-score: exploring the relations between machine learning metrics. Towar. Data Sci. https://towardsdatascience.com/a-look-at-precision-recall-and-f1-score-36b5fd0dd3ec

DCDynUnet: Deep Supervision Attention Context for Brain Segmentation

Vo Thanh Hoang Son[2], Tran Nguyen Quynh Tram[2], Nguyen Quoc Huy[3](✉), and Do Nhu Tai[1](✉)

[1] University of Economics Ho Chi Minh City—UEH, Ho Chi Minh City, Vietnam
taidn@ueh.edu.vn
[2] HCMC University of Foreign Language-Information Technology, Ho Chi Minh City, Vietnam
tramtnq@huflit.edu.vn
[3] Sai Gon University, Ho Chi Minh City, Vietnam
nqhuy@sgu.edu.vn

Abstract. Brain segmentation is considered a gold standard in the analysis of brain diseases. However, this task requires much time, effort, and highly professional skills. Many research studies applying artificial intelligence and deep learning have been developed to assist doctors in brain segmentation to solve this task automatically. In this paper, our primary goal is to propose a deep learning model using attention and deep supervision mechanisms to improve the performance of brain segmentation. We will use an attention block in the deep supervision mechanism proposed by us. This attention block helps to identify standard features between the steps in the decoder stage and the output features of the network. This helps to enhance these standard features and achieve good performance. In addition, we also conducted ablation studies to evaluate the effectiveness of the proposed fusion method. Experimental results have shown that our performance is quite good, with the highest results being 96.06% of CSF, 93.36% of WM, and 91.94% of GM of the iSeg2019 dataset.

Keywords: Brain Segmentation · Attention Mechanism · Deep Supervision mechanism

1 Introduction

The rapid development of medical imaging techniques has led to the creation of a vast amount of medical data. This has paved the way for the strong growth of machine-learning methods in medicine. Medical image segmentation is an application that aims to clarify anatomical structures in images. It often plays a vital role in diagnosis, as computer-aided and artificial intelligence-powered segmentation can help improve diagnosis accuracy [1].

Brain segmentation is a specific problem in the field of medical image segmentation. This problem provides a non-invasive tool to quantify brain structure based on structural MRI images obtained in clinical settings. Manual description of brain structure has long

P. T. Nghia et al. (Eds.): ICTA 2023, LNNS 847, pp. 158–165, 2023.
https://doi.org/10.1007/978-3-031-49529-8_17

been considered the "gold standard." However, manual segmentation is time-consuming and resource-intensive, making it impractical to deploy at scale. As a result, deep learning algorithms have been sought to reduce manual efforts [2]. This problem helps to correctly identify the three main types of brain components: gray matter, white matter, and cerebrospinal fluid (CSF). Figure 1 shows an overview of the processing steps of the problem. The input image is the channels of the MRI image. The output will return a 3D image containing 3-pixel values, representing grey matter, white matter, and CSF, respectively.

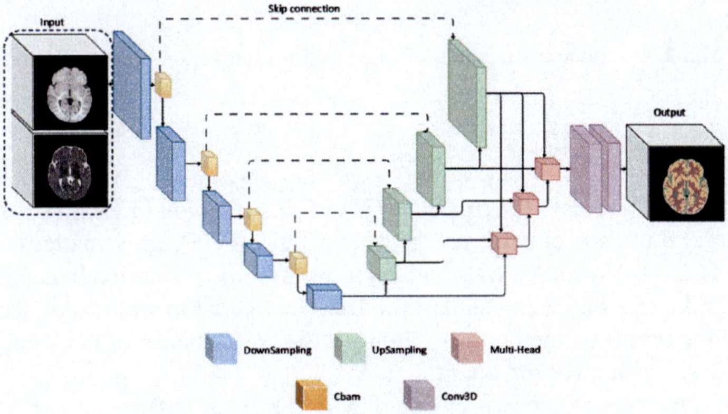

Fig. 1. Overview of DCDynUnet model

In recent years, the methods based on convolutional neural networks (CNNs) have provided significant insights into efficient and feasible ways to segment the brain. Methods such as V-Net, SegResnet, and SynthSeg have shown promising results on various brain segmentation tasks. In addition, multimodal fusion is a widely used method in medical image analysis because it can provide richer object features, which help improve object detection and segmentation results [3].

Our work will apply the multi-modal fusion method at the feature and decision levels. Besides, we propose a method called DCDynUnet (Deep supervision Context DynUnet), which is based on the DynUnet model, using deep supervision and attention mechanisms. More than that, we will ablation study the fusion method between the proposed models with the main goal of improving the accuracy of the results.

The paper will be organized into four main sections: introduction, proposed method, experiment, and conclusion. The opening will provide an overview of the form and its objectives. The proposed method section will explain in detail the proposed model. The experiment section will present specific experimental strategies and results. The conclusion will summarize the entire research process in the paper.

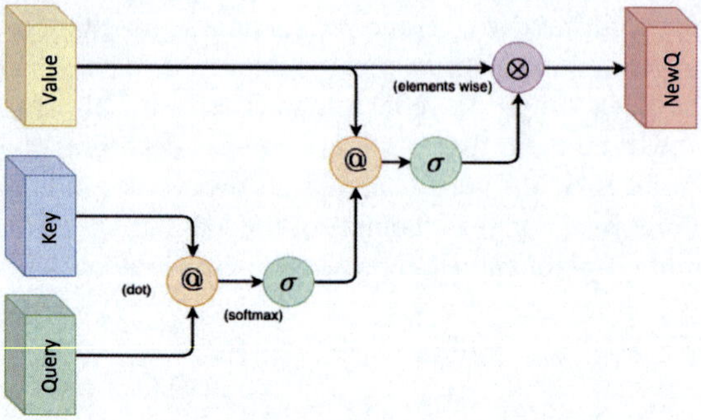

Fig. 2. Architecture of the Multi-head deep supervision block (MHDS)

2 Proposed Method

The main problem addressed in this paper is the classification of each voxel in an MRI volume into one of four classes: cerebrospinal fluid (CSF), grey matter (GM), white matter (WM) and background. In this chapter, we propose a solution that applies a deep supervisor and attention mechanism to the DynUnet [4, 5] model, called DCDynUnet. The goal of the proposed method is to improve the performance of the DynUnet model by using a deep supervisor and attention block.

The overall process is shown in Fig. 1 with the input data being two MRI blocks showing brain image $I \in R^{w \times h \times d \times 2}$. The goal is to obtain the input value of 3 MRI blocks $P \in R^{w \times h \times d \times 3}$. With each block corresponding to each classification including cerebrospinal fluid (CSF), gray matter (GM), white matter (WM), and the background (BG). The basic encoder phase is like conventional UNet models. However, we apply 3D CBam [6] blocks right after each encoding phase of the model. The goal is to take advantage of the CBam block to enhance the feature synthesis process. More than that, our proposed decoder method effectively integrates the Multi-Head Attention Block with the Deep supervisor method. Thereby helping to perfect the decoder process and improve the resulting performance.

We were inspired by the Multi-head attention block from [7] to create the multi-head deep supervision block shown in Fig. 2. The input of this block will be 3 different feature maps, which will be called Key, Query, and Value abbreviated as K, Q and V. Q is the feature extracted from the decoder block at the corresponding stage, K is the feature aggregated from the previous stage, and V is the feature at the last decoder stage of the model. The synthesis method of multi-head deep supervision is shown in Fig. 2 where the operator (@) represents the matrix multiplication between 2 feature maps, and the operator (⊗) the elementwise multiply.

This block will perform a matrix multiplication between K and Q, which helps to synthesize standard information from the two feature maps between two adjacent steps. Then, the results of the matrix multiplication will be multiplied by V. This helps to fuse the new features with the features in the last layer. In the next step, the elements of the standard features will be multiplied by V, which will enhance the standard features for

V. Note that after each stage of using matrix multiplication, a SoftMax layer will be used to maintain the confidence threshold between 0 and 1.

3 Experiment

3.1 Experiments Setup

Dataset. Our experiment used the iSeg2019 shown in Fig. 3, a 6-month infant brain MRI Segmentation dataset. The dataset was introduced in the MICCAI Grand Challenge 2019 [8]. This dataset consists of MRI images of infants at around six months of age. Each record includes two MRI modalities, T1-weight and T2-weight, and a label map. The label map contains four values, namely Background, cerebrospinal fluid (CSF), grey matter (GM), and white matter (WM). The training dataset consists of 10 and 13 samples for the testing dataset. However, the testing dataset does not include the label map. Therefore, during the experiment, we will split the training dataset into five folds with a training and testing data ratio of 8:2.

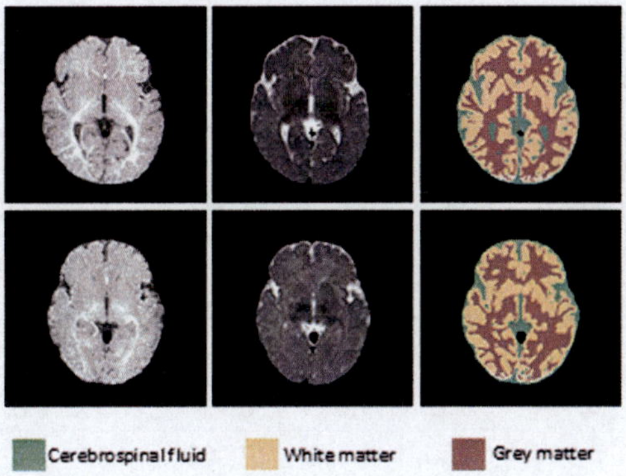

Fig. 3. Examples of samples from iSeg-2019 included T1w modality, T2w modality, and the ground truth.

Data augmentation: The input data will be augmented using the random flip method and adjusted randomly within 10% intensity. In addition, the input will be cut into small batches of size 128 × 128 × 128. Similarly, the ground truth data will also be made accordingly to match and normalized as one-hot encoder to facilitate the classification process of the model.

The number of epochs was limited to 100 epochs. We trained the model using Adam optimization, with an initial learning rate of 0.001. The learning rate schedule was Cosine Annealing, with a minimum learning rate of 0.00001. The experiments ran on the NVIDIA Tesla P100 16GB hardware and used Pytorch and Monai frameworks.

$$DiceScore = \frac{2\sum_{i=1}^{N} p_i g_i + \epsilon}{\sum_{i=1}^{N} p_i^2 + \sum_{i=1}^{N} g_i^2 + \epsilon} \tag{1}$$

Loss and Evaluation Metric. We used the dice loss for training. Mean dice score [8] was used for the result evaluation. The formular of Dice score was repesent in (1). Where p_1 is the predicted i-th pixel, g_i is the grouth truth value at the i-th position too. Epsilon will help the formula not to fall into undefined cases. In addition, to provide the most objective view of our method, we will use additional metrics MHD - Mean Hausdorff Distance and ASD - Average surface Distance [9]. The purpose of these two metrics is to evaluate the Hausdorff and surface distance between the ground truth and the prediction.

3.2 Result

At the experimental stage, we first experimented with the training of modern deep learning models as reference results with the proposed methods in this paper. These models include Segresnet, V-Net, SwinUnetR and DynUnet [5, 8, 10, 11]. According to the results recorded from Table 1, it shows that the DynUnet model when using the Cbam block, the results increased by about 1% on the classes. Then, when the model uses our proposed method, it increased by about 2% on each class. This shows that our method has brought quite good performance.

Table 1. Results on conventional networks and the model DynUnet with/without DS (Deep supervision) or CBam and Our's DCDynUnet.

Model	Dice Score ↑			MHD ↓			ASD ↓		
	CSF	WM	GM	CSF	WM	GM	CSF	WM	GM
Segresnet	0,935	0,906	0,891	9,718	7,121	7,476	0,166	0,336	0,388
V-Net	0,924	0,735	0,869	10,11	11,35	92,40	0,196	0,423	0,470
SwinUnet	0,938	0,894	0,875	10,01	6,918	7,508	0,140	0,357	0,421
DynUnet	0,943	0,914	0,903	9,629	6,935	**7,303**	0,126	0,293	0,333
DynUnet + DS	0,944	0,914	0,900	**9,603**	7,002	7,323	0,126	0,291	0,333
DynUnet + CBam	0,954	0,925	0,909	9,723	6,982	7,374	0,124	0,300	0,328
DCDynUnet	**0,960**	**0,933**	**0,918**	9,689	**6,865**	7,430	**0,122**	**0,286**	**0,317**

After the experiment proved the effectiveness of the proposed method, we continued to experiment with fusion to improve the efficiency of our methods using the fusion method. We used the 3 most basic fusion methods: late fusion, early fusion, and joint fusion in [3] with results showing in Table 2.

According to the results in Table 2, the most effective method is Early fusion with 2 combined models: DynUnet and DCDynUnet (Ours). The best results achieved are 96% on the CSF class, 93.3% on the WM class, and 91.9% on the GM class.

Figure 4 shows the segmentation results obtained by the deep learning models, including DynUnet, DCDynUnet, and the early fusion method between the DynUnet

Table 2. Ablation studio between merge methods and models.

Model 1	Model 2	Fusion method	Dice Score		
			CSF	WM	GM
DynUnet	DynUnet + DS	late	0.9572	0.9299	0.9154
		early	0.9575	0.9305	0.9157
		joint	0,9573	0,9303	0,9156
DynUnet + DS	DynUnet + DS	late	0,9536	0,9245	0,9093
		early	0.9538	0.9257	0.9102
		joint	0.9535	0.9251	0.9096
DynUnet	DCDynUnet	late	0.9585	0.9315	0.9171
		early	**0.9606**	**0.9336**	**0.9194**
		joint	0.9600	0.9332	0.9190

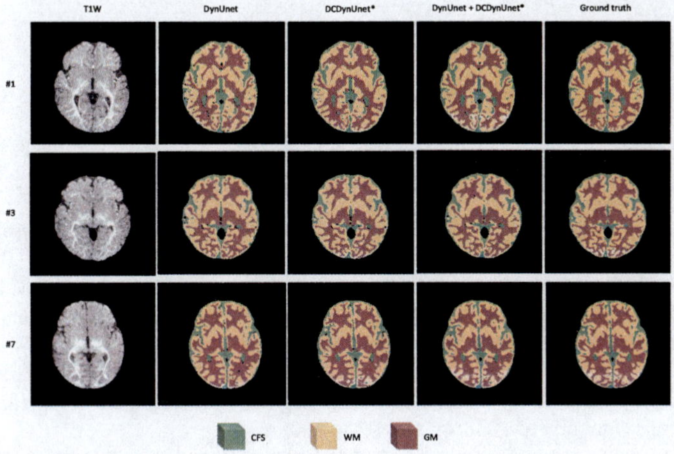

Fig. 4. Predictable outcomes from models versus ground truth

and DCDynUnet models. We can see that in our methods, the number of black pixels has been minimized compared to the original DynUnet method.

4 Conclusion

In this paper, we propose a model called DCDynUnet, which applies the attention and deep supervision mechanisms to the DynUnet network. Our goal is to use the CBAM block to summarize information in the encoding stage and use the Multi-Head supervision method to select and enhance common information in the decoding stage to solve the brain tumor segmentation problem.

In this approach, CBAM will play a role in contributing to feature extraction for the encoder. It will help the model to recognize and focus correctly on the necessary areas. Then, in the decoder part, our multi-head attention block will be attached immediately. Through this, it will learn the common information between feature maps between steps. This will help the model to distinguish the components of the brain more clearly.

The results are quite good when achieving 95,9% on CFS, 93,2% on WM, and 91.8% on GM. In addition, we also use the fusion method to improve the performance of the methods. The best results are achieved when early fusion of the DynUnet and DCDynUnet models.

Our method has achieved relatively good performance, however, the use of all CBAM and Deep supervision blocks may not be optimal. In the future, we will explore how to select the correct stage to attach the CBAM block as well as MHDS in the most reasonable way into the DynUnet model. In addition, we also hope to build a general standard so that we can use our MHDS into similar U-shape networks.

Acknowledgement. This research was funded by University of Economics Ho Chi Minh City, Vietnam.

References

1. Wang, R., Lei, T., Cui, R., Zhang, B., Meng, H., Nandi, A.K.: Medical image segmentation using deep learning: a survey. IET Image Process. **16**(5), 1243–1267 (2022). https://doi.org/10.1049/ipr2.12419
2. Huo, Y., et al.: 3D whole brain segmentation using spatially localized atlas network tiles. Neuroimage **194**, 105–119 (2019). https://doi.org/10.1016/j.neuroimage.2019.03.041
3. Huang, S.-C., Pareek, A., Seyyedi, S., Banerjee, I., Lungren, M.P.: Fusion of medical imaging and electronic health records using deep learning: a systematic review and implementation guidelines. Npj Digit. Med. **3**(1), 136 (2020). https://doi.org/10.1038/s41746-020-00341-z
4. Isensee, F., Jäger, P.F., Kohl, S.A.A., Petersen, J., Maier-Hein, K.H.: Automated design of deep learning methods for biomedical image segmentation. Nat. Methods **18**(2), 203–211 (2021). https://doi.org/10.1038/s41592-020-01008-z
5. Ranzini, M.B.M., Fidon, L., Ourselin, S., Modat, M., Vercauteren, T.: MONAIfbs: MONAI-based fetal brain MRI deep learning segmentation (2021). https://doi.org/10.48550/arXiv.2103.13314
6. Woo, S., Park, J., Lee, J.-Y., Kweon, I.S.: Cbam: convolutional block attention module. In: Proceedings of the European Conference on Computer Vision (ECCV), pp. 3–19 (2018)
7. Vaswani, A., et al.: Attention is all you need. Adv. Neural Inf. Process. Syst. **30** (2017)
8. Milletari, F., Navab, N., Ahmadi, S.-A.: V-net: fully convolutional neural networks for volumetric medical image segmentation. In: 2016 Fourth International Conference on 3D Vision (3DV). Ieee, pp. 565–571 (2016)
9. Heimann, T., et al.: Comparison and evaluation of methods for liver segmentation from CT datasets. IEEE Trans. Med. Imaging **28**(8), 1251–1265 (2009). https://doi.org/10.1109/TMI.2009.2013851

10. Myronenko, A.: 3D MRI brain tumor segmentation using autoencoder regularization. In: Crimi, A., Bakas, S., Kuijf, H., Keyvan, F., Reyes, M., van Walsum, T. (eds.) Brainlesion: Glioma, Multiple Sclerosis, Stroke and Traumatic Brain Injuries. LNCS, vol. 11384, pp. 311–320. Springer, Cham (2019). https://doi.org/10.1007/978-3-030-11726-9_28
11. Hatamizadeh, A., Nath, V., Tang, Y., Yang, D., Roth, H.R., Xu, D.: Swin UNETR: Swin transformers for semantic segmentation of brain tumors in MRI images. In: Brainlesion: Glioma, Multiple Sclerosis, Stroke and Traumatic Brain Injuries: 7th International Workshop, BrainLes 2021, Held in Conjunction with MICCAI 2021, Virtual Event, September 27, 2021, Revised Selected Papers, Part I. Springer, pp. 272–284 (2022)

Enhancing Wildfire Detection Using Semi-supervised Fuzzy Clustering on Satellite Imagery

Phung The Huan[1](✉), Hoang Thi Canh[1], Vu Duc Thai[1], Do Huy Khoi[1], and Le Truong Giang[2]

[1] Thai Nguyen University of Information and Communication Technology, Thainguyen, Vietnam
{pthuan,htcanh,vdthai,dhkhoi}@ictu.edu.vn
[2] Hanoi University of Industry, Hanoi, Vietnam
letruonggiang@haui.edu.vn

Abstract. The current concern over wildfires stems from their devastating impact on ecosystems, wildlife and human lives. These fires have adverse effects on the environment, resulting in the destruction of vegetation cover, loss of biodiversity and environmental degradation. Detecting wildfires in a timely manner allows for prompt response, mitigating environmental damage and ensuring the safety of individuals, particularly those residing near forested areas. There are various methods available for wildfire detection, including the use of sensor networks, employing artificial intelligence and machine learning for fire prediction and satellite image observation, which is a prominent focus of researchers. In this study, we introduce several fuzzy clustering methods and their applications for detecting wildfire in satellite images. The experimental results demonstrate the effectiveness of these methods in identifying wildfire areas within satellite imagery. Notably, the Picture Fuzzy Clustering method exhibits high clustering efficiency and provides clear images for wildfire detection. This approach proves valuable for disaster prevention and wildfire detection, enabling swift decision-making and implementation of solutions to mitigate potential harm to human lives, properties and ecosystems.

Keywords: Clustering · Fuzzy semi-supervised clustering · Picture fuzzy clustering · Satellite · Wildfire

1 Introduction

In image processing, the image segmentation process plays a crucial role. Image segmentation is aimed at analyzing an image and dividing it into regions with similar characteristics based on boundaries or connected regions. Image segmentation partitions a digital image into multiple segments to accurately determine the objects belonging to image regions and the background region. Alternatively, image segmentation is the process of assigning labels to all image points with the requirement that points with the same

P. T. Nghia et al. (Eds.): ICTA 2023, LNNS 847, pp. 166–175, 2023.
https://doi.org/10.1007/978-3-031-49529-8_18

label belong to the same specific image region. Image segmentation is the most important and complex process in image processing, serving subsequent object recognition steps. In general, image segmentation methods can be classified into three fundamental trends: (i) Using image processing techniques, including thresholding methods, edge-based methods and region-based methods [1]. (ii) Using clustering techniques, such as K-means [2] and Fuzzy C-Means (FCM) [3]. (iii) Using soft computing techniques [4] and deep learning-based techniques [1]. These trends represent different approaches to image segmentation and each approach has its own advantages and disadvantages.

The process of grouping all data points into different clusters, where data points that are more similar are assigned to the same cluster, is called data clustering [5]. Clustering methods can be divided into two types: hard clustering and fuzzy clustering. In hard clustering, a data point is assigned to only one cluster with certainty, whereas in fuzzy clustering, a data point can belong to multiple clusters with membership values indicating the degree of belongingness. To improve the quality of clustering, fuzzy clusters are supplemented with additional predefined information such as membership degrees and pairwise constraints, making them fuzzy semi-supervised clusters. Due to the additional information, fuzzy semi-supervised clustering methods generally achieve better clustering accuracy than hard clustering and fuzzy clustering methods alone. The advantages of the fuzzy clustering method compared to hard clustering methods like K-means, DBSCAN, etc., are its flexibility and noise handling capability. Furthermore, clustering methods can be enhanced by using advanced fuzzy clustering techniques [6–8]. Nowadays, data clustering finds extensive practical applications. Some typical applications include marketing for customer segmentation to identify and classify potential customers, predict customer behavior [9], biology for grouping plant species based on their characteristics [10], library management to track readers and books [11], insurance [12], finance [13], document classification [14], web user profiling [15], dental X-ray segmentation [16] and medical image segmentation [17]. The satellite images were collected from satellites orbiting the Earth at an altitude of approximately 35.78 km above sea level, so the image quality is affected by the capture angle, environmental elements like clouds, fog, weather conditions, etc. Particularly in satellite image processing, there are significant applications such as traffic classification [18], post-disaster assessment [19], ship detection from satellite image [20], satellite image classification for drought, flood and wildfire prediction [21, 22], etc.

In this paper, we introduce some approaches in fuzzy clustering and their application in wildfire detection from satellite images. The remaining of the paper: Sect. 2 introduces several fuzzy clustering methods. In Sect. 3, we discuss the application of clustering methods in wildfire detection. Section 4 presents the conclusion and contributions of the research, along with suggesting future research directions.

2 Clustering Techniques

2.1 Fuzzy C-Means (FCM) Method

FCM (Fuzzy C-Means) is the most famous fuzzy clustering method proposed by Bezdek [3]. This method is based on the idea of minimizing the distance between data points and cluster centers. The objective function is constructed as follows:

$$J_{(u,v)} = \sum_{a=1}^{N}\sum_{b=1}^{C} u_{ab}^{m} d_{ab}^{2} \rightarrow Min \tag{1}$$

With the constraints:

$$\sum_{b=1}^{C} u_{ab} = 1; \quad u_{ab} \in [0, 1], \ \forall a = 1, .., N \tag{2}$$

where $X = \{X_1, X_2, ..., X_k, ..., X_N\}$: the dataset consisting of N elements, C: the number of clusters, $V = \{V_1, V_2, ..., V_b, ..., V_C\}$: the set of cluster centers, u_{ab}: the membership degree of a data point a to cluster b, d_{ab}: the distance from data point a to cluster center b, m: the fuzzy coefficient. The cluster centers V_b and membership degrees u_{ab} are determined using the Lagrange multiplier method for the objective function (1) and the constraints (2) as follows:

$$V_b = \frac{\sum_{a=1}^{N} u_{ab}^{m} X_a}{\sum_{a=1}^{N} u_{ab}^{m}} \tag{3}$$

$$u_{ab} = \frac{1}{\sum_{k=1}^{C} \left(\frac{d_{ab}}{d_{ak}}\right)^{\frac{2}{m-1}}} \tag{4}$$

2.2 Semi-supervised Fuzzy C-Means (SSFCM) Method

The SSFCM method was proposed in a fuzzy clustering research study [23]. In this method, the objective function of SSFCM consists of two components: the unsupervised learning component and the supervised learning component. The objective function of this method is defined as follows:

$$J_{(u,d)} = \sum_{a=1}^{N}\sum_{b=1}^{C} u_{ab}^{m} d_{ab}^{2} + \alpha \sum_{a=1}^{N}\sum_{b=1}^{C} (u_{ab} - f_{ab} b_b)^{m} d_{ab}^{2} \rightarrow Min \tag{5}$$

where $X = \{X_1, X_2, ..., X_k, ..., X_N\}$ the dataset consisting of N elements, C: the number of clusters, $V = \{V_1, V_2, ..., V_b, ..., V_C\}$: the set of cluster centers, u_{ab}: the membership degree of a data point a to cluster b, d_{ab}: the distance from data point a to cluster center b, m: the fuzzy coefficient, the membership degree f_{ab} assigned to labeled data point a in cluster b. α: the parameter used to adjust the balance between the supervised and unsupervised components, the parameter δ_a is used to differentiate between labeled and unlabeled data elements.

$$f_{\mathrm{ab}} = \begin{cases} 1 \text{ if element } \mathrm{X}_k \text{ is labeled} \\ 0 \; otherwise \end{cases} \tag{6}$$

$$b_{\mathrm{k}} = \begin{cases} 1 \text{ if element } \mathrm{X}_k \text{ is labeled} \\ 0 \; otherwise \end{cases} \tag{7}$$

The values of cluster centers V_b and membership degrees u_{ab} are determined as follows:

$$V_b = \frac{\sum_{a=1}^{N} u_{ab}^2 x_a + \alpha \sum_{a=1}^{N} (u_{ab} - f_{ab}\delta_a)^2 X_a}{\sum_{a=1}^{N} u_{ab}^2 + \alpha \sum_{a=1}^{N} (u_{ab} - f_{ab}\delta_a)^2 X_a},$$
$$\forall b = 1, \ldots, C \tag{8}$$

$$u_{ab} = \frac{1}{1+\alpha} \left\{ \frac{1 + \alpha\left(1 - b_b \sum_{b=1}^{C} f_{ab}\right)}{\sum_{j=1}^{C} \frac{d_{ab}^2}{d_{bj}^2}} + \alpha f_{ab} b_k \right\},$$
$$\forall a = 1, .., N, \; b = 1, \ldots, C \tag{9}$$

2.3 Fuzzy Clustering on Advance Fuzzy Set

To enhance clustering performance, in addition to techniques that modify components of the objective function, another approach being researched is the application of advanced fuzzy clustering techniques. These techniques leverage advanced fuzzy sets such as Type-1 fuzzy sets, Type-2 fuzzy sets, intuitionistic fuzzy sets and picture fuzzy sets. Below are some definitions:

The traditional fuzzy set was defined by Zadeh in 1965 as follows:

Definition 1. [24] A fuzzy set is defined as follows:

$$\dot{A} = \left\{ \left(\alpha, \mu_{\dot{A}}(\alpha) \right) | \forall \alpha \in X, \mu_{\dot{A}}(\alpha) \in [0, 1] \right\} \tag{10}$$

where X is a non-empty universe of discourse and $\mu_{\dot{A}}(\alpha)$ is the membership function of $\alpha \in X$.

The picture fuzzy set is proposed by Bui Cong Cuong based on the extension of Zadeh's fuzzy set.

Definition 2. [25] A picture fuzzy set is defined as follows:

$$A = \{(\alpha, \mu_A(\alpha), \eta_A(\alpha), \gamma_A(\alpha)) | \alpha \in X\} \tag{11}$$

where X is a non-empty universe of discourse, $\mu_A(\alpha)$ is the positive degree, $\eta_A(\alpha)$ is the neutral degree and $\gamma_A(\alpha)$ is the negative degree of each element $\alpha \in X$.

With the constraints:

$$\begin{gathered} 0 \leq_A (\alpha), \eta_A(\alpha), \gamma_A(\alpha) \leq 1, \\ 0 \leq \mu_A(\alpha) + \eta_A(\alpha) + \gamma_A(\alpha) \leq 1 \end{gathered} \tag{12}$$

The refusal degree of an element is calculated as follows:

$$\xi_A(\alpha) = 1 - (\mu_A(\alpha) + \eta_A(\alpha) + \gamma_A(\alpha)) \tag{13}$$

If $\xi_A(\alpha) = \eta_A(\alpha) = 0$ then picture fuzzy set is transformed into a traditional fuzzy set.

2.4 Picture Fuzzy Clustering (FC-PFS) Method

In a recent study, the authors [26] proposed a clustering method based on panoramic fuzzy set called FC-PFS. The objective function of this method is defined as follows:

$$J_{(u,\eta,\xi,d)} = \sum_{a=1}^{N}\sum_{b=1}^{C} (\mu_{ab}(2 - \xi_{ab}))^m d_{ab}^2 + \sum_{a=1}^{N}\sum_{b=1}^{C} \eta_{ab}(\log \eta_{ab} + \xi_{ab}) \rightarrow Min \tag{14}$$

with the constraints:

$$0 \leq \mu_{ab}, \eta_{ab}, \xi_{ab} \leq 1, \quad 1 \leq a \leq N, \ 1 \leq b \leq C, 0 \leq \mu_{ab} + \eta_{ab} + \xi_{ab} \leq 1 \tag{15}$$

$$\sum_{b=1}^{C} \mu_{ab}(2 - \xi_{ab}) = 1, 1 \leq a \leq N \tag{16}$$

$$\sum_{b=1}^{C} (\eta_{ab} + \frac{1}{C}\xi_{ab}) = 1, 1 \leq a \leq N \tag{17}$$

Using the Lagrange multiplier method, the authors have determined the centroid values and membership degrees as follows.

The values of cluster centers:

$$v_b = \frac{\sum_{a=1}^{N} (\mu_{ab}(2 - \xi_{ab}))^m X_a}{\sum_{b=1}^{N} (\mu_{ab}(2 - \xi_{ab}))^m} \tag{18}$$

The membership degrees:

$$\mu_{ab} = \frac{1}{(2 - \xi_{ab}) \sum_{k=1}^{C} \left(\frac{d_{ab}^2}{d_{ak}^2} \right)^{1/(m-1)}} \tag{19}$$

$$\eta_{ab} = \frac{\exp(-\xi_{ab})}{\sum_{k=1}^{C} \exp(-\xi_{ak})} \left(1 - \frac{1}{C} \sum_{k=1}^{C} \xi_{ak} \right) \tag{20}$$

The refusal degree:

$$\xi_{ab} = 1 - (\mu_{ab} + \eta_{ab}) - (1 - (\mu_{ab} + \eta_{ab})^{\delta})^{1/\delta} \tag{21}$$

In which the refusal exponent coefficient is denoted as δ with $\delta \in (0, 1)$.

2.5 The Flowchart of the Algorithm

The general algorithm flowchart is shown as follows (see Fig. 1):

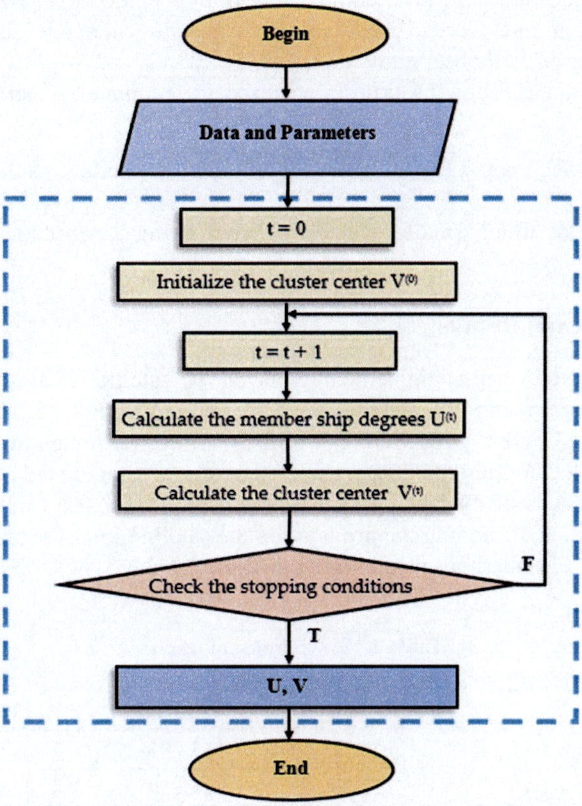

Fig. 1. The flowchart of the fuzzy clustering algorithm.

In general, fuzzy clustering algorithms have the following conceptual idea: The input consists of datasets and parameters and the output is the cluster centers V and the membership matrix U.

The first step of fuzzy clustering algorithms is to input the data into a matrix X consisting of N elements. Then, the initial values such as the number of clusters C, fuzzy coefficient m, maximum number of iterations $MaxStep > 0$ and minimum expected error *epsilon* are determined. Next, the initial cluster centers are randomly initialized. The next step involves computing the membership matrix and the cluster center matrix. The algorithm checks the stopping condition, which is if the number of iterations is greater than the threshold *MaxStep* or if the distance between the cluster centers of two consecutive iterations is smaller than *epsilon*. If the stopping condition is satisfied, the algorithm stops; otherwise, it continues to execute the loop.

3 Application to Wildfire Detection from Satellite Images

3.1 Environmental Setup

The FCM, SSFCM and FC-PFS methods were implemented on an HP computer with 16GB RAM and an Intel Core i5 processor. The experimental images data was obtained from NASA: https://earthobservatory.nasa.gov/topic/fires.

In this paper, the following criteria are used to compare the performance of the algorithms:

(i) Clustering Accuracy [27]: Higher values indicate better results, measured in percentage (%).
(ii) Computational time: Smaller values indicate better performance, measured in seconds (s).

3.2 Environmental Results

We conducted experimental implementations on 20 satellite images in the wildfire dataset. Our program ran 20 times for each image and initialized labels for 20% of randomly selected pixels (wildfire/non-wildfire), while the remaining 80% of pixels were left unlabeled. Within the 20% of labeled pixels, we conducted experiments with different percentages of mislabeled labels, from 0%, 10%, 20% to 30%. Finally, we compiled the results of running the program on the satellite image dataset for the FCM, SSFCM and FC-PFS methods as presented in Table 1 below:

Table 1. Environmental results

Method	FCM	SSFCM	FC-PFS
Clustering accuracy (%)	68,12	73,92	85,18
Computational time (s)	112,25	140,58	189,62

According to Table 1, considering the Clustering accuracy criterion, the FC-PFS method exhibits the highest clustering accuracy, followed by the SSFCM method and finally the FCM method with the lowest clustering accuracy. The clustering accuracy ratio of FC-PFS to FCM is 1.25 and to SSFCM is 1.15.

Considering the Computational time, the FCM method demonstrates the best algorithm running time, followed by the SSFCM method and finally the FC-PFS method. The algorithm running time ratio of FCM to SSFCM is 0.80 and to FC-PFS is 0.59.

The results obtained after applying the FCM, SSFCM and FC-PFS techniques to three original images (image **1**, image **2** and image **3**) out of a total of 20 experimental images are presented in Figs. 2, 3 and 4 as follows:

In the original images, the wildfire regions are areas that appear red or brown, resembling the color of fire or smoke. In the resulting images, the wildfire regions are identified as areas that are colored red. In satellite images of wildfires, there may be various objects

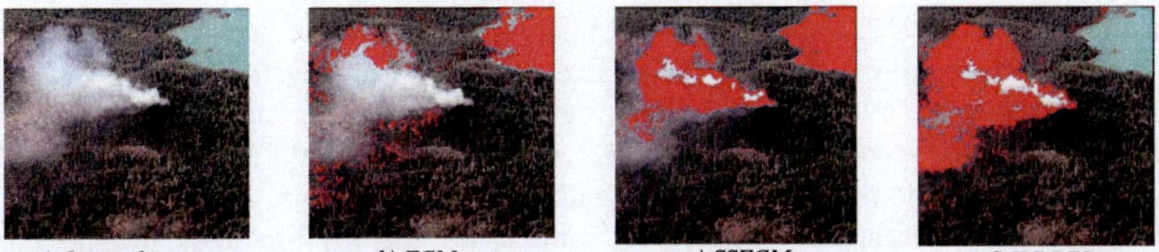

a) Original image *b) FCM* *c) SSFCM* *d) FC-PFS*

Fig. 2. The results of wildfire detection for image **1**

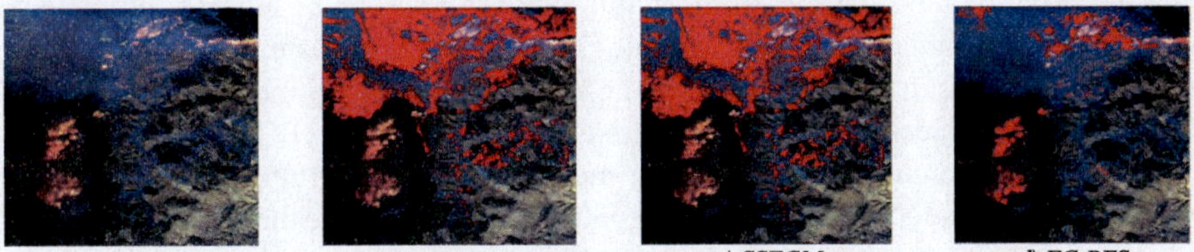

a) Original image *b) FCM* *c) SSFCM* *d) FC-PFS*

Fig. 3. The results of wildfire detection for image **2**

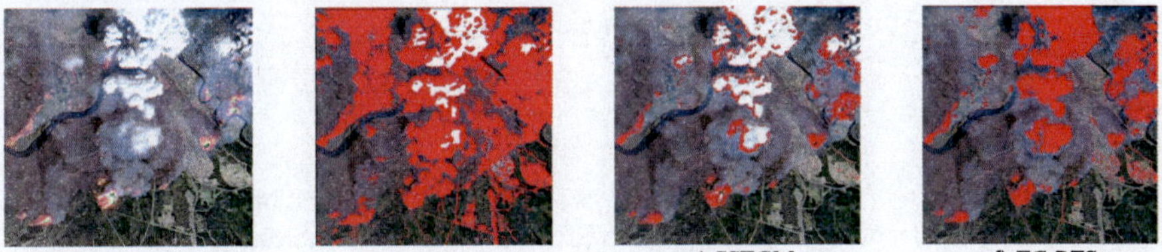

a) Original image *b) FCM* *c) SSFCM* *d) FC-PFS*

Fig. 4. The results of wildfire detection for image **3**

present, including wildfire regions (corresponding to positive membership), green vegetation areas (corresponding to negative membership), rocky terrain (corresponding to neutral membership), along with some other noise (corresponding to refusal membership). An object in the image can belong to one of these four categories with different membership degrees.

For image **1** (Fig. 2), in the original image, there is a smoke plume rising from the forest, which may indicate the presence of a wildfire. However, there is also a water area in the image with colors relatively similar to the smoke. Using the FCM clustering technique, a portion of the smoke is segmented, but it mistakenly includes the water area as part of the wildfire. With the SSFCM technique, the entire smoke area is accurately segmented, but it also misidentifies the water area. With the FC-PFS method, the smoke area is accurately segmented without mistakenly including the water area. For image **2** (Fig. 3), in the original image, there are two fire areas with characteristic fire colors. The colors of the mountains and forests in the image also have little difference. Using the FCM clustering technique, a portion of the fire is segmented, but it mistakenly includes non-fire areas as part of the wildfire. The SSFCM technique yields similar results to the FCM technique. With the FC-PFS method, the fire areas are accurately segmented with fewer misclassifications of non-fire areas. For image **3** (Fig. 4), similar results are obtained as with the other images, where the FC-PFS clustering method still yields the highest effectiveness compared to the FCM and SSFCM methods. The FC-PFS method can accurately segment the fire areas, including smoke and flame.

In general, it can be observed that in some cases, the wildfire areas and the water bodies in satellite images have certain similarities, especially in terms of color, making them difficult to distinguish (both appear as bright white). This can be clearly seen in the illustrations: many wildfire areas were misidentified, particularly those determined using the FCM method. However, the actual wildfire areas were successfully detected, enabling timely evaluation of disaster mitigation and post-disaster recovery solutions. As shown in the result images, it can be noted that the FC-PFS method provides more accurate wildfire detection results compared to the FCM and SSFCM methods.

The FC-PFS picture fuzzy clustering method yields better clustering results compared to the FCM and SSFCM methods, especially in cases where the data contains noise, such as satellite images with mixed clouds, fog and confusing colors. In this method, each data element can be assigned to one of the four attributes of the picture fuzzy set: positive degree, negative degree, neutral degree and refusal degree to belong to a cluster. Unlike traditional fuzzy sets, incorporating these additional attributes in picture fuzzy sets can improve the clustering performance.

References

1. Minaee, S., Boykov, Y., Porikli, F., Plaza, A., Kehtarnavaz, N., Terzopoulos, D.: Image segmentation using deep learning: a survey. IEEE Trans. Pattern Anal. Mach. Intell.Intell. **44**(7), 3523–3542 (2021)
2. Likas, A., Vlassis, N., Verbeek, J.J.: The global k-means clustering algorithm. Pattern Recogn.Recogn. **36**(2), 451–461 (2003)
3. Bezdek, J.C., Ehrlich, R., Full, W.: FCM: the fuzzy c-means clustering algorithm. Comput. Geosci.. Geosci. **10**(2–3), 191–203 (1984)
4. Senthilkumaran, N., Rajesh, R.: Image segmentation-a survey of soft computing approaches. In: 2009 International Conference on Advances in Recent Technologies in Communication and Computing, pp. 844–846 (2009)
5. Omran, M.G., Engelbrecht, A.P., Salman, A.: An overview of clustering methods. Intell. Data Anal. **11**(6), 583–605 (2007)
6. Huan, P.T., Thong, P.H., Tuan, T.M., Hop, D.T., Thai, V.D., Minh, N.H., Son, L.H., et al.: TS3FCM: trusted safe semi-supervised fuzzy clustering method for data partition with high confidence. Multimed. Tools Appl. **81**(9), 12567–12598 (2022)
7. Thong, P.H., Smarandache, F., Tuan, T.M., Ngan, T.T., Thai, V.D., Giang, N.L., Son, L.H.: Picture-neutrosophic trusted safe semi-supervised fuzzy clustering for noisy data. Comput. Syst. Sci. Eng. **46**(2) (2023)
8. Tuan, T.M., et al.: A new approach for semi-supervised fuzzy clustering with multiple fuzzifiers. Int. J. Fuzzy Syst. **24**(8), 3688–3701 (2022)
9. Ramasubbareddy, S., Srinivas, T.A.S., Govinda, K., Manivannan, S.S.: Comparative study of clustering techniques in market segmentation. In: Innovations in Computer Science and Engineering, Springer, Singapore, pp. 117–125 (2022)
10. Wu, Y., Guo, Y., Xiao, Y., Lao, S.: AAE-SC: a scRNA-seq clustering framework based on adversarial autoencoder. IEEE Access **8**, 178962–178975 (2020)

11. Santoso, J.T., Jumini, S., Bhawika, G.W., Susilo, D., Wibowo, D., Rahim, R.: Unsupervised Data Mining Technique for Clustering Library in Indonesia. Library Philosophy and Practice, pp. 1–9 (2021)
12. Farzanfar, E., Delafrooz, N.: Determining the customer lifetime value based on the benefit clustering in the insurance industry. Indian J. Sci. Technol. **9**(1), 1–8 (2016)
13. Majumdar, S., Laha, A.K.: Clustering and classification of time series using topological data analysis with applications to finance. Expert Syst. Appl. **162**, 113868 (2020)
14. Curiskis, S.A., Drake, B., Osborn, T.R., Kennedy, P.J.: An evaluation of document clustering and topic modelling in two online social networks: Twitter and Reddit. Inf. Process. Manag.Manag. **57**(2), 102034 (2020)
15. Knorr, W., Pytharoulis, I., Petropoulos, G.P., Gobron, N.: Combined use of weather forecasting and satellite remote sensing information for fire risk, fire and fire impact monitoring. Comput. Ecol. Softw. **1**(2), 112 (2011)
16. Narayana, G.S., Kolli, K.: Fuzzy K-means clustering with fast density peak clustering on multivariate kernel estimator with evolutionary multimodal optimization clusters on a large dataset. Multimed. Tools Appl. **80**(3), 4769–4787 (2021)
17. Zhao, K., et al.: View-collaborative fuzzy soft subspace clustering for automatic medical image segmentation. Multimed. Tools Appl. **79**(13), 9523–9542 (2020)
18. Cao, L., Wang, C., Li, J.: Vehicle detection from highway satellite images via transfer learning. Inf. Sci. **366**, 177–187 (2016)
19. Li, Z., Li, Y., Lu, W., Huang, J.: Crowdsourcing logistics pricing optimization model based on DBSCAN clustering algorithm. IEEE Access **8**, 92615–92626 (2020)
20. Tuan, T.M., Thong, P.H., Ngan, T.T.: An improvement of trusted safe semi-supervised fuzzy clustering method with multiple fuzzifiers. J. Comput. Sci. Cybern. **38**(1), 47–61 (2022)
21. Chen, B., Jiang, T., Chen, L.: Weblog Fuzzy Clustering Algorithm based on Convolutional Neural Network. Microprocessors and Microsystems, p. 103420 (2020)
22. Sinha, D., Kumari, R., Tripathi, S.: Semisupervised classification based clustering approach in WSN for forest fire detection. Wireless Pers. Commun.Commun. **109**, 2561–2605 (2019)
23. Pedrycz, W., Waletzky, J.: Fuzzy clustering with partial supervision. IEEE Trans. Syst. Man Cybern. Part B (Cybernetics) **27**(5), 787–795 (1997)
24. Zadeh, L.A.: Fuzzy Sets, Information and Control, vol. 8, pp. 338–353 (1965)
25. Cuong, B.C., Kreinovich, V.: Picture fuzzy sets. J. Comput. Sci. Cybern. **30**(4), 409–420 (2014)
26. Thong, P.H., Son, L.H.: Picture fuzzy clustering: a new computational intelligence method. Soft. Comput.Comput. **20**(9), 3549–3562 (2016)
27. Gan, H., Fan, Y., Luo, Z., Huang, R., Yang, Z.: Confidence-weighted safe semi-supervised clustering. Eng. Appl. Artif. Intell.Artif. Intell. **81**, 107–116 (2019)

Evaluating the Performance of Some Deep Learning Model for the Problem of Emotion Recognition Based on EEG Signal

Thuong Duong Thi Mai(✉) and Trung-Nghia Phung

University of Information and Communication Technology, Thai Nguyen, Vietnam
dtmthuong@ictu.edu.vn

Abstract. Using deep learning architectures to recognize human emotions from electroencephalography (EEG) signals, thereby deploying practical applications, is a problem that is currently of great interest to many scientists. In which, depending on the application, the EEG signal can be received from devices with many electrodes (32, 64) or small (14, 2). For each application, to ensure good recognition efficiency, determining the appropriate deep learning architecture as well as the corresponding input features is an important issue that needs to be solved. In this paper, based on the DEAP emotion database, we evaluate the performance of several different deep learning architectures (CNN, BiLSTM, a combination of CNN and BiLSTM) that input is different features (FFT, Welch) extracted from 32-electrode to 14-electrode EEG signals. The experimental results obtained show that the CNN network with FFT features has the best performance with the highest average recognition accuracy and the lowest average loss value. However, the CNN network architecture with Welch features gives the best stability when switching from 32 electrodes to 14 electrodes.

Keywords: Human Emotion Recognition · EEG · Deap Learning · DEAP Dataset · CNN network

1 Introduction

Human emotions are psychological and physiological states generated by conscious human processes. They are closely related to human behavior. In most cases, emotions reflect people's actual thoughts and behaviors caused by psychological conditions [1]. Therefore, enabling machines to understand and interpret human emotional states will greatly contribute to realizing intelligent human-machine interaction systems [2]. Moreover, research on emotion recognition technology can further advance various disciplines, including computer science, psychology, neuroscience, medicine, and crime investigation [3].

There are many ways for computers to understand human emotions, such as extracting affective features from text and facial expressions [2]. However, since brain activity plays an essential role in creating and changing emotions, using EEG signals collected from the brain can help uncover information related to real-time emotional changes

P. T. Nghia et al. (Eds.): ICTA 2023, LNNS 847, pp. 176–185, 2023.
https://doi.org/10.1007/978-3-031-49529-8_19

in people objectively. Especially in recent years, with the development of dry electrode technology, non-invasive electrode devices have been widely used for EEG signal acquisition [3]. This makes emotion recognition from EEG signals an intriguing research topic for scientists worldwide [1, 2].

As depicted in Fig. 1, building an EEG-based emotion recognition system involves: emotional stimulation, EEG signal collection, EEG signal preprocessing, feature extraction, and training a recognition model [4].

The performance of an emotion recognition model depends on various factors, including EEG signal collection equipment, feature extraction methods, and emotion classification techniques.

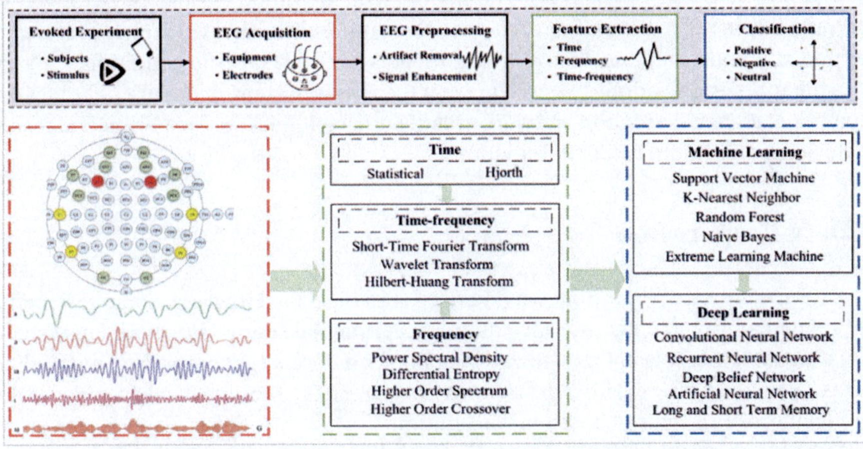

Fig. 1. Implementation steps of the emotion recognition problem [4]

Regarding equipment for collecting EEG signals, there are two main types: devices with many EEG electrodes (e.g. Biosemi ActiveTwo, ESI NeuroScan System) and devices with a small number of EEG electrodes (e.g. Emotiv EPOC) [4, 5]. Devices that collect EEG data using many electrodes are often expensive and inconvenient for portability. In contrast, devices with few electrodes like the Emotiv EPOC are wearable, portable, wireless, and inexpensive, but have lower accuracy [5]. Therefore, when implementing practical applications, it is necessary to develop emotion recognition models that work well even with a small number of EEG electrodes.

Model performance also heavily depends on EEG signal input characteristics. Various feature extraction and selection methods have been proposed from the time, frequency, or time-frequency domains [6]. Commonly used features are frequency domain attributes analyzed via FFT and Welch transforms [1, 3].

Regarding emotion classification techniques, previous EEG-based emotion recognition methods were mainly developed using traditional machine learning techniques [2]. Recently, inspired by emerging deep learning capabilities, various advanced deep learning models such as CNNs and LSTMs have been increasingly utilized for EEG emotion recognition [2, 3].

The above issues have led to research on deep learning models for emotion recognition often being conducted using large, well-known EEG datasets like DEAP, SEED, DREAMER, and MPED. Among them, DEAP is the most widely used, accounting for 51% of research projects [2]. Using the 32 EEG channel DEAP dataset, we can discard some electrode data and retain only data from electrodes corresponding to low-channel devices like Emotiv Epoch. This allows us to conveniently use the DEAP dataset to evaluate deep learning model performance for devices with varying numbers of electrodes.

Based on this analysis, in this paper, we will evaluate the performance of some deep learning models (CNN, CNN-BiLSTM hybrid) for EEG-based emotion recognition. To enable comparison with other publications, we utilize the DEAP dataset in two scenarios - with 32 channels and 14 channels (compatible with the Emotiv Epoch device). We will also vary the input features for each model, using either FFT or Welch features, to determine the most suitable deep learning models for each specific condition.

In the following sections, we will present the research steps in detail, including data preparation, feature extraction, network architecture configuration, experimental results, conclusions, and future research directions.

2 Data Preparation

To perform emotion recognition, we first need suitable data. However, some researchers have limited ability to construct specialized experimental setups. Most are interested in evaluating their methods on recognized benchmark datasets for comparison with related studies. Hence, there is demand for open-source EEG emotion databases developed specifically for EEG-based emotion recognition.

We chose to use the DEAP dataset [7, 8] in this investigation. In this dataset, physiological signals from 32 participants (16 men, 16 women) were recorded using a 40-channel acquisition system including 32 EEG electrodes and 8 additional physiological signal channels. EEG signals were collected while participants watched 40 one-minute videos. After each video, participants rated their emotional experience from 1 to 9 on four scales: Arousal, Valence, Dominance, and Liking. These ratings are used as labels corresponding to the EEG data from that video. This process repeated for each video.

We utilized a preprocessed version of the raw DEAP data provided by the dataset authors, divided into data and labels. The data contains a $40 \times 40 \times 8064$ matrix (trials x videos x channels). The labels are a 40×4 matrix (trials x emotions) with one column per emotion scale. In this preprocessed data, signals are downsampled to 128 Hz and bandpass filtered from 4 to 40.5 Hz. Data from the first 3 s of each trial is discarded.

From the preprocessed DEAP dataset, we constructed two sample sets. The first set utilizes the 32 EEG channels, as channels 33–40 contain EOG and other non-EEG measurements like temperature and blood pressure unrelated to this study. For the second set, we only used data from the 14 electrodes corresponding to the Emotiv Epoch channel configuration.

3 Feature Extraction

The main purpose of feature extraction for EEG-based emotion recognition is to obtain information that can effectively reflect individual emotional states. This information is then used in emotion classification algorithms. Recognition accuracy depends largely on the extracted features. Therefore, extracting EEG features related to emotional states is critical. EEG feature analysis is typically performed in the time domain, frequency domain, or both.

Frequency domain features have been more successful for automatic EEG-based emotion recognition compared to time domain features [3, 5, 9]. Frequency domain analysis converts the EEG signal from the time domain to the frequency domain to evaluate and extract features. The EEG signal is often divided into sub-bands, and features like power spectral density (PSD), differential entropy (DE), and higher order spectra (HOS) are extracted. This study selected two common frequency analysis methods: Fast Fourier Transform (FFT) and Welch's method.

3.1 FFT Feature Extraction

Fast Fourier Transform (FFT) is an efficient algorithm for discrete Fourier transform (DFT) computation. The steps to perform FFT in this study include:

- Initialize parameters: EEG channel number (32), FFT window size (256 samples, 2 s), step size (16 samples), EEG frequency bands (4, 8, 12, 16, 25, 45 Hz).
- Divide the time domain data into short segments (windows) of length 256 samples (window_size) for computational efficiency. Move the window through the data with a step size of 16 samples.
- For each segment, FFT converts the time domain data to the frequency domain. The result is a series of complex numbers representing different fundamental frequencies in the original data. The real part represents the amplitude and imaginary part represents the phase shift of each frequency.

The FFT spectrogram shows the strength of each frequency component, which is useful for analysis of audio, signals, and EEG data, as it distinguishes different frequency components [10].

3.2 Welch Feature Extraction

The Welch method estimates the Power Spectral Density (PSD) of a signal. PSD indicates the energy distribution across frequencies. The Welch method divides the signal into overlapping segments, computes the PSD of each segment, and averages the PSD estimates. Segments are windowed before PSD calculation, commonly using Hanning or Hamming windows [11]. The steps in this study are:

- Initialize parameters (same as FFT)
- Divide the signal into m overlapping segments by taking segments from start to start + window_size, with step_size controlling overlap.
- Apply a Hamming window to each segment to reduce spectral leakage.

- Perform Fast Fourier Transform on each windowed segment.
- Average the PSDs of all segments to produce the final estimate.

Welch averaging reduces variability versus direct FFT, providing a more robust PSD estimate. However, averaging loses insight into time-localized frequency changes. Careful selection of window size and overlap is needed. The Welch method requires more computations and resources than direct FFT.

4 Network Architecture Selection

Deep learning network architecture depends greatly on the type of data as well as input characteristics. Therefore, depending on the data set and input characteristics, we need to choose the appropriate network architecture.

With the DEAP dataset, it is a multidimensional dataset. Each EEG signal is a multidimensional data (data from 32 or 14 channels), reflecting the activity of many different brain regions. This creates an important spatial feature. CNN network architecture is very suitable to handle these spatial features.

Next, the EEG signal contains specific frequency bands, such as Alpha, Beta, Gamma, Delta, Theta, that can be related to the user's emotions. Converting the signal from the time domain to the frequency domain helps reveal these frequency characteristics.

When the EEG signal is divided into small segments (to find frequency characteristics), each signal segment has a temporal relationship with other segments. BiLSTM can learn patterns and changes over time from these signal segments. Furthermore, the term "bi" in BiLSTM means "two-way", meaning it considers both past and future information of each data point. This helps improve performance in many tasks.

Therefore, we chose to experiment with two types of network architectures (CNN and combining CNN with BiLSTM). The parameters of this network architecture are determined based on experiments combined with hyperparameter tuning technique, on the Google Colab Pro + environment [12].

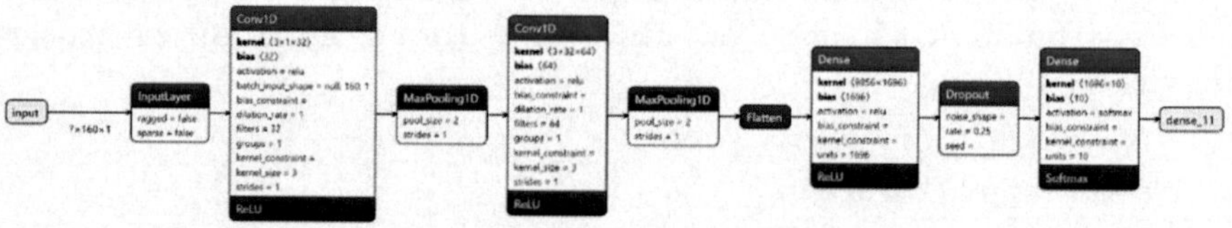

Fig. 2. Proposed CNN network architecture

Figure 2 specifically describes the CNN network architecture with the parameters obtained when using 32 EEG channels. This architecture includes: Two Conv1D layers that learn local and global features from EEG signal data extracted through FFT. Interleaved between the two convolutional layers are 2 MaxPooling1D layers, which help reduce the size of features while avoiding overfitting and increasing computation speed. Next, a Dense layer with 1696 units and ReLU activation function acts as the classification layer. A Dropout (0.2) Layer eliminates randomness to reduce overfitting. Finally, the fully connected layer has 10 classification layers corresponding to the rating scale for each emotion and uses the Softmax activation function to calculate the probability for each class.

In the case of using 14 EEG channels, the network architecture remains the same, with only 70 input features instead of 160.

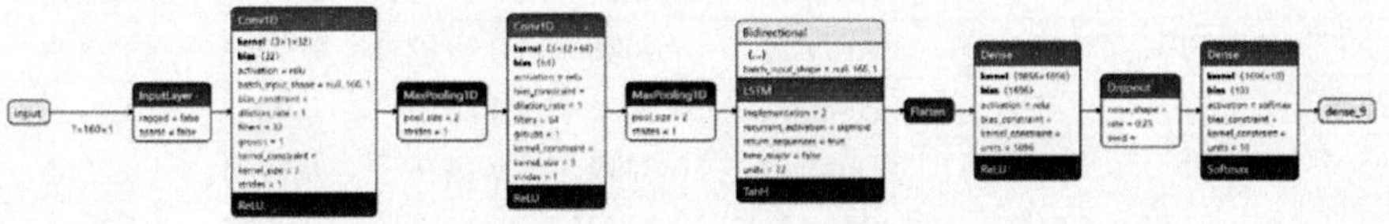

Fig. 3. Proposed CNN and BiLSTM network architecture

Since EEG signals are often highly complex and informative, BiLSTM can capture associations between data points. Therefore, the combined CNN and BiLSTM network architecture and its parameters when using 32 EEG channels are also described in Fig. 3. In this architecture, the first convolutional layer plays the role of learning basic data features, while the second convolutional layer learns more complex features. The two MaxPooling layers still play a role in reducing dimensionality, increasing efficiency and preventing overfitting. The added BiLSTM layer aims to process time series information and capture longer relationships in the data.

5 Some Experimental Results

Using the open-source Tensorflow library and conducting experiments on the Google Colab Pro+ environment, we have evaluated the performance of the proposed model architectures when input features change. To compare with results of previous publications, we use two measures - recognition accuracy (Acc) and loss function (Loss). Table 1 specifically describes the values obtained for deep learning models using 32 and 14 channels for each label.

From Table 1, for 32-channel data, the CNN model with FFT features gives the best Acc results for all labels and the lowest Loss overall. However, for 14-channel data, the CNN model with Welch features has the best recognition result for each label. This is noteworthy when implementing practical emotion recognition where output identifies separate labels.

Table 2 shows the average model performance. Experiments show the CNN + FFT model has the highest mean accuracy (0.9542) and lowest mean loss (0.1786) for 32 channels. With 14 channels, this model still has the highest accuracy (0.9246) and lowest loss (0.2792).

Evaluating a model requires assessing both accuracy and loss. Next, we evaluate how accuracy and loss values change between 32 and 14 channels. We calculate the absolute difference in accuracy (ΔAcc) and loss (ΔLoss), and the total absolute deviation (ΔTotal = ΔAcc + ΔLoss). ΔTotal provides an overview of differences when reducing channels from 32 to 14, evaluating accuracy and loss together. ΔTotal also assesses model stability with fewer channels - an important aspect of adaptability and reliability.

Table 1. Performance results of deep learning models with 32 and 14 channel EEG data

Model	Feature	Label	Acc 32	Loss 32	Acc 14	Loss 14
CNN	FFT	Arousal	**0,9497**	0,1926	0,9241	0,2826
CNN	Welch	Arousal	0,9415	0,1836	**0,9307**	0,2655
CNN + BiLSTM	FFT	Arousal	0,9326	0,2440	0,9231	0,2792
CNN + BiLSTM	Welch	Arousal	0,9448	0,2212	0,9203	0,2893
CNN	FFT	Domain	**0,9632**	0,1636	0,9068	0,2925
CNN	Welch	Domain	0,9492	0,1656	**0,9116**	0,2768
CNN + BiLSTM	FFT	Domain	0,9370	0,2245	0,9075	0,2845
CNN + BiLSTM	Welch	Domain	0,9273	0,2437	0,9006	0,2931
CNN	FFT	Like	**0,9529**	0,1864	0,9146	0,3048
CNN	Welch	Like	0,9390	0,1972	**0,9200**	0,2881
CNN + BiLSTM	FFT	Like	0,9306	0,2488	0,9137	0,3085
CNN + BiLSTM	Welch	Like	0,9348	0,2258	0,9163	0,3084
CNN	FFT	Valence	**0,9511**	0,1719	0,8920	0,3490
CNN	Welch	Valence	0,9457	0,1878	**0,9014**	0,3328
CNN + BiLSTM	FFT	Valence	0,9347	0,2403	0,8956	0,3401
CNN + BiLSTM	Welch	Valence	0,9454	0,2152	0,9008	0,3349

Table 2. Average performance of deep learning models

Model	Feature	Mean Acc_32	Mean Loss_32	Mean Acc_14	Mean Loss_14
CNN	FFT	**0.9542**	**0.1786**	**0.9246**	**0.2792**
CNN	Welch	0.9438	0.1836	0.9066	0.2867
CNN + BiLSTM	FFT	0.9337	0.2394	0.9161	0.3025
CNN + BiLSTM	Welch	0.9381	0.2265	0.8975	0.3392

Table 3. Evaluation results based on absolute difference

Model	Feature	Δ Acc	Δ Loss	Δ Total
CNN	FFT	0.0297	0.1005	0.1302
CNN	Welch	0.0372	0.1032	0.1404
CNN + BiLSTM	FFT	0.0176	0.0630	0.0806
CNN + BiLSTM	Welch	0.0406	0.1127	0.1533

Table 2 shows the CNN + FFT model is best by mean. However, Table 3 shows this model has the highest total deviation of 0.1302, meaning significant difference from 32 to 14 channels. The CNN + BiLSTM + FFT model has the lowest deviation (0.0806), giving it the best stability.

Next, we compare the CNN + FFT model to published works using 32 and 14 DEAP channels. As Table 4 shows, our model has higher accuracy than prior 32-channel studies classifying 4 (or 2) labels. We found limited 14-channel studies, but Table 5 shows good CNN-FFT results.

Table 4. Comparison with some published works using 32 channels

Model	Feature	Acc 32				Mean Acc 32
		Arousal	Valence	Dominance	Liking	
BiDCNN [13]	DE	94.72	94.38	NA	NA	NA
LSTM [14]	FFT, PSD	NA	NA	NA	NA	91.38
CNN [15]	Entropy & HFD	NA	NA	NA	NA	95.20
CNN-BiLSTM [16]	DE	94.86	94.02	NA	NA	94.44
Merged LSTM [17]	DWT	83.85	84.89	84.37	80.72	83.46
CNN (Propose)	FFT	94.97	95.11	96.32	95.29	**95.42**

Table 5. Comparison with some published works using 14 channels

Model	Feature	Acc 14				Mean Acc 14
		Arousal	Valence	Dominance	Liking	
DBN-GCs [18]	Raw	75.92	76.83	NA	NA	NA
CNN (Propose)	FFT	93.07	90.14	91.16	92.00	**92.46**

6 Conclusion

Applying deep learning models for emotion recognition based on EEG signals is a reasonable research direction with many potential practical applications. However, finding the optimal model is not simple, as performance depends on factors including application type, EEG data properties, deep learning network architecture and parameters, and input characteristics.

By investigating CNN and CNN-BiLSTM models on the DEAP dataset with two feature types (FFT, Welch) from 32 or 14 electrodes, we have the following conclusions:

For EEG-based emotion recognition applications with multiple output labels, a CNN model with FFT inputs is recommended, as this architecture provides the best average performance across labels.

For classifying a single label with 14 electrodes, a CNN model with Welch inputs should be considered.

To maintain stability when reducing from 32 to 14 channels, the CNN model with Welch inputs is recommended, as this model has the lowest total absolute deviation.

References

1. Yu, C., Wang, M.: Survey of Emotion Recognition Methods Using EEG Cognitive Robotics, vol. 2, pp. 132–146 (2022). ISSN 2667-2413https://doi.org/10.1016/j.cogr.2022.06.001
2. Houssein, E.H., Hammad, A., Ali, A.A.: Human emotion recognition from EEG-based brain–computer interface using machine learning: a comprehensive review. Neural Comput. Appl.Comput. Appl. **34**, 12527–12557 (2022). https://doi.org/10.1007/s00521-022-07292-4
3. Wang, X., Ren, Y., Luo, Z., He, W., Hong, J., Huang, Y.: Deep learning-based EEG emotion recognition: current trends and future perspectives. Front. Psychol. **27**(14), 1126994 (2023). https://doi.org/10.3389/fpsyg.2023.1126994.PMID:36923142;PMCID:PMC10009917
4. Wang, J., Wang, M.: Review of the emotional feature extraction and classification using EEG signals. Cogn. Robot., 1 (2021).https://doi.org/10.1016/j.cogr.2021.04.001
5. Vempati, R., Sharma, L.: A systematic review on automated human emotion recognition using electroencephalogram signals and artificial intelligence. Results Eng. **18**, 101027 (2023). https://doi.org/10.1016/j.rineng.2023.101027
6. Lin, W., Li, C.: Review of studies on emotion recognition and judgment based on physiological signals. Appl. Sci. **13**(4), 2573 (2023). https://doi.org/10.3390/app13042573
7. Akter, S., Prodhan, R.A., Pias, T.S., Eisenberg, D., Fernandez, J.F.: M1M2: deep-learning-based real-time emotion recognition from neural activity. Sensors **22**(21), 8467 (2022). https://doi.org/10.3390/s22218467
8. Koelstra, S., et al.: DEAP: a database for emotion analysis using physiological signals. IEEE Trans. Affect. Comput.Comput. **3**, 18–31 (2011). https://doi.org/10.1109/T-AFFC.2011.15
9. Abdulrahman, A., Baykara, M.: A comprehensive review for emotion detection based on EEG signals: challenges, applications, and open issues. Traitement du Signal **38**, 1189–1200 (2021)
10. Murugappan, M., Murugappan, S.: Human emotion recognition through short time Electroencephalogram (EEG) signals using Fast Fourier Transform (FFT). In: 2013 IEEE 9th International Colloquium on Signal Processing and its Applications, pp. 289–294 (2013)
11. Garg, N., Sharma, K.: Feature Extraction for Emotion Recognition: A Review. Emotion Recognition—Recent Advances, New Perspectives and Applications. IntechOpen (2023). https://doi.org/10.5772/intechopen.109740
12. Mezzah, S., Tari, A.: Practical hyperparameters tuning of convolutional neural networks for EEG emotional features classification. Intell. Syst. Appl. **18**, 200212 (2023). ISSN 2667-3053https://doi.org/10.1016/j.iswa.2023.200212
13. Yoon, H.J., Chung, S.Y.: EEG-based emotion estimation using Bayesian weighted-log-posterior function and perceptron convergence algorithm. Comput. Biol. Med. **43**(12), 2230–2237 (2013). ISSN 0010-4825, https://doi.org/10.1016/j.compbiomed.2013.10.017
14. Huang, D., Chen, S., Liu, C., Zheng, L., Tian, Z., Jiang, D.: Differences first in asymmetric brain: a bi-hemisphere discrepancy convolutional neural network for EEG emotion recognition. Neurocomputing **448**, 140–151 (2021). https://doi.org/10.1016/j.neucom.2021.03.105

15. Sakalle, A., Tomar, P., Bhardwaj, H., Acharya, D., Bhardwaj, A.: A LSTM based deep learning network for recognizing emotions using wireless brainwave driven system. Expert Syst. Appl. **173**, 114516 (2021). https://doi.org/10.1016/j.eswa.2020.114516
16. Cui, H., Liu, A., Zhang, X., Xiang, C., Wang, K., Chen, X.: EEG-based emotion recognition using an end-to-end regional-asymmetric convolutional neural network. Knowl.-Based Syst. Based Syst. **205**, 106243 (2020). https://doi.org/10.1016/j.knosys.2020.106243
17. Cui, F., Wang, R., Ding, W., Chen, Y., Huang, L.: A novel DE-CNN-BiLSTM multi-fusion model for EEG emotion recognition. Mathematics **10**(4), 582 (2022). https://doi.org/10.3390/math10040582
18. Garg, A., Kapoor, A., Bedi, A., Sunkaria, R.: Merged LSTM Model for Emotion Classification Using EEG Signals, pp. 139–143 (2019). https://doi.org/10.1109/ICDSE47409.2019.8971484

Feature Reduction for Interpretability of Neuro-Fuzzy Classifier

Thanh-Tung Nguyen[1], Dang Thi Hien[2], and Tuan-Linh Nguyen[2](✉)

[1] Thai Nguyen University of Information and Communication Technology, Thainguyen, Vietnam
nttung@ictu.edu.vn
[2] Thai Nguyen University of Technology, Thainguyen, Vietnam
{danghien,ntlinh}@tnut.edu.vn

Abstract. The growing use of machine learning in healthcare demands models that not only predict well but clearly explain their reasoning. Neuro-fuzzy classifiers combine neural networks and fuzzy logic for accurate yet interpretable modeling. However, high-dimensional medical data challenges comprehension. This study presents techniques to distill neuro-fuzzy classifiers down to their most essential components for enhanced interpretability. Recursive feature elimination is employed, iteratively removing the least contributory features according to classifier weights. On a breast cancer prognosis task, these methods significantly reduce features to the minimal set necessary for strong accuracy. The pruned neuro-fuzzy classifier demonstrates competitive prediction with increased transparency through an intuitive rule-based structure relying on only the most critical inputs. Feature reduction thus holds promise for optimizing the balance between predictive capacity and comprehensive explanations in medicine. However, challenges remain in scaling these methods to even higher-dimensional data like medical images while retaining accuracy. Future work should explore the effects of different criteria for ranking feature importance beyond classifier weights. Overall, this study provides an initial demonstration of how feature reduction techniques can produce simplified neuro-fuzzy classifiers more amenable to clinical implementation, but continued research is needed to extend these approaches to complex real-world medical data. Distilling machine learning models down to their most essential components will be critical for trust and adoption of artificial intelligence in high-stakes healthcare applications.

Keywords: Feature reduction · Neuro-fuzzy classifier · Breast cancer · Interpretable

1 Introduction

Advances in medicine have made large amounts of patient data available to medical practitioners, enabling research into predictive models for early cancer diagnosis and prognosis. Predicting clinical outcomes from complex medical data remains challenging but is important for enhancing treatment decisions and managing patient care [1]. As

P. T. Nghia et al. (Eds.): ICTA 2023, LNNS 847, pp. 186–193, 2023.
https://doi.org/10.1007/978-3-031-49529-8_20

a result, there has been significant research applying machine learning tools in artificial intelligence to detect meaningful patterns in data for the early prediction of cancer. Traditional machine learning algorithms rely on pre-selected features to build models, lacking the ability to provide an explicit representation of the knowledge gained. However, developing explicit knowledge from data could improve decision making and patient management.

Machine learning methods, especially deep learning approaches, have shown great promise for medical applications due to their ability to process large datasets and automatically learn complex features. Deep learning is a type of machine learning based on artificial neural networks with multiple layers that can learn hierarchical representations of data [2]. In medicine, due to large workloads, deep learning has emerged as a powerful machine learning approach with superior diagnostic performance compared to other models. This allows for improved diagnosis, classification and prediction of clinical outcomes [3]. Deep learning forms a hierarchical representation of data through layer-by-layer pattern detection, effectively characterizing complex training information [4]. While deep neural networks efficiently represent problems internally, it remains unclear how this representation distributes across layers and arises from learning [5]. This lack of transparency in training presents critical issues for healthcare, where explaining predicted patient outcomes based on characteristics is essential. Explainability, interpretability and transparency have thus become central concerns for deep learning in medicine in recent years [6, 7]. Explicit linking of inputs to outcomes is needed to build trust in critical applications and support explainable clinical decision making.

Rule extraction approaches like decision lists, decision trees and simple IF-THEN rules can reveal hidden network knowledge and provide more comprehensible explanations of decision processes [8]. Fuzzy logic deals with uncertainties and ambiguities in raw data, benefiting from the ability of rule-based systems to incorporate subjective expert information [9]. This improves classifications or changes system behavior. Both fuzzy and neural systems effectively represent data. Adaptive neuro fuzzy inference systems (ANFIS) combine artificial neural networks and fuzzy inference systems to approximate non-linear problems using adaptive if-then statements [10]. ANFIS automatically learns fuzzy membership functions and derives rules from large training datasets [11]. It works well for applications requiring extensive knowledge bases to accommodate various scenarios not fully captured by domain experts. While ANFIS has improved inference accuracy, interpreting thousands of fuzzy rules and parameters for high-dimensional problems remains challenging. Optimizing fuzzy systems involves balancing two conflicting goals: interpretability and accuracy. Simultaneously optimizing for both interpretability and accuracy is expected to further enhance system performance.

This paper introduces a neuro-fuzzy classifier (NFC) to better balance predictive performance and model interpretability. The proposed NFC leverages the rule-based structure of adaptive neuro fuzzy inference system (ANFIS) to create a classifier that can provide explanations for its classifications. To further improve the model, an evolutionary algorithm is used to optimally permute input features. This optimization determines the best groupings of mutually relevant features between subsets. Grouping relevant features together aims to maximize relevance within feature subsets, while minimizing irrelevant features between them. This process can improve overall model performance. The performance of the proposed NFC is evaluated on a breast cancer classification problem. After training, input feature size is further reduced through recursive feature elimination. This greatly reduces model complexity by removing non-critical features. Not all input variables are important for prediction.

2 Neuro-Fuzzy Classifier

The neuro-fuzzy classifier integrates an adaptive neuro-fuzzy inference system with a softmax layer applied to the expanded of consequence part of the neuro-fuzzy network, allowing it to perform multi-class classification tasks. As shown in Fig. 1, the consequence portion of the NF classifier contains multiple consequence units, with each unit receiving the same output from the premise part. Individual units then compute an output value corresponding to one class. Therefore, the number of consequence units scales directly with the number of classes the model is designed to classify.

The softmax layer serves to normalize the neuro-fuzzy (NF) outputs into a proper probability distribution using an exponential function. It calculates the probability that an input belongs to each class based on the outputs of the underlying NF network. By coupling an NF with a softmax layer in this way, the resulting NF classifier can predict the class probability distribution for inputs rather than just a single class. This multi-class capability expands the applicability of the classifier beyond simple two-class problems. Integrating a softmax layer is important because it allows the NF classifier to work with problems involving more than two target classes. Without the softmax normalization, the raw NF outputs would not represent a valid probability distribution over classes. The softmax helps bridge the gap between the numerical NF outputs and probabilistic class predictions. It therefore represents a key component that transforms the NF into a multi-class classifier suitable for complex classification tasks with numerous target classes. This integrated structure means the proposed NFC can similarly be applied to solve general multi-class problems beyond basic binary classification.

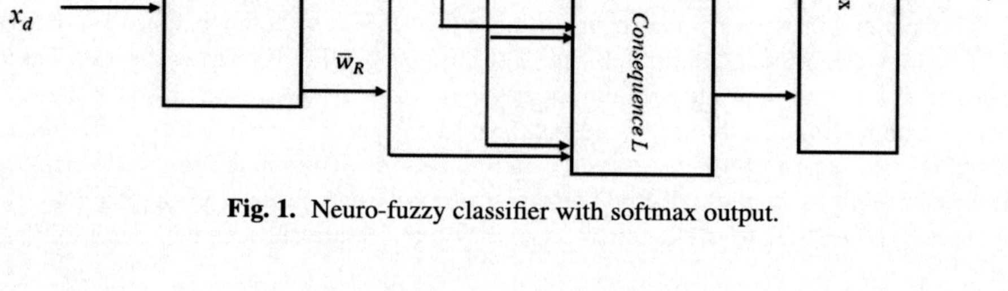

Fig. 1. Neuro-fuzzy classifier with softmax output.

3 Feature Reduction by Recursive Feature Elimination and Causality Analysis

Explanation involves ascribing causal responsibility, even when the true underlying causes are unknown. To understand why a system selects one feature over others, we must cite a causal difference between the selected feature and the absence of that feature in the non-selected histories. This process of selecting a subset of potential causes to provide as an explanation is known as causal selection.

Determining causality from observational data can be achieved using the Linear Non-Gaussian Additive Model (LiNGAM) algorithm, a form of Independent Component Analysis. LiNGAM consistently estimates the underlying causal network if its assumptions hold and sufficient data is available. This paper leverages DirectLiNGAM [12], an extension of LiNGAM [13] that can be applied even when there are more variables than observations. This is crucial as the original LiNGAM would fail in such high-dimensional settings.

However, not all input features may be relevant for the target classification task. Irrelevant features can sometimes degrade a classifier's performance. Therefore, the number of features is significantly reduced using a combination of Recursive Feature Elimination (RFE) and causality-based selection. RFE ranks feature sets according to a causal measurement, removing those contributing least to classification. Causality selection identifies the most influential features based on their causal effects. Together, RFE and causality selection effectively prune irrelevant features to improve model efficacy while retaining causally important predictors. This helps ensure only the most relevant features are used at each layer for optimal prediction.

Gaining insight into the input features is crucial for explaining its predictions. To select the most important features at input layer, a combination of RFE and causality analysis is performed. First, RFE ranks the feature sets from input layer based on their contributions to the classifier. Features ranked lowest are iteratively eliminated to reduce dimensionality.

Causality analysis is then used to determine which features have a direct causal relationship to the target classifying labels of benign or malignant. This involves applying the DirectLiNGAM algorithm to identify causal directionality between features and cancer classes from observational data. Features found to causally influence cancer labels are retained.

4 Experiment results

We evaluate the performance of the proposed neuro-fuzzy classifier on the task of breast cancer diagnostic classification. The dataset used is the Wisconsin Diagnostic Breast Cancer (WDBC) dataset from the UCI Machine Learning Repository [14], which contains 569 breast mass specimens. For each sample, there are 30 tumor features along with a subject ID number and a binary class label indicating whether the tumor is benign or malignant.

The 30 features were computed from digital images obtained through fine needle aspirate (FNA) of breast masses. Specifically, 10 morphological attributes of the cell nuclei were measured, including radius, texture, perimeter, area, smoothness, compactness, concavity, concave points, symmetry, and fractal dimension. The measurements were then processed to derive average, standard error, and extremes of these attributes across the specimen sample.

For each specimen, mean, standard error, and extreme values were calculated from the 10 basic cell attribute measurements. This resulted in 30 features describing characteristics like the mean radius and radius error, as well as descriptors of the distribution of values like tumor radius maximum and radius minimum. Together, these features form a complex but informative representation of each breast tumor specimen that can aid diagnostic classification. By evaluating NFC on this benchmark WDBC dataset, we aim to demonstrate its ability to effectively interpret and learn from high-dimensional biomedical data for an important healthcare application.

The traditional machine learning classifier such as Support Vector Machine (SVM) and Multiple Layer Perceptron (MLP) are employed for comparing the performance of the NFC. To ensure fair model comparisons, hyperparameters for each classifier were carefully optimized. For the support vector machine (SVM), a grid search procedure was used to select the optimal hyperparameters, which were determined to be: kernel = 'poly', degree = 3, C = 0.35, coef0 = 0.125, and gamma = 0.0625. These values define the polynomial kernel function with a degree of 3 and help control model complexity and scaling. The MLP parameters were chosen empirically based on standard settings commonly used for classification tasks. Specifically, the networks contained dense layers with dropout regularization.

Hyperparameters of the NFC like the number of fuzzy rules and learning rate were tuned to achieve the best performance on validation data. This validation-based approach helped ensure each model was optimized independently prior to performance evaluation and comparison on the test set.

To obtain robust and statistically significant performance comparisons, each prediction model was trained and tested using 10-fold cross-validation repeated over 10 iterations. This involved randomly splitting the dataset into 10 equal partitions, using 9 folds for training and 1 fold for testing in each iteration. The models were retrained from scratch 10 times to ensure every data point took part in both the training and test sets.

The average test accuracy and associated variance over the 10 iterations were then calculated for each model and recorded. As summarized in Table 1, the average F-scores achieved were: 0.918 for the support vector machine (SVM), 0.971 for the MLP, and 0.989 for the proposed NFC. Compare with the state-of-the-art from [15], our proposed AH-ANFIS model has a comparable performance.

These results clearly demonstrate that the ANFIS-related models generally outperformed the traditional SVM and MLP classifiers in terms of cancer classification accuracy. This could be attributed to NFC's ability to learn fuzzy rules that capture complex cancer-related patterns in an interpretable manner.

Figure 2 depicts the results of critically analyzing the breast cancer datasets to identify the most important predictive features. For breast cancer prediction, the DirectLiNGAM algorithm was applied to calculate a causality score representing each variable's contribution to the overall inference performance. Variables with a score of zero do not causally impact prediction and can be neglected. Causality scoring revealed the top

Table 1. Comparison of average test accuracy

	SVM	MLP	NFC	SOTA [15]
Average	0.918	0.971	0.989	0.993
STD	0.0095	0.0126	0.0105	–

predictive features as: *Radius Maximum, Symmetry Standard Error, Perimeter Mean, Texture Mean, Smoothness Maximum, Texture Maximum, Area Maximum, Concavity Maximum, Area Standard Error, Perimeter Maximum, Concavity Mean, Texture Standard Error, Concavity Standard Error,* and *Symmetry Maximum.* These image-derived texture and shape measurements from breast tissue samples causally impact the model's diagnostic performance. By quantifying each variable's causal importance, the proposed method provides insight and justification for why certain clinical or imaging biomarkers should be prioritized over others for optimizing predictive accuracy.

Index	20	1	4	5	9	28	21	19	3	2	25	13	22	30	24
Feature	Fractal dimension SE	Radius mean	Area mean	Smoothness mean	Symmetry mean	Concave points max	Radius max	Symmetry SE	Perimeter mean	Texture mean	Smoothness max	Perimeter SE	Texture max	Fractal dimension max	Area max
Abbr.	FSE	RdM	AM	SmM	SM	CpMx	RMx	SSE	PM	TM	SmM	PSE	TMx	FMx	AMx
Score	0	0	0	0	0	0	0.65	0.17	0.41	0.15	0.33	0	0.85	0	0.38
Selected	0	0	0	0	0	0	1	1	1	1	1	0	1	0	1

Index	11	27	8	14	6	26	15	23	7	16	12	17	29	10	18
Feature	Radius SE	Concavity max	Concave points mean	Area SE	Compactness mean	Compactness max	Smoothness SE	Perimeter max	Concavity mean	Compactness SE	Texture SE	Concavity SE	Symmetry max	Fractal dimension mean	Concave points SE
Abbr.	RSE	CnMx	CpM	ASE	CmM	CmMx	SmS	PMx	CnM	CmSE	TSE	CSE	SMx	FdM	CpSE
Score	0	0.67	0	0.43	0	0	0	0.53	0.24	0	0.16	0.31	0.19	0	0
Selected	0	1	0	1	0	0	0	1	1	0	1	1	1	0	0

Fig. 2. Critical analysis of breast cancer input features result

To further streamline the model and enhance interpretability, recursive feature elimination (RFE) was combined with the causal feature analysis. RFE works by recursively removing the feature that least improves prediction performance, based on the weights learned by the fitted model, until a desired number of features remains. By applying RFE to the causal feature selection results, the number of clinically relevant input features was substantially reduced. Specifically, when applied to the breast cancer prediction problem involving 30 original inputs, RFE narrowed the feature set down from the 14 variables identified through causality scoring. This helped prune less impactful features and simplify the model. The combined approach of causal analysis followed by RFE ultimately selected a highly optimized set of just 14 critical features that most directly and causally influenced the predictive performance of the proposed neuro-fuzzy classifier.

Reducing the dimensionality this way resulted in a more streamlined and interpretable model focused only on the smallest set of features necessary to achieve strong classification accuracy. It demonstrates how integrating causal reasoning with recursive selection techniques can distill predictive models down to their most essential components.

5 Conclusion

This study developed and evaluated techniques for refining neuro-fuzzy classifiers to focus only on the most important features for enhanced interpretability. By applying an evolutionary algorithm optimization and recursive feature elimination to hierarchically structured neuro-fuzzy models, the most predictive input variables were identified from high-dimensional medical datasets. On a breast cancer diagnostic problem, these feature selection methods significantly reduced the input dimensionality while maintaining strong predictive performance. The pruned classifier demonstrated classification accuracy comparable to more complex editions relying on far fewer features. By distilling models down to their essence through recursive feature elimination guided by causal influence, neuro-fuzzy systems can achieve a superior balance between predictive capacity and transparent interpretability for domain experts. For critical applications in healthcare, this improved transparency facilitates comprehensive explanations and builds user trust in model predictions. Moving forward, integrating feature selection approaches like those presented offers promising opportunities to optimize machine learning models for both accuracy and interpretability in medicine.

Acknowledgement. This work was supported by Thainguyen University of Technology.

References

1. Kourou, K., Exarchos, T.P., Exarchos, K.P., Karamouzis, M.V., Fotiadis, D.I.: Machine learning applications in cancer prognosis and prediction. Comput. Struct. Biotechnol. J. **13**, 8–17 (2015)
2. Chen, J.H., Asch, S.M.: Machine learning and prediction in medicine—beyond the peak of inflated expectations. N. Engl. J. Med. **376**(26), 2507 (2017)
3. Nichols, J.A., Herbert Chan, H.W., Baker, M.A.B.: Machine learning: applications of artificial intelligence to imaging and diagnosis. Biophys. Rev. **11**(1), 111–118 (2019)
4. Wei-Ning, W., Ying-Lin, Y., Sheng-Ming, J.: Image retrieval by emotional semantics: a study of emotional space and feature extraction. In: 2006 IEEE International Conference on Systems, Man and Cybernetics, vol. 4, pp. 3534–3539 (2006)
5. Grégoire, M.: On layer-wise representations in deep neural networks. Ph.D. dissertation, Technische Universität, Berlin (2013)
6. Paredes, S., Henriques, J., Rocha, T., Valente, F., Carvalho, P., Morais, J.: Interpretable and personalized CVD risk assessment: innovative approaches. In: European Journal of Clinical Investigation, vol. 51, p. 97 (2021)
7. Antoniadi, A.M. et al.: Current challenges and future opportunities for XAI in machine learning-based clinical decision support systems: a systematic review. Appl. Sci. **11**(11) (2021)
8. Petkovic, D., Ab Hamid, S.H., Cojbasic, Z., Pavlovic, N.T.: Adapting project management method and ANFIS strategy for variables selection and analyzing wind turbine wake effect (Retraction of Vol. 74, P. 463, 2014). Springer One New York Plaza, Suite 4600, New York, NY, United States (2020)
9. Guillaume, S.: Designing fuzzy inference systems from data: an interpretability-oriented review. IEEE Trans. Fuzzy Syst. **9**(3), 426–443 (2001)

10. Jang, J.S.R.: ANFIS: adaptive-network-based fuzzy inference system. IEEE Trans. Syst. Man Cybern. **23**(3), 665–685 (1993)
11. Petković, D., Pavlović, N.T., Ćojbašić, Ž: Wind farm efficiency by adaptive neuro-fuzzy strategy. Int. J. Electr. Power Energy Syst. **81**, 215–221 (2016)
12. Shimizu, S. et al.: DirectLiNGAM: A direct method for learning a linear non-Gaussian structural equation model. J. Mach. Learn. Res. **12**, 1225–1248 (2011)
13. Shimizu, S., Hoyer, P.O., Hyvärinen, A., Kerminen, A.: A linear non-gaussian acyclic model for causal discovery. J. Mach. Learn. Res. **7**, 2003–2030 (2006)
14. Bache, K., Lichman, M.: UCI Machine Learning Repository (2013)
15. Devarriya, D., Gulati, C., Mansharamani, V., Sakalle, A., Bhardwaj, A.: Unbalanced breast cancer data classification using novel fitness functions in genetic programming. Expert Syst. Appl. **140**, 112866 (2020)

Improved Accuracy of Path System on Creating Intelligence Base

Don T. Tran[1], Vinh Q. Nguyen[1], Cuong V. Nguyen[3], Dong L. T. Tran[4], Hoang T. Tran[4], and N. D. Anh[2](✉)

[1] Dept. of Radio- Electronic Equipment, Academy of Military Sciences and Technology, Hanoi 100000, Vietnam

[2] The University of Fire Fighting & Prevention, Hanoi 100000, Vietnam
anhpchp@yahoo.co.uk

[3] Thai Nguyen University of Information and Communication Technology, Thai Nguyen 240000, Vietnam

[4] Center of Electrical Engineering, Duy Tan University, Da Nang 550000, Vietnam

Abstract. Artificial Intelligence (AI) is a branch of computer science, created by humans to help computers to automate intelligent behaviors and have human-like intelligence. Currently, AI is a spearhead technology with potential applications in many areas of social life. In particular, in the military, AI is being developed more and more perfectly and started to be applied in practice. This paper proposes a method of applying an artificial neural network to compensate for the lost GNSS signal in the loosely coupled structure to improve the accuracy of the positioning and navigation process for a class of flying devices. The obtained results proved the correctness of the indicated method.

Keywords: Inertial Navigation Systems · Global Navigation Satellite Systems · Artificial Intelligence · Neural Networks · Kalman Filter

1 Introduction

Navigation Satellite System (GNSS) has been successfully used in various fields over the past decade. GNSS is based on various information from satellite positions in orbit to locate objects in real time. An important part of GNSS is that their accuracy does not degrade with time. However, GNSS is only accurate when it receives signals from more than three satellites. This limitation can affect the accuracy of the system. Another limitation of GNSS is that it uses the energy of radio waves because of its parametric navigation and is therefore susceptible to interference and tampering by enemy forces [1, 12]. On the other hand, an inertial navigation system (INS) combines inertial measurement sensors (IMUs) and algorithms to calculate its position, velocity, and posture. The IMU consists of three gyroscopes and accelerometers placed orthogonal to each other to measure angular velocity and acceleration, respectively. The signals from these devices are processed to obtain the necessary navigational parameters (position, velocity, and posture) by integrating the raw inertial measurements of the IMU. The integration

P. T. Nghia et al. (Eds.): ICTA 2023, LNNS 847, pp. 194–205, 2023.
https://doi.org/10.1007/978-3-031-49529-8_21

process is realized using a set of kinematic equations and is performed using the navigation tool [4, 12]. Generally, there are two main types of IMUs: Base and Baseless, each with its advantages and disadvantages. The INS cannot function as a standalone navigation system because the sensors have drifted over time. Often scientists use INS in combination with GNSS to increase the accuracy of the system [12]. Many fusion algorithms are used to integrate INS/GPS data. The traditionally used fusion techniques are Bayesian filtering methods such as Kalman Filter (KF) or Extended Kalman Filter (EKF) [12]. With precise dynamic models and random models of GPS and INS errors, KF can generate very accurate georeferenced solutions as long as there is constant access to GPS signals. If a GPS loss occurs, the KF will operate in predictive mode and correct the INS measurements based on the system failure model. KF is computationally efficient and especially useful for real-time applications. However, KF has limitations: The main disadvantage associated with using KF for INS/GNSS integration is the need for an accurate random model for each sensor failure [11]. Although navigation-level and tactical-level systems with random errors can be appropriately modeled, modeling the errors of low-level tactical and MEMS-based IMUs is always challenging. In addition, there are some significant limitations of KF, such as sensor dependency and observability issues. The limitations of KF above lead to the need for AI-integrated INS/GPS research.

The field of AI is increasingly noticed in the development of future technology, especially with modern computer technology in hardware and software. AI has been proven to be a successful and effective tool for providing solutions to some technical and scientific problems that cannot be solutions to some technical and scientific problems that cannot be properly solved by conventional techniques [14]. Targets of AI technologies, including artificial neural networks (ANNs) [13], neural fuzzy systems [15], alarm computing [16], expert systems [17], and genetic algorithms (GA) [18], etc., are to provide some intelligence and robustness in the complex and uncertain systems similar to those seen in natural biological species[19].

2 Inertial Navigation System

2.1 Inertial Navigation Algorithm

Coordinate system conversion methods:

The article uses reference systems, inertial reference systems (I-frames), earth reference systems (e-frames), geographic reference systems (n-frames), and object reference systems (b-frames). Because the accelerometer in the INS system uses a Strapdown structure attached to the aircraft, moving with the aircraft, the measured acceleration vector has a constant change in direction and direction, making it difficult to accurately determine the speed and position of the aircraft, which leads to the conversion of the acceleration vector measured in the b-frame system to the e-frame or n-frame coordinate system. Convert from b-frame system to n-frame system [1]:

$$\overrightarrow{f^n} = C_b^n \overrightarrow{f^b} \tag{1}$$

For the n-frame navigation coordinate system. This is the coordinate system often referred to in our daily life as the north, east, and down direction. It will be denoted by the superscript n. It's determined by fitting at a tangent plane to a fixed point of the geodetic reference ellipse. This point will be the origin of the coordinate system and the x-axis points towards the true north, the y-axis to the east and the z-axis completes the right-hand coordinate system by pointing towards the earth's interior So, just knowing the coordinates of the IMU block, we can easily determine the orientation of the n-frame system in the space created by the e-frame system according to the following formula [1–5].

$$C_n^e = \begin{pmatrix} -\sin\varphi\cos\lambda & -\sin\lambda & -\cos\varphi\cos\lambda \\ -\sin\varphi\sin\lambda & \cos\lambda & -\cos\varphi\sin\lambda \\ \cos\varphi & 0 & -\sin\varphi \end{pmatrix} \tag{2}$$

In there: λ: Longitude of the object (rad); φ: Latitude of the object (rad). The vector conversion from coordinate system B to coordinate system C is usually used mainly by three methods [2–6] that is, the DCM direction cosine matrix method(direction cosine matrix), Euler angle method, Quaternion method are detailed in [3–6].

Inertial navigation algorithm:

The object position in the n-frame navigation coordinate system is represented as curve coordinates [4, 5]:

$$r^n = (\varphi\,\lambda\,h)^T \tag{3}$$

Velocity in the n-frame navigation coordinate system is determined by the formula [3, 4]:

$$v^n = \begin{bmatrix} v_N \\ v_E \\ v_D \end{bmatrix} = \begin{bmatrix} M+h & 0 & 0 \\ 0 & (N+h)\cos\varphi & 0 \\ 0 & 0 & 0 \end{bmatrix} \begin{bmatrix} \dot\varphi \\ \dot\lambda \\ \dot h \end{bmatrix} \tag{4}$$

Therefore, the time derivative of the position coordinates [2–4]:

$$\dot r^n = \begin{bmatrix} \dot\varphi \\ \dot\lambda \\ \dot h \end{bmatrix} = \begin{bmatrix} \frac{1}{M+h} & 0 & 0 \\ 0 & \frac{1}{(N+h)\cos\phi} & 0 \\ 0 & 0 & -1 \end{bmatrix} \begin{bmatrix} v_N \\ v_E \\ v_D \end{bmatrix} \tag{5}$$

We have (4) $v^n = C_e^n \dot r^e$. Substitute and $\dot r^e = C_i^e\left(\dot r^i - \Omega_{ie}^i r^i\right)$ derive the formula, we have the result:

$$\dot v^n = C_i^n \ddot r^i + C_i^n(\Omega_{ni}^i - \Omega_{ie}^i)\dot{\underline{r}}^i - C_i^n \Omega_{ni}^i \Omega_{ie}^i \underline{r}^i \tag{6}$$

where Ω_{ie}^i is the rotational speed of the Earth's mussel. The transformation process (6) obtains the formula (2) (4):

$$\dot v^n = C_b^n f^b - (2\omega_{ie}^n + \omega_{en}^n) \times v^n + g^n \tag{7}$$

We have:

$$\dot{C}_b^n = C_b^n \Omega_{nb}^b = C_b^n(\Omega_{ib}^b - \Omega_{in}^b) \tag{8}$$

Combining the formulas (5), (7) and (8) deduce the system of equations to determine the direction angle, velocity, and position of the object in the n-frame system with the following form [2–4]:

$$\begin{bmatrix} \dot{r}^n \\ \dot{v}^n \\ C_b^n \end{bmatrix} = \begin{bmatrix} B^{-1}v^n \\ C_b^n f^b - (2\omega_{ie}^n + \omega_{en}^n) \times v^n + g^n \\ C_b^n(\Omega_{ib}^b - \Omega_{in}^b) \end{bmatrix} \tag{9}$$

with: $B^{-1} = \begin{bmatrix} \frac{1}{M+h} & 0 & 0 \\ 0 & \frac{1}{(N+h)\cos\phi} & 0 \\ 0 & 0 & -1 \end{bmatrix}$

Equation (7) can be rewritten as a matrix [3]:

$$\begin{bmatrix} \dot{v}_N^n \\ \dot{v}_E^n \\ \dot{v}_D^n \end{bmatrix} = \begin{bmatrix} f_x^n \\ f_y^n \\ f_z^n \end{bmatrix} + D_{3x3} \begin{bmatrix} v_N^n \\ v_E^n \\ v_D^n \end{bmatrix} + \begin{bmatrix} 0 \\ -g \\ 0 \end{bmatrix} \tag{10}$$

With:

$$D_{3x3} = \begin{bmatrix} 02\omega_{iez}^n + \omega_{enz}^n - \left(2\omega_{iey}^n + \omega_{eny}^n\right) \\ -(2\omega_{iez}^n + \omega_{enz}^n)02\omega_{\text{iex}}^n + \omega_{enx}^n \\ -\left(2\omega_{iey}^n + \omega_{eny}^n\right) - (2\omega_{iex}^n + \omega_{enx}^n)0 \end{bmatrix}$$

The angular velocity in the n-frame system relative to the I-frame projected onto the n-frame system ω_{in}^n is determined as follows [1–5]:

$$\omega_{in}^n = \omega_{ie}^n + \omega_{en}^n \tag{11}$$

With: $\omega_{ie}^n = \left[\omega_{ie}\cos\varphi \;\omega_{ie}\sin\varphi \;0\right]^T$; $\omega_{en}^n = \left[\frac{v_D^n}{N+h} \;\frac{v_D^n}{N+h}tg\varphi \; -\frac{v_N^n}{M+h}\right]^T$ Derivatives for latitude and longitude have:

$$\dot{\lambda} = \frac{v_N}{(N+h)\cos\varphi} \tag{12}$$

$$\dot{\varphi} = \frac{v_N}{M+h} \tag{13}$$

Integrating (12) and (13) we get the longitude and latitude of the aircraft [4]:

$$\lambda = \int_0^t \frac{v_N}{(N+h)\cos\varphi} + \lambda_0 \tag{14}$$

$$\varphi = \int_0^t \frac{v_N}{M+h} + \varphi_0 \tag{15}$$

M radius of curvature of the shear plane formed by the z-axis and the north (meridian) axis; N radius of curvature of the shear plane formed by the z-axis and the east axis. The diagram of the navigation algorithm in the n-frame system (Fig. 1):

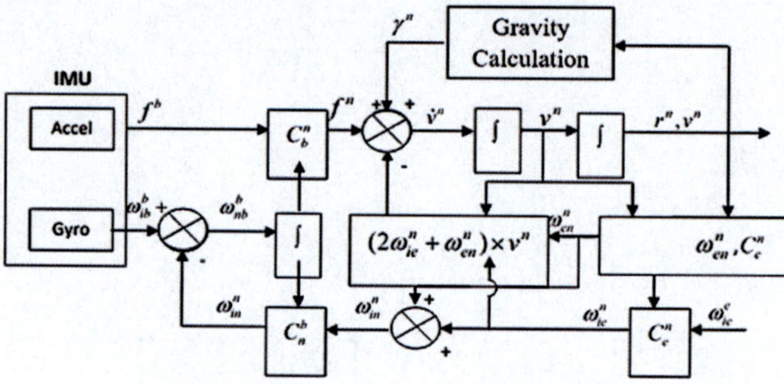

Fig. 1. Diagram of the inertial navigation algorithm

2.2 INS error model

The error parameters are determined according to the equations below [4].

$$\hat{r}^n = r^n + \delta r^n \tag{16}$$

$$\hat{v}^n = v^n + \delta v^n \tag{17}$$

$$\hat{C}_b^n = (I - E^n)C_b^n \tag{18}$$

$$\gamma^n = g^n + \delta g^n \tag{19}$$

where: γ^n is the standard gravity vector; E^n is the symmetry matrix (or vector product) of the directional errors (Matrix of the errors of the directional angle of the aircraft).

$$E^n = \begin{bmatrix} 0 & -\delta\psi & \delta\theta \\ \delta\psi & 0 & -\delta\phi \\ -\delta\theta & 0 & \delta\phi \end{bmatrix} \tag{20}$$

with $\eta = [\delta\phi \, \delta\theta \, \delta\psi]$ is called the directional error vector. From formula (18) using the orthogonality of the matrix C_b^n: $E^n = I - \hat{C}_b^n C_b^{nT}$ (20). Take the derivative on both sides of 2.20 and transform we get [4]:

$$\dot{E}^n = -C_b^n\left(\delta\omega_{ib}^b - \delta\omega_{in}^b\right) \tag{21}$$

Transforming (21) we get the vector derivative of directional error [4, 7]:

$$\dot{\eta}^n = -\omega_{in}^n \eta^n + \delta\omega_{in}^n - C_b^n \delta\omega_{ib}^b \tag{22}$$

where ^ and δ: predicted value and error respectively. From the above formulas, through the calculation and transformation process, we have the error model of the position and speed of the INS system as follows [3–5]:

$$\begin{bmatrix} \delta\dot{r}^n \\ \delta\dot{v}^n \\ \dot{\eta}^n \end{bmatrix} = \begin{bmatrix} \delta v^n \\ F_1\delta r^n + F_2\delta v^n + f^n\eta^n + C_b^n\delta f^b \\ F_3\delta r^n + F_4\delta v^n - \omega_{in}^n\eta^n - C_b^n\delta\omega_{ib}^b \end{bmatrix} \tag{23}$$

In there: $R = \sqrt{M \cdot N}$

$$F_1 = \begin{bmatrix} F_{111} & F_{121} & F_{131} \\ F_{112} & F_{122} & F_{132} \\ F_{113} & F_{123} & F_{133} \end{bmatrix}; F_2 = \begin{bmatrix} F_{211} & F_{221} & F_{231} \\ F_{212} & F_{222} & F_{232} \\ F_{213} & F_{223} & F_{233} \end{bmatrix}$$

$$F_3 = \begin{bmatrix} F_{311} & F_{321} & F_{331} \\ F_{312} & F_{322} & F_{332} \\ F_{313} & F_{323} & F_{333} \end{bmatrix}; F_4 = \begin{bmatrix} F_{411} & F_{421} & F_{431} \\ F_{412} & F_{422} & F_{432} \\ F_{413} & F_{423} & F_{433} \end{bmatrix}$$

$$F_{111} = -2v_E\omega_e\cos\varphi - \frac{v_E^2}{(N+h)\cos^2\varphi};$$

$$F_{112} = 2\omega_e(v_N\cos\varphi - v_D\sin\varphi) + \frac{v_E v_N}{(N+h)\cos^2\varphi}$$

$$F_{113} = 2v_E\omega_e\sin\varphi;\ F_{131} = \frac{-v_N v_D}{(M+h)^2} + \frac{v_E^2\tan\varphi}{(N+h)^2}$$

$$F_{132} = \frac{-v_E v_N}{(N+h)^2} - \frac{v_N v_E\tan\varphi}{(N+h)^2};\ F_{133} = \frac{v_E^2}{(N+h)^2} + \frac{v_N^2}{(M+h)^2} - \frac{2\gamma}{(R+h)}$$

$$F_{211} = \frac{v_D}{M+h};\ F_{212} = 2\omega_e\sin\varphi + \frac{v_E\tan\varphi}{N+h};\ F_{213} = -2\frac{v_N}{M+h}$$

$$F_{221} = -2\omega_e\sin\varphi - 2\frac{v_E\tan\varphi}{N+h};\ F_{222} = \frac{v_D + v_N\tan\varphi}{N+h};$$

$$F_{223} = -2\omega_e\cos\varphi - 2\frac{v_E}{N+h}$$

$$F_{231} = \frac{v_N}{M+h};\ F_{232} = 2\omega_e\cos\varphi + \frac{v_E}{N+h};\ F_{311} = -\omega_e\sin f;$$

$$F_{313} = -\omega_e\cos f - \frac{v_E}{(N+h)\cos^2 f}$$

$$F_{331} = \frac{-v_E}{(N+h)^2};\ F_{332} = \frac{v_N}{(M+h)^2};\ F_{333} = \frac{v_E\tan f}{(N+h)^2};$$

$$F_{412} = \frac{-1}{M+h};\ F_{421} = \frac{1}{N+h};\ F_{423} = \frac{-\tan\varphi}{N+h}$$

$$F_{233} = F_{312} = F_{321} = F_{322} = F_{323} = F_{411} = F_{413} = F_{422} = 0;$$

$$F_{431} = F_{432} = F_{433} = F_{121} = F_{122} = F_{123} = 0$$

3 Inertial Navigation System

3.1 Integrated INS/GNSS System Based on Kalman Filter

The Kalman filter is a set of mathematical equations that describe an efficient recursive computation method that allows the state of a process to be estimated such that the mean-variance of the deviation (between the true and the estimated value) is minimal. For linear Kalman filters are often applied to estimate the state of a discrete linear process in time. However, the limitation of the linear Kalman filter is that it optimally works when the system is linear, and the noise acting on the system is Gaussian noise [1]. In practice, the system is often nonlinear, so applying a linear Kalman filter will cause large errors. There are many methods to linearize the system, such as applying the first-order Taylor series. The filter is then called an extended Kalman filter (EKF). The model of applying the Kalman filter combining the INS system and G NSS system is described as shown in Fig. 2, [1, 5, 12]:

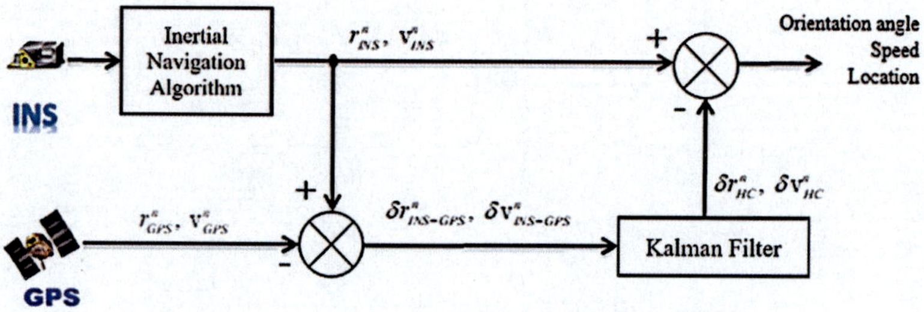

Fig. 2. Integrated INS/GPS system using Kalman filter

The Kalman filter is built to estimate the position and velocity error of the INS system, with the observed signal being the difference between the position and velocity obtained from the GPS r^n_{GPS}, v^n_{GPS} and the position and velocity determined from the INS., r^n_{INS}, v^n_{INS}. The output information of the Kalman filter includes the results of the error estimate of the position and velocity of the aircraft δr^n_{HC}, δv^n_{HC}. Taking the results of the velocity and position determined from the INS system minus the estimated error values through the Kalman filter, the exact position, and velocity of the aircraft will be obtained. With the structure as shown in Fig. 2, the INS and GPS systems operate independently and are the basis for an artificial neural network (ANN) that compensates for GPS signals in case of a loss of GPS signal in a certain period. The error model of the system (24) is built from Eq. (9), determining the direction, velocity, and position of the object in the n-frame system. The error model of the INS system is described in terms of state space [5, 6]:

$$\delta \dot{x}(t) = F(t)\delta x(t) + G(t)u_c(t) \tag{24}$$

In there: $\delta x(t) = \left[\delta r^n\, \delta v^n\, \eta\, \delta f^b\, \delta \omega_{ib}^b\right]$: State vector describing the error of the INS system $F(t) = \begin{bmatrix} 0_{3x3}\, I_{3x3}\, 0_{3x3}\, 0_{3x3}\, 0_{3x3} \\ F_1\, F_2 f^n\, C_b^n\, 0_{3x3} \\ F_3\, F_4\, -\omega_{in}^n\, 0_{3x3}\, -C_b^n \\ 0_{3x3}\, 0_{3x3}\, 0_{3x3}\, 0_{3x3}\, 0_{3x3} \\ 0_{3x3}\, 0_{3x3}\, 0_{3x3}\, 0_{3x3}\, 0_{3x3} \end{bmatrix}$.

$u_c(t) = \left[u_{acc}\ \mathrm{u}_{gyros}\right]^T$: Vector represents the random error component in the measurement results of the accelerometer and angular velocity sensor $G(t) = \begin{bmatrix} 0_{3x3} & C_b^n & 0_{3x3} & 0_{3x3} & 0_{3x3} \\ 0_{3x3} & 0_{3x3} & -C_b^n & 0_{3x3} & 0_{3x3} \end{bmatrix}^T$. Discrete Eq. (23) to get (24):

$$\delta x_{k+1} = \Psi_k \delta x_k + u_{d,k} \tag{25}$$

In which: $\Psi_k = e^{\int_{t_k}^{k+1} F(t)dt}$. The observation equation is built based on the difference between the position and velocity determined by GPS and the position and velocity from the INS system. The observation equation has the form [1]:

$$\delta y_k = H_k \delta x_k + \mathrm{w}_{d,k} \tag{26}$$

With: $\delta y_k = \left[\, r_{GPS}^n[\mathrm{k}] - r_{INS}^n[k]\ v_{GPS}^n[\mathrm{k}] - \mathrm{v}_{INS}^n[k]\ 0_{3x1}\ 0_{3x1}\ 0_{3x1}\right]^T$.

3.2 Inertial Navigation System using AI

The Kalman filter plays an important role in the INS/GNSS combined navigation information processing system, where measurement updates are used to correct and limit errors to the navigation parameters Street. However, the Kalman filter only gives an optimal result of the output navigation errors if and only if the navigation information signal of both the GNSS system and the INS system is always available, in the absence of measurement updates, so the integrated navigation information processing INS/GNSS will reduce accuracy in the absence of measurement updates from GNSS. It is a fact that GNSS signals will be completely or partially lost when moving through tunnels or around obstacles, or harsh electromagnetic interference environments also degrade GNSS signal quality; A special thing is that GNSS uses radio energy to get navigation information, so in the combat environment it tends to be jammed and tampered with by enemy forces. From the above problems, it is necessary to compensate for the GNSS signal when it is lost or interfered with, and the choice of an artificial neural network ANN is one of the solutions to improve navigation accuracy (Figs. 3 and 4).

Initially, signals from INS and GPS were used by the author to train the network. The input parameter set used to train the network is the position and velocity vectors of the INS system at the sampling points, and the position and velocity vector values of the GNSS the next time will be taken $\{r_{GPS}^n, v_{GPS}^n\}$ as The target set of the network training process, the author assumes a 3-layer MLP perceptron or a DLNN deep learning neural network so that the values $\left\{\left(r_{ANN}^n - r_{GPS}^n\right)^2; \left(\delta v_{ANN}^n - \delta v_{GPS}^n\right)^2\right\}$ reach the optimal value, now we have a neural network with appropriate weights and biases. A class of aviators

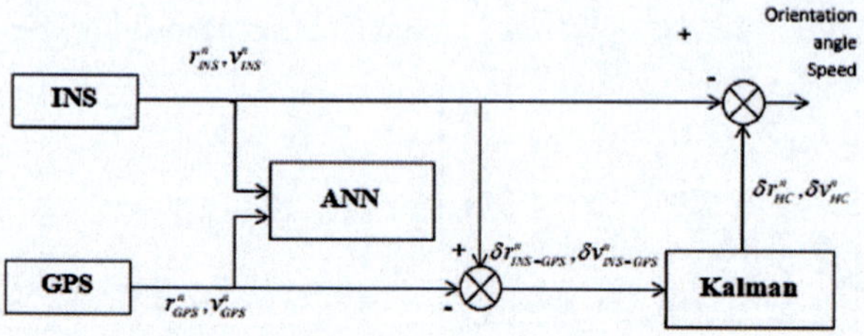

Fig. 3. ANN training process when GPS or dataset information is available

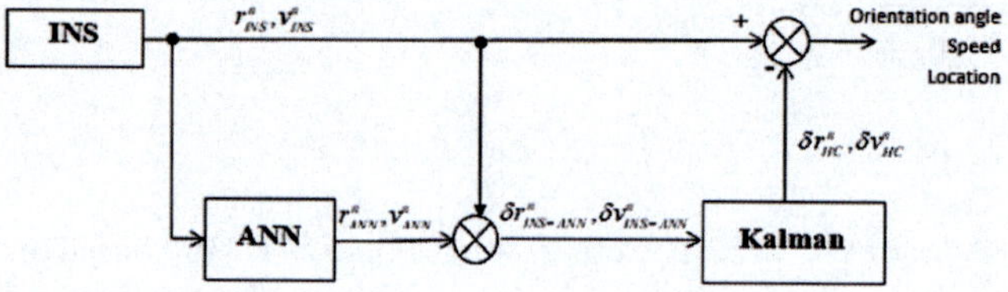

Fig. 4. Compensation for GPS signal by ANN

with constant speed or small offsets allows the measured values from the accelerometer and angular velocity to always be within a limited range, which affects the process of generating target parameters and prediction of ANN does not occur in force majeure cases [8]. We see that a NN can be trained for a certain function by modifying the connection weights between the elements to give the desired output. The neural network is calibrated based on the comparison between the output and the desired target until the output matches the target or the given training target is met. Then this network is called a supervised learning network (Fig. 5).

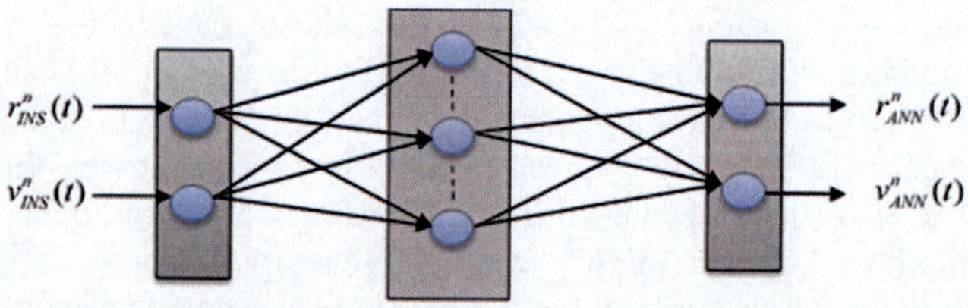

Fig. 5. Artificial neural network model

4 Simulation Results

With an input parameter of a moving object type according to the document [12]:

The results in Fig. 6 show that, for some types of non-stop and low-speed flying devices, enough input data for ANN will select a suitable artificial neural network model to compensate for the GNSS signal. If in case of loss of the GNSS signal causes a large

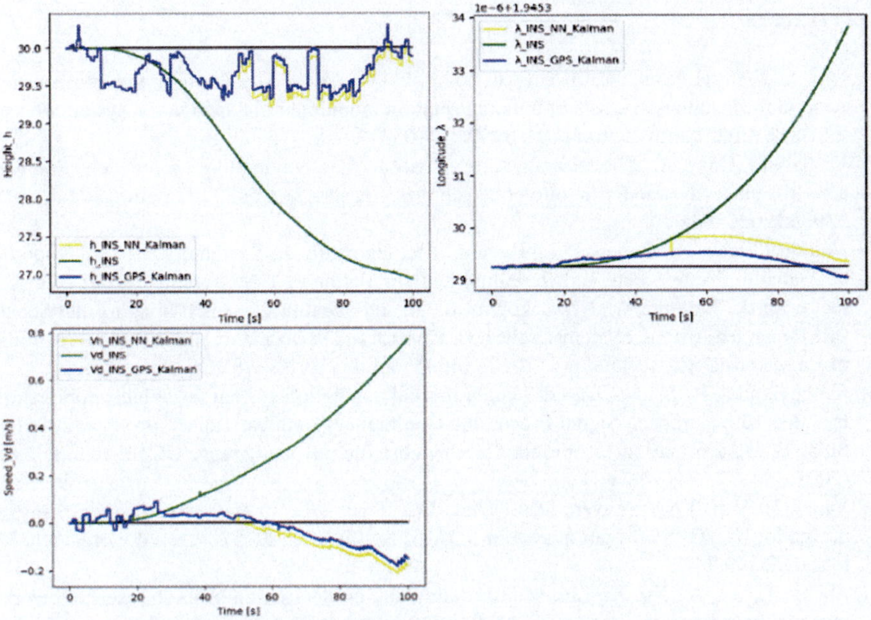

Fig. 6. Speed, longitude and height in cases with and without GPS

error, in case ANN compensates the GNSS signal with the allowed standard value, the signal after correction will follow the signal when both INS and GNSS systems are fully available. The results show that in the case of GNSS signal loss where the pure INS signal is not corrected from the error value of the Kalman filter, due to the gyroscope drift leads to a large error of the navigation result, but in the case of The GNSS signal is compensated by the ANN allowing the output to follow over 95% of the result when the GNSS signal is present.

5 Conclusion

Artificial Neural Networks allow a new approach to applications to improve positioning and navigation accuracy using MEMS sensors, which allows applications with high accuracy and applications in advanced navigation fields. This paper proposes a method of applying an artificial neural network to compensate for the lost GNSS signal to improve the accuracy of the positioning and navigation process. The obtained results proved the correctness of the indicated method.

References

1. Niu, M., Ma, H., Sun, X., Huang, T., Song, K.: A new self-calibration and compensation method for installation errors of uniaxial rotation module inertial navigation system. Sensors **22**, 3812 (2022). https://doi.org/10.3390/s22103812
2. Khatab, E., Onsy, A., Abouelfarag, A.: Evaluation of 3D vulnerable objects' detection using a multi-sensors system for autonomous vehicles. Sensors **22**, 1663 (2022). https://doi.org/10.3390/s22041663
3. Muraleedharan, N., Cohen, D.S., Estrela, V.V., Hemanth, J., Saotome, O., Nikolakopoulos, G., Sabatini, R., Wackett, L.: Modelling and simulation of UAV systems. In: Estrela, V.V., Hemanth, J., Saotome, O., Nikolakopoulos, G., and Sabatini, R. (eds.): Imaging and Sensing for Unmanned Aircraft Systems. Volume 1: Control and Performance, pp. 101–121. Institution of Engineering and Technology (2020). https://doi.org/10.1049/PBCE120F_ch5
4. Skog, Händel, P.: A low-cost GPS aided inertial navigation system for vehicle applications. In: 2005 13th European Signal Processing Conference, Antalya, Turkey, pp. 1–4 (2005)
5. Shin, E.-H.: Estimation techniques for low-cost inertial navigation. UCGE report, 20219 (2005)
6. Van, T.P., Van, T.N., Nguyen, D.A., Duc, T.C., Duc, T.T.: 15-state extended Kalman filter design for INS/GPS navigation system. JOACE **3**, 109–114 (2015). https://doi.org/10.12720/joace.3.2.109-114
7. Zheng, T., Xu, A., Xu, X., Liu, M.: Modeling and compensation of inertial sensor errors in measurement systems. Electronics **12**, 2458 (2023). https://doi.org/10.3390/electronics12112458
8. Skog, I., Händel, P.: Calibration of a MEMS inertial measurement unit. In: XVII IMEKO World Congress, pp. 1–6 (2006)
9. Mok, E., Retscher, G., Wen, C.: Initial test on the use of GPS and sensor data of modern smartphones for vehicle tracking in dense high rise environments. In: 2012 Ubiquitous Positioning, Indoor Navigation, and Location Based Service (UPINLBS). pp. 1–7. IEEE, Helsinki, Finland (2012). https://doi.org/10.1109/UPINLBS.2012.6409789
10. Guerrero-Castellanos, J.F., Madrigal-Sastre, H., Durand, S., Marchand, N., Guerrero-Sanchez, W.F., Salmeron, B.B.: Design and implementation of an attitude and heading reference system (AHRS). In: 2011 8th International Conference on Electrical Engineering, Computing Science and Automatic Control, pp. 1–5. IEEE, Merida City, Mexico (2011). https://doi.org/10.1109/ICEEE.2011.6106610
11. Vinh, N.Q.: INS/GPS integration system using street return algorithm and compass sensor. Procedia Comput. Sci. **103**, 475–482 (2017). https://doi.org/10.1016/j.procs.2017.01.030
12. Vinh, N.Q., Thoa, P.D.: Improving high quality in combination processing the high-measurement signals. In: Satapathy, S.C., Bhateja, V., Nguyen, B.L., Nguyen, N.G., and Le, D.-N. (eds.) Frontiers in Intelligent Computing: Theory and Applications, pp. 328–340. Springer Singapore, Singapore (2020). https://doi.org/10.1007/978-981-32-9186-7_34
13. Kubat, M.: Neural networks: a comprehensive foundation by Simon Haykin, Macmillan, 1994, ISBN 0–02–352781–7. Knowl. Eng. Rev. **13**(4):409–412 (1999)
14. Jaakkola, H., Thalheim, B., Henno, J.: About the essence of intelligence—will artificial intelligence (ever) cover human intelligence? In: Tropmann-Frick, M., Jaakkola, H., Thalheim, B., Kiyoki, Y., Yoshida, N. (eds.) Frontiers in Artificial Intelligence and Applications. IOS Press (2022). https://doi.org/10.3233/FAIA210475
15. Karaboga, D., Kaya, E.: Adaptive network based fuzzy inference system (ANFIS) training approaches: a comprehensive survey. Artif. Intell. Rev. Intell. Rev. **52**, 2263–2293 (2019). https://doi.org/10.1007/s10462-017-9610-2

16. Tamilselvi, S.: Introduction to evolutionary algorithms. In: Ventura, S., María Luna, J., María Moyano, J. (eds.) Genetic Algorithms. IntechOpen (2022). https://doi.org/10.5772/intechopen.104198
17. Rajan, H.: An Experiential Introduction to Principles of Programming Languages. MIT Press (2022)
18. Introduction to Genetic Algorithms. Springer Berlin Heidelberg, Berlin, Heidelberg (2008). https://doi.org/10.1007/978-3-540-73190-0
19. Tetko, I.V., Kůrková, V., Karpov, P., Theis, F. (eds.): Artificial Neural Networks and Machine Learning—ICANN 2019: Image Processing: 28th International Conference on Artificial Neural Networks, Munich, Germany, September 17–19, 2019, Proceedings, Part III. Springer International Publishing, Cham (2019). https://doi.org/10.1007/978-3-030-30508-6

The Vehicle Routing Problem with Drones for Fresh Agricultural Products

Quang Huy Vuong[1], Trung-Nghia Phung[2], and Minh-Trien Pham[3](✉)

[1] Hanoi University, Hanoi, Vietnam
huyvq@hanu.edu.vn
[2] Thai Nguyen University of Information and Communication Technology, Z115, 250000 Thai Nguyen, Vietnam
ptnghia@ictu.edu.vn
[3] VNU University of Engineering and Technology, Hanoi, Vietnam
trienpm@vnu.edu.vn

Abstract. The delivery of fresh agricultural products is a significant challenge due to their perishability. This paper addresses a real-life delivery of fresh agricultural products vehicle routing problem with drones (VRPD). The model takes into account the impact of perishability as part of the overall distribution costs and explores the feasibility of using drones for delivery. The model is constructed using mixed-integer linear programming optimization, to minimize the quality deterioration of agricultural products and the completion time for all customers. For a large number of customers, we proposed a heuristic approach by utilizing an improved Large Neighborhood Search algorithm with an adaptive repairing parameter. To validate the proposed algorithm, various datasets have been tested. The results demonstrate that the proposed algorithm reduces delivery times and the deterioration value of products significantly. Therefore, it provides an effective decision-making reference for logistics distribution enterprises.

Keywords: Vehicle Routing Problem · Drone Delivery · Fresh Agricultural · Products

1 Introduction

Delivery plays a critical role in fresh product logistics. In recent years, the management of fresh product supply chains has received much interest from both academics and practitioners. In particular, the process of transporting fresh agricultural products plays an important role in the supply chain and raises many problems for distribution companies. The biggest difficulty is preserving the nutritional characteristics of fresh agricultural products during transportation due to their perishability. Therefore, different operating strategies are needed compared to the cost-saving perspective of the traditional delivery strategy. In general, there are two types of agri-food supply chains: fresh produce agri-food supply chains and packaged products agri-food supply chains [1]. In this paper, we focus on fresh agricultural products' last-mile delivery.

P. T. Nghia et al. (Eds.): ICTA 2023, LNNS 847, pp. 206–214, 2023.
https://doi.org/10.1007/978-3-031-49529-8_22

To ensure the quality of the product, the cost of storage or damage is often added to the objective function of the problem. In [2], Zhang et al. have shown that controlling product quality throughout the supply chain requires a focus on both time and temperature during transit. Product quality is expressed as a function of time and temperature during transport and storage. Tabu search algorithm is used to optimize the supply chain with the goal of minimizing transportation and storage costs. Rong et al. [3] modeled a cost function based on the temperature range and the number of perishable products during transportation. Osvald and Stirn [4] formulated the VRP with Time Windows (VRPTW) problem considering the perishability of agricultural products. This model considers the negative effect of perishability throughout the distribution by approximating the loss in product quality as a function of time. Chen et al. [5] studied a multi-objective model consisting of a function that minimizes distribution costs and a function that maximizes the freshness of delivered foods. The problem is based on the model of the VRPTW problem with the assumption that when the vehicles come from the warehouse, the freshness of the products is at the maximum.

The contributions of this paper are as follows. First, a modeling framework with trucks and drones for the delivery of fresh agricultural products is presented. The objective of this model is to minimize the quality deterioration value of agricultural products and the completion time. Second, a heuristic approach to solve the problem is proposed. Finally, we conduct the numerical results to assess the effectiveness of the proposed model and algorithm.

2 Materials and Methods

2.1 Vehicle Routing Problem with Drones for Agricultural Products

The Vehicle Routing Problem with Drones for Agricultural Products (VRPD-AP) problem is defined on a directed graph G = (V, A), in where the set of nodes V = {0, 1,..., n} represents the depot (node 0) and the set of customers C = {1,..., n}. A set of k trucks $K = \{1, 2,..., k\}$ depart from the depot, serve an arbitrary set of customers, and return to the depot. A set of drones $D = \{1, 2,..., d\}$ serve customers directly from the depot and back to the depot, with a limit of serving just one customer per trip. Each customer is served only once by truck or drone.

Let t_{ij} be the cost of the truck moving from node i to node j, while the cost required for the drone to serve customer i is denoted $t_{i'}$, which includes the total cost of flying to the customer and flying back to the depot. Each vehicle has a capacity constraint. Let Q_t be the maximum payload each truck can carry per trip, Q_d represents the maximum payload of each drone. Each customer node i (i = 1, 2, 3, ..., n) has a non-negative transport request value Q_i representing the amount of requested cargo to be served. Assume that at the depot the value $Q_0 = 0$.

The trucks and drones depart from the depot at time 0, the objective of the problem is to minimize the total cost of delivery to all customers and the total quality deterioration value of the agricultural products over time. The value of damage over time will be defined in the following section.

2.2 Total Quality Deterioration Value

The loss of quality during transportation of perishable products is a significant cost for companies. Considering the properties related to the perishability of agricultural products during transport, a model of total quality deterioration value (*QDV*) is proposed. Based on the different degradation levels over time and the amount of transport of the agricultural product, the *QDV* is calculated according to the following equation:

$$QDV = QDV_t + QDV_d \tag{1}$$

Subject to:

$$QDV_t = \sum\nolimits_{i \in C} \sum\nolimits_{k \in K} D_i . \left[q(0) - q\left(t_i^k\right) \right] . y_i^k \tag{2}$$

$$QDV_d = \sum\nolimits_{i \in C} \sum\nolimits_{d \in D} D_i . \left[q(0) - q\left(td_i^d\right) \right] . yd_i^d \tag{3}$$

where, QDV_t and QDV_d represent the deterioration value of products served by traditional trucks and drones, respectively. The demand of customer *i*th is represented by the parameter D_i; q(t) represents the quality of the agricultural product after a period t with a deterioration factor φ; t_i^k and td_i^d represent the time taken by the truck and drone to reach customer node *i*, respectively; y_i^k equals 1 when node *i* is visited by truck k and 0 otherwise, while yd_i^d equals 1 when node *i* served by drone *d* and 0 otherwise. Thus, we can represent the quality of the agricultural product delivered to customer node *i* of the truck at time t_i^k is $q\left(t_i^k\right)$. Similarly, if customer *i* is served by a drone, the quality of the product at time td_i^d is $q\left(td_i^d\right)$. Equations (4) show the deterioration value as a linear function of time:

$$q(t) = q_0 - \varphi . t \tag{4}$$

2.3 Mathematical Model

Notation:

Set and indices: $K = \{1, 2, \ldots, k\}$—a set of trucks; $D = \{1, 2, \ldots, d\}$—a set of drones; C $= \{1, 2, \ldots, n\}$—a set of customers; $N = \{0, 1, \ldots, n + 1\}$—a set of all nodes (include the depot 0 and a copy of depot n + 1); $C_0 = \{0, 1, \ldots, n\}$—a set of customer nodes and a depot; $C_+ = \{1, 2, \ldots, n + 1\}$—a set of customer nodes and a copy of depot; $Q = \{Q_1, Q_2, \ldots, Q_n\}$—a set of customer's demand; *i, j*—the index for customer nodes; *k*—the index for trucks; *d*—the index for drones.

Parameters: α—total transportation cost; t_{ij}—transportation cost of a truck moving from node *i* to *j*; $t_{i'}$—transportation cost of a drone serving a customer node *i*; Q_t—maximum capacity of a truck; Q_d—maximum capacity of a drone; Q_i—demand of a customer node *i*; *QDV*—quality deterioration value.

Decision variables: x_{ij}^k—a binary variable equal to 1 when the route between node *i* and node *j* is traveled by truck k, equal to 0 otherwise; y_i^k—a binary variable equal to 1 when node i is visited by truck k, equal to 0 otherwise; yd_i^d—a binary variable equal to 1 when node i is served by drone d, equal to 0 otherwise; t_i^k—arrival time of truck k at node i; td_i^d—arrival time of drone d at node i; u_i—auxiliary variables.

MILP Model:

The model is constructed with mixed integer linear programming (MILP). The objective function for the delivery of fresh agricultural products can be described as Eq. (5) and the model is formulated as follows:

$$\text{Min}F = \theta.QDV + (1-\theta).\alpha \tag{5}$$

$$\alpha \geq \sum_{k\in K}\sum_{i\in C_0}\sum_{j\in C_+} t_{ij}x_{ij}^k + \sum_{d\in D}\sum_{i\in C} t_i^{'}y_i^d \tag{6}$$

$$\alpha \geq \sum_{k\in K} t_{n+1}^k + \sum_{d\in D}\sum_{i\in C} td_i^d \tag{7}$$

$$QDV \geq \text{QDV}_t + \text{QDV}_d \tag{8}$$

$$\sum_{k\in K} y_i^k + \sum_{d\in D} yd_i^d = 1;\ \forall i\in C \tag{9}$$

$$\sum_{i\in C_+} x_{0i}^k = 1;\ \forall k\in K \tag{10}$$

$$\sum_{i\in C_+} x_{i,n+1}^k = 1;\ \forall k\in K \tag{11}$$

$$\sum_{j\in C_+} x_{ij}^k = \sum_{j\in C_0} x_{ij}^k = y_i^k;\ \forall i\in C, k\in K \tag{12}$$

$$\sum_{i\in C} D_i y_i^k \leq Q_t;\ \forall k\in K \tag{13}$$

$$\sum_{i\in C} D_i yd_i^d \leq Q_d;\ \forall d\in D \tag{14}$$

$$t_j^k \geq t_i^k + t_{ij} - M\left(1 - x_{ij}^k\right);\ \forall i\in C_0, j\in C_+, k\in K \tag{15}$$

$$t_i^d \geq t_i^{'} - M\left(1 - y_i^d\right);\ \forall i\in \text{C}, d\in D \tag{16}$$

$$u_i^k - u_j^k + Q_t x_{ij}^k \leq Q_t - D_j;\ \forall i\in N, j\in N, k\in K \tag{17}$$

$$D_i \leq u_i^k \leq Q_t;\ \forall i\in N, k\in K \tag{18}$$

$$x_{ij}^k, y_i^k, y_i^d \in \{0,1\}, t_i^k, t_i^d \geq 0;\ \forall i\in \text{N}, j\in N, k\in K, d\in D \tag{19}$$

The objective function (5) minimizes the total delivery time of both vehicles and the total deterioration over time of the agricultural products; (6) and (7) define the lower bound of the total delivery time of the two types of vehicles; (8) defines the total deterioration value of agricultural products transported by all trucks and drones; (9) ensures that each customer is served by either a truck or a drone; (10) and (11) require each truck in the fleet to depart and return at the depot; (12) ensure the flow conservation condition of each truck at node *i*, ensuring that if each truck goes to a node, it must also leave that customer node; (13) and (14) limit the capacity of the vehicles that each truck or drone is not allowed to carry more than a given maximum payload; (15) and (16) in order to define arrival times serving all customer nodes of each truck and drone; (17) and (18) work as a subtour elimination ensuring that there are no subtours in the truck's routes. Finally, constraint (19) defines the type and domain of the variables and parameters.

2.4 Large Neighborhood Search Algorithm for VRPD-AP

In this section, we adapt the Large Neighborhood Search (LNS) algorithm, previously applied to various VRP variations, to address the VRPD problem. Given the specific constraints involved in our proposed model for the delivery of perishable agricultural products via trucks and drones, we introduce modifications to the LNS algorithm's operations. These adjustments are necessary due to the unique constraints of our problem, which differ from the base local search operators commonly used.

Algorithm 1: LNS-VRPD-AP

$S_{VRPD\text{-}AP} \leftarrow$ InitialSolution()
$S_{best} \leftarrow S_{VRPD\text{-}AP}$, $S_{current} \leftarrow S_{VRPD\text{-}AP}$
$t = 0, i = 0$
repeat
    while $i \leq i_{max}$ **do**:
        Call Destroy phase with destroy methods $d() \in \Omega^-$ and Repair
        phase with repair methods $r() \in \Omega^+$:
        $S_{temp} \leftarrow r(d(S_{current}))$
        $S_{temp} \leftarrow$ LocalSearch(S_{temp})
    if $cost(S_{temp}) < cost(S_{current})$ **then**:
        $S_{current} \leftarrow S^*_{temp}$
        $i = 0$
    else:
        $i = i + 1$
if $cost(S_{current}) < cost(S_{best})$ **then**:
    $S_{best} = S_{current}$
else:
    $S_{current} \leftarrow$ Perturbate($S_{current}$)
    $i = 0$
until $t \geq t_{max}$
return S_{best}

The search is to find a better solution in the search space. Destroy and repair operators are described in detail in the next sections.

Initial Solution: At each step, customer node i assigned to truck k or drone d.

Destroy Method: At each iteration, the LNS-VRPD-AP destroys part of the current solution by removing some customer nodes from the solution.

Truck nodes removal—D1. Removing β customers from routes served by trucks excludes the depot node.

Drone nodes removal—D2. Removing β customers from the customer sets served by the drones excludes the depot node.

Random removal—D3. Randomly remove each customer node in the current solution until β customers are eliminated. Now the set $C_{current}$ represents the set of all customers served by both vehicles.

Worst removal—D4. Removing the customer nodes that cause a high cost in the solution until all customers are removed. The set $C_{current}$ represents the set of all customers served by both vehicles.

Repair Methods: After the destroy phase, a set of removed customer nodes U is generated. The purpose of the repair phase is to re-insert the customer nodes in the set U with strategies to improve the cost of the solution compared to the original solution as follows: (1) *Random greedy insertion—R1.* This operator is intended to explore the search space. At each loop, a customer node i is randomly selected to insert at the best position p in the truck or drone route; (2) *Greedy insertion—R2.* This operator is a basic construction heuristic operator. Let $\Delta v_{i,p}$ be the cost value of the current solution when inserting customer node i in set U at position p (position p is the order of service in truck or drone routes). We calculate the value $f_{i,p} = \min_{p \in K,D} \{\Delta v_{i,p}\}$ which representing the lowest cost change at positions p. Then customer node i will be removed from the set U and inserted in the best position p which increases the cost least; (3) *Deep greedy insertion—R3.* The deep greedy insertion operator works differently from the basic greedy insertion operator R2. Instead of inserting the first customer i in the set U into the current solution, this operator traverses all customer nodes i $\in$ U, if i satisfy $\min_{i \in U}\{\Delta v_{i,p}\} \forall p \in K, D$, it will be inserted in the best position p in the current solution.

In the LNS algorithm, the repair methods take up most of the computation time. This comes from the fact that repair methods need to traverse and reschedule the solution at each customer node insertion step, while destroy methods only remove customer nodes from the solution at each iteration. Searching for larger neighborhoods leads to better local solutions quality, but is more time-consuming. Therefore, there needs to be a mechanism to balance the computation time and the algorithm's ability to explore deeper search space. Here, a repair method selection mechanism is proposed to solve this problem. To find out the performance of the proposed adaptive repair method selection mechanism, the repair method selection mechanism are evaluated by comparing the solutions obtained with other scenarios as in Table 1. As a result, for large-scale problems, adaptive parameters showed better performance. It can be explained that this strategy tends to explore the search space in the early search steps to diversify the solution, then fine-tuning it in later iterations, and at the same time avoid the local optimization by applying rotation mechanism of selecting repair operators when there is no improvement in the objective function value after 3 iterations.

Table 1. Comparison of different repair operator selection scenarios.

Instance	R = Adaptive		Combination of R1, R2, R3		R = R3	
	Cost	Time (s)	Cost	Time (s)	Cost	Time (s)
P-n16-k2	188.18	50.10	194.18	54.10	188.18	71.10
P-n17-k2	199.43	52.51	201.45	56.51	194.13	75.01
P-n18-k2	204.22	56.12	210.30	66.17	198.20	78.98
P-n19-k2	226.16	60.18	228.01	69.18	236.19	101.12
Average	204.49	54.72	208.48	61.49	204.17	81.55

3 Results and Discussion

Simulation Scenarios and Parameters Settings: In this section, simulation results based on different data sets along with a case study are given to analyze the MILP model and proposed algorithm. The input datasets based on the standard CVRP dataset introduced by Augerat et al. and are publicly available in the online library [6]. For the proposed algorithm, parameters are set as follows: $t_{max} = 300$, $i_{max} = 100$. The parameters are selected through testing before running the algorithm. The value of θ varies in the range [0, 1], and the deterioration factor φ is set to 0.2. For MILP models, the Gurobi 9.5 solver is used. All simulations use Python 3.7 on an Intel Core-i7 8700 computer, 32GB of RAM running on the Linux operating system, Ubuntu 20.04.

Comparison of proposed LNS algorithm and MILP VRPD-AP model: From Table 2, it shown that the proposed LNS algorithm performing well on small data sets in terms of calculation time criteria. However, for the large-size of customers, Table 3 shows that the best cost values over 10 runs of LNS VRPD-AP are all equivalent to MILP-VRPD. The GAP value of the proposed algorithm shows the capability of finding solution that approximates the best value of the MILP model. In terms of computation time, the average computation time of the proposed heuristic algorithm is approximately 9 times lower than the average computation time of MILP-VRPD.

Table 2. Comparison between proposed LNS algorithm and MILP VRPD-AP for small size instance.

Instance	LNS VRPD-AP			MILP VRPD-AP	
	Cost	GAP %	Time (s)	Cost	Time (s)
P-n8-k2	114.96	0.0	26.10	114.96	0.19
P-n9-k2	138.39	0.0	27.55	138.39	0.30
P-n10-k2	150.10	0.0	27.13	150.10	1.15
P-n8-k4	161.00	1.47	26.05	163.37	0.13
P-n9-k4	207.59	0.0	26.66	207.59	0.10
P-n10-k4	223.62	0.0	26.89	223.62	0.21

Table 3. Comparison between proposed LNS and MILP VRPD-AP for large size instance.

Instance	MILP-VRPD		LNS VRPD-AP			
	Cost	Time (s)	GAP %	Best	Average	Time (s)
P-n16-k2	188.18	137.82	0.39	188.18	188.91	58.10
P-n17-k2	188.53	129.50	2.26	188.53	192.80	58.50
P-n18-k2	196.64	359.09	2.6	196.64	201,74	60.12
P-n19-k2	204.55	1511.67	3.25	206.79	211.19	60.08
Average	194.47	534.52	1.62	195.04	197.63	59.2

As a result, the adaptive repairing parameter in the LNS algorithm plays an important role in improving the efficiency of the proposed heuristic algorithm by reducing the computation time.

4 Conclusion

In this paper, a model and solution for the delivery of fresh agricultural products using a combination of trucks and drones has been proposed. In addition to modeling the problem of taking advantage of incorporating drones into the transportation system, the model also considers the value of deterioration in the quality of agricultural products during the delivery process with the aim of minimizing delivery time and loss of product quality. A modified LNS algorithm is proposed to solve the problem on large data sets. Computational scenarios on many different data sets have been considered. As the results, the proposed model can effectively reduce the degree of spoilage of agricultural products during delivery. In addition, different scenarios were tested for analysis to consider the trade-off between minimizing the deterioration value of the products and the cost of the entire transportation process. Finally, this study can be used as a reference for companies involved in improving the efficiency of agricultural product distribution.

References

1. Onggo, B.S., Panadero, J., Corlu, C.G., Juan, A.A.: Agri-food supply chains with stochastic demands: a multi-period inventory routing problem with perishable products. Simul. Model. Pract. Theory Pract. Theory **97**, 101970 (2019). https://doi.org/10.1016/J.SIMPAT.2019.101970
2. Zhang, G., Habenicht, W., Spie, W.E.L.: Improving the structure of deep frozen and chilled food chain with tabu search procedure. J. Food Eng. **60**(1), 67–79 (2003). https://doi.org/10.1016/S0260-8774(03)00019-0
3. Rong, A., Akkerman, R., Grunow, M.: An optimization approach for managing fresh food quality throughout the supply chain. Int. J. Prod. Econ. **131**(1), 421–429 (2011). https://doi.org/10.1016/j.ijpe.2009.11.026
4. Osvald, A., Stirn, L.Z.: A vehicle routing algorithm for the distribution of fresh vegetables and similar perishable food. J. Food Eng. **85**(2), 285–295 (2008). https://doi.org/10.1016/j.jfoodeng.2007.07.008
5. Chen, H.K., Hsueh, C.F., Chang, M.S.: Production scheduling and vehicle routing with time windows for perishable food products. Comput. Oper. Res.. Oper. Res. **36**(7), 2311–2319 (2009). https://doi.org/10.1016/j.cor.2008.09.010
6. “CVRPLIB—All Instances.” Accessed: 27 Jul 2023. [Online]. Available: http://vrp.atd-lab.inf.puc-rio.br/index.php/en/

ViT-SigNet: Combining Deep CNN and Vision Transformer for Enhanced Signature Verification

Ton Do Thanh[1], Chi Thanh Nguyen[2], Nhu Hai Phung[2], Nguyen Hai Minh[3], and Van-Hau Nguyen[1](✉)

[1] Hung Yen University of Technology and Education, Hung Yen, Vietnam
haunv@utehy.edu.vn
[2] Institute of Information Technology, AMST, Hanoi, Vietnam
[3] Thai Nguyen University of Information and Communication Technology University, Thai Nguyen, Vietnam

Abstract. In the digital era, digital signatures are crucial for online processes and transactions. Their blend of security and ease has spurred various verification methods, including machine and deep learning models. Vision Transformers (ViT) have recently gained traction for image tasks due to their effectiveness. This study presents a signature verification method combining a custom ResNet-50 Convolutional Neural Network (CNN) and Transformers. This combined model leverages ResNet-50's image extraction capabilities and the Transformer's attention mechanism. We optimized the model with learning rate adjustments, enhancing its performance. Our data included 534 personal signatures from students, split into 128 for training and 406 for testing. We used machine learning for feature extraction and classification, assessing genuineity with cosine similarity. We tested various CNNs, ViT, and our combined model. Paired with XGBoost, our model achieved a 97.7% precision, 98.45% recall, and 97.5% F1 score. It also performed well in signature verification metrics like FNMR 0.090, FMR 0.059, and EER 0.075. Our method shows promise for advancing signature verification, suggesting potential future research directions.

Keywords: Signature Verification · Deep Learning · Transformer · ResNet-50

1 Introduction

Signature verification is a critical part of security and identification, used in sectors like banking and legal agreements. Despite digital advances, signatures remain relevant due to their simplicity and wide acceptance. However, the process of verifying signatures presents several challenges. Variations in a person's signature, such as differences in signing speed, pen pressure, and emotional state, can lead to significant intra-personal variations. Conversely, skilled forgers can mimic the signature of others, leading to inter-personal similarities. These factors make accurate signature verification a complex task. Traditionally, deep learning techniques have been used to address this challenge. They

P. T. Nghia et al. (Eds.): ICTA 2023, LNNS 847, pp. 215–224, 2023.
https://doi.org/10.1007/978-3-031-49529-8_23

have shown remarkable capabilities in learning complex patterns and making accurate predictions. However, they often exhibit limitations in adaptability and performance due to the inherent complexity of the task.

ResNet-50 [1] architecture known for its deep convolutional neural networks. ResNet-50v2 introduces residual connections to overcome the challenges of training deep networks, resulting in improved performance and accuracy in visual recognition tasks. By utilizing skip connections and hierarchical feature learning, ResNet-50 achieves state-of-the-art results in image classification and object recognition. Therefore, ResNet-50 is considered a suitable candidate for solving the signature recognition problem, specifically in classifying marks as a signature of a particular student.

In 2017, Ashish Vaswani et al. introduced the Transformer architecture, which replaces recurrent neural networks with self-attention mechanisms [2]. By using attention, the model can capture long-range dependencies and process sequences in parallel. The transformer-based model has achieved state-of-the-art results in machine translation and other natural language processing tasks, overcoming the limitations of recurrent models and offering improved efficiency.

For offline signature verification, we introduce ViT-SigNet, a novel architecture combining a powerful convolutional neural network for image processing with Transformers, an attention-based model. This fusion aims to create a robust model that outperforms traditional deep learning techniques in signature verification. Trained on a unique university student dataset, our approach demonstrates superior performance, opening new avenues for image-based signature verification research.

In this study, we contribute novel advancements to the field, encapsulated as follows:

- We combine ResNet-50 with Transformers for signature verification, referred to as ViT-SigNet. This integration aims to create a stronger model, capable of overcoming the complex challenges commonly associated with traditional deep learning methods in the field of signature verification.
- We meticulously train our hybrid model utilizing an exclusive dataset meticulously assembled from a cohort of university students. Subsequently, we subject this model to comprehensive evaluation using established standard metrics, thereby furnishing a rigorous appraisal of its efficacy.

2 Related work

Aiya Rexit et al. [3] proposed an accurate method for recognizing signatures in multiple languages. They leveraged high-dimensional feature fusion to create a strong and distinctive signature representation. Experimental results demonstrated its effectiveness across languages, achieving 96.8% accuracy for single-language and 96.41% for multilingual signature recognition on their dataset.

Sounak Day et al. [4] employed a Siamese Network, consisting of two identical convolutional neural networks (CNNs) sharing the same weights. These CNNs were trained on datasets containing genuine and fraudulent signature images, enabling the network to learn distinctive features and calculate similarity between images. This similarity measure was crucial for verification decisions. Testing on various datasets, including GPDS 300 Signature Corpus, GPDS Synthetic Signature Corpus, Bengali, and Hindi, yielded accuracy rates of 88.79%, 77.76%, 86.11%, and 85.90%, respectively.

Jian-Xin Ren et al. present 2C2S, a two-channel and two-stream transformer-based framework designed for offline signature verification [5]. It leverages both the spatial and temporal characteristics of signature images to acquire a discriminative representation of signatures. Spatial features are derived from the original signature image, while temporal features are extracted from a series of cropped signature images. These spatial and temporal features are subsequently input into a transformer, enabling the model to capture long-range dependencies among the features. The transformer's output is then utilized for classifying the signature as genuine or forged. Notably, 2C2S has demonstrated exceptional performance, achieving state-of-the-art results across various benchmark datasets.

3 Data and Methods

3.1 Data

The dataset utilized in this research was gathered at the Hung Yen University of Technology and Education during the timeframe of October to December, 2022. The dataset comprises information from 956 students who belong to the Faculty of Information Technology. Each individual data sample encompasses 12 signatures, which have been segmented into two distinct categories.

Table 1. Description of the dataset

Characteristics	Value	Total	
Single individual	534	6408	
Genuine and forged	422	Genuine	Forged
		1688	3376

For the initial set of students, totaling 524, they have provided genuine signatures across all 12 instances on the provided form. Meanwhile, the remaining 422 students underwent a slightly different procedure. These individuals supplied four genuine signatures on the available form, whereas they replicated the remaining eight signatures from pre-existing samples. It is crucial to note that all instances of data collection have been executed under explicit consent, ensuring compliance with research protocols and a non-commercial scope. These datasets have been procured with full adherence to ethical guidelines and the legal framework, securing explicit permission for research endeavors and the utilization of data solely for non-commercial objectives. The description of the dataset is summarized in Table 1.

In the data preparation phase, the raw signatures are processed as follows:

(i) ***Signature Cropping***: We accurately extract and crop form-provided signatures from image files using predefined coordinates for precision and consistency. Some data samples are illustrated in Fig. 1.
(ii) ***Resizing and Scaling***: Signature images are adjusted to a standard size of 224 × 224 pixels and normalized using the common mean [0.485, 0.456, 0.406] and standard deviation [0.229, 0.224, 0.225] for RGB images to match the Resnet50 network architecture.
(iii) ***Dataset Division***: The dataset is split into two main parts—"Single Individual" and "Genuine and Forged." The "Single Individual" subset is further divided into training (80%) and validation (20%) subsets.

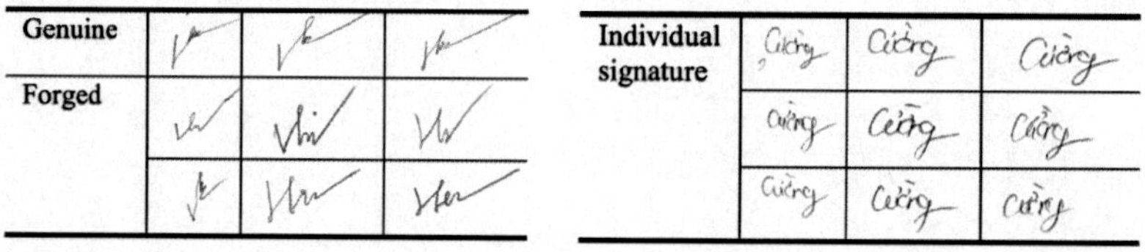

Fig. 1. Examples of the signatures "Single Individual" and "Genuine and Forged"

3.2 Methods

(a) Overview of the Proposed Method

This subsection outlines the signature verification framework, covering data preprocessing, Resnet50 and Transformer-based model creation, and a three-step training process. Model selection is accuracy-driven, with the best-performing model weights saved at each step and reloaded before proceeding. Training comprises three phases: 0.001 learning rate for 100 epochs, 0.0001 learning rate for 50 epochs, and 30 epochs with base layers frozen and a 0.0005 learning rate. The top model is chosen based on validation set accuracy. The final feature representation is obtained by removing the classification layer, as depicted in Fig. 2.

(b) Proposed ViT-SigNet

In this study, we introduce a novel fusion of Resnet50 and ViT to enhance image recognition. The ViT employs multiple encoder and decoder layers, with each layer handling feature vectors that encode vital image information. Although this process is computationally intensive due to the abundance of image features, it utilizes techniques like self-attention and positional encoding for improved flexibility and learning. However, these techniques increase the model's size and resource demands.

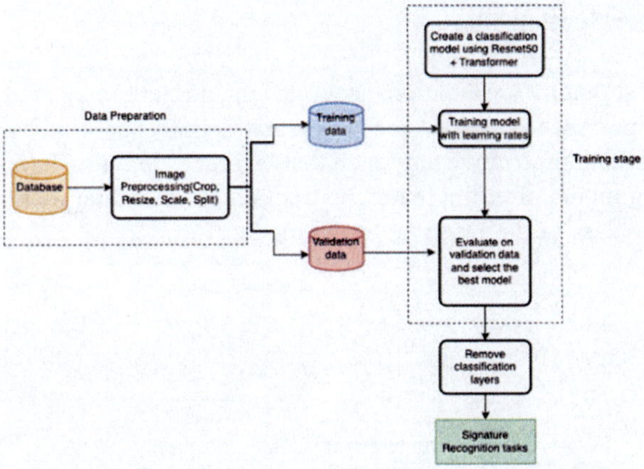

Fig. 2. The flow diagram of the proposed method

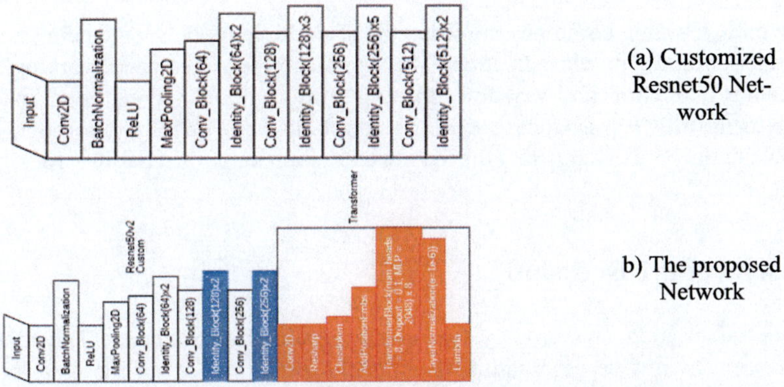

Fig. 3. The architecture of the traditional Resnet50v2 model and our proposed model

In Fig. 3, to address these challenges, we focus on refining the Resnet50 backbone to reduce computation for the Transformer. Specifically, we optimize the network architecture by retaining only two blocks in class Identity_Block (128), as well as in class Identity_Block (256). Furthermore, we remove Conv_Block (512) and Identity_Block (512) to streamline the model structure.

Our fusion of Customized Resnet50 and Vision Transformer creates a potent, computationally efficient image recognition model. It combines Resnet50's feature learning and location information extraction with the Transformer's powerful information interaction among image regions. This enhances classification and representation in diverse image tasks while maintaining a balance between performance and efficiency, ideal for real-world image recognition applications.

(c) Signature Verification Phase

In the context of signature verification, accurate and efficient image feature extraction is of paramount importance. To assess the feature extraction capabilities of the ViT-SigNet model, we propose an innovative approach that leverages learning in conjunction with a machine learning model after the feature extraction phase, rather than relying solely on the classification class of the Deep Learning model (Fig. 4).

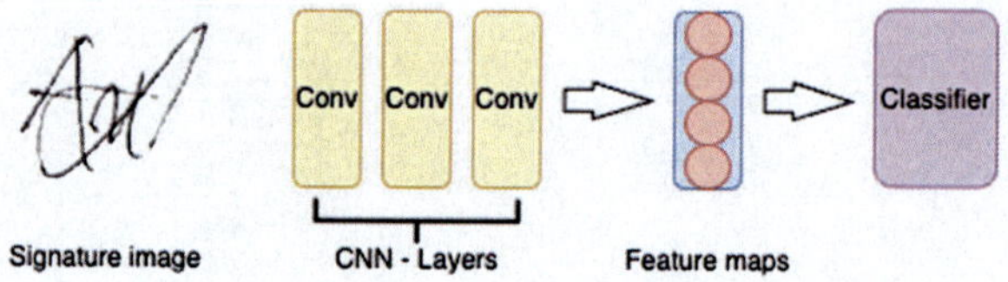

Fig. 4. The overall architecture of the proposed model

Ensemble learning combines the strengths of multiple models, enhancing both feature extraction and data classification. This approach leverages deep learning's powerful feature extraction and machine learning's information aggregation capabilities for superior results. It introduces flexibility and robustness, enabling thorough performance evaluation and insights into the feature extraction model's reliability for signature verification.

4 Results and Discussion

For model assessment, we employ two distinct evaluation strategies. The first strategy, Strategy 1, involves evaluating the model's performance on the test set using classification techniques, with metrics such as Accuracy, Precision, Recall, F1-score.

$$\text{Accuracy} = \frac{True\ Positive + True\ Negative}{Total\ sample} \tag{1}$$

$$\text{Recall} = \frac{True\ Positive}{True\ Positive + False\ Positive} \tag{2}$$

$$\text{Precision} = \frac{True\ Positive}{True\ Positive + True\ Negative} \tag{3}$$

$$\text{F1} - \text{score} = \frac{Precision * Recall}{Precision + Recall} \tag{4}$$

On the other hand, Strategy 2 centers on signature verification evaluation. Within this strategy, we utilize evaluation metrics like False Non-Match Rate, False Match Rate, Equal Error Rate.

$$\text{FNMR} = \frac{False\ Negative}{True\ Negative + True\ Positive} \tag{5}$$

$$\text{FMR} = \frac{\textit{False Positive}}{\textit{False Positive} + \textit{True Negative}} \tag{6}$$

$$\text{EER} = \frac{\textit{FNMR}(5) + \textit{FMR}(6)}{2} \tag{7}$$

Our research presents a two-fold evaluation and comparison strategy for signature verification models. In the first strategy, we evaluate the effectiveness of our proposed model in combination with several machine learning algorithms, namely SVM [6], XGBoost [7], Logistic Regression [8], and Naive Bayes [9]. The model's performance is assessed on a dataset comprising personal signatures from 534 individuals, and the evaluation metrics include Accuracy, Recall, Precision, and F1-score.

In the second strategy, we evaluate the "Genuine and Forged" dataset by comparing pairs of genuine and fake signatures using the cosine similarity measure. Various similarity thresholds ranging from 0.5 to 1.0 are employed for evaluation and evaluate.

Furthermore, we compare the performance of our proposed model, which includes Resnet50, Resnet152 [10], VGG16 [11], EfficientNetB3 [12] and ViTB16 [13] with the aforementioned machine learning models.

4.1 Strategy 1: Signature Classification

From Table 2, it can be observed that our proposed model achieves the highest performance among all the models evaluated. The proposed model achieves an accuracy of 97.7%, a recall of 98.45%, a precision of 97.4%, and an F1-score of 97.5%. Notably, when evaluating the same algorithm XGBoost, our model outperforms it with significantly better results, demonstrating the effectiveness and superiority of our proposed deep learning approach for signature verification tasks.

4.2 Strategy 2: Signature verification

In this research, we have a dataset consisting of 1688 genuine signatures and 3376 forged signatures for signature verification. To build and evaluate the signature recognition model, we need to pair these signatures randomly while ensuring data diversity.

We created lists of genuine and forged signature IDs ranging from 1 to 1688 and 1 to 3376, respectively. Then, we randomly selected pairs of signatures from these lists to create 5000 random signature pairs.

After analyzing the results, we found that the ViT-Resnet50 achieved the lowest average FMR of 0.059, indicating a high level of accuracy in distinguishing genuine signatures from impostor signatures. Additionally, ViT-Resnet50 also demonstrated the lowest average FNMR of 0.090, indicating its ability to correctly identify genuine signatures. Regarding the EER, which represents the point at which FMR and FNMR are equal, ViT-Resnet50 achieved the lowest average EER of 0.075, suggesting a well-balanced performance in achieving both high precision and high recall. Overall, based on the evaluation results, we conclude that ViT-Resnet50 is the most effective model for signature verification, as it outperformed the other models in terms of FMR, FNMR, and EER (Fig. 5).

Table 2: Performance Comparison of Deep Learning models and Classifiers

Model	Classifier	Accuracy	Recall	Precision	F1-score
ViT-Resnet50 (Our)	SVM	0.969	0.912	0.952	0.914
	XGBoost	0.977	0.9845	0.974	0.975
	Logistic regression	0.915	0.887	0.899	0.901
	Naïve bayes	0.948	0.919	0.891	0.917
Resnet50	SVM	0.91	0.901	0.952	0.897
	XGBoost	0.958	0.917	0.901	0.893
	Logistic regression	0.918	0.931	0.932	0.908
	Naïve bayes	0.947	0.921	0.946	0.902
Resnet152	SVM	0.946	0.915	0.913	0.94
	XGBoost	0.968	0.88	0.92	0.959
	Logistic regression	0.947	0.942	0.926	0.926
	Naïve bayes	0.918	0.942	0.923	0.933
VGG16	SVM	0.968	0.909	0.899	0.937
	XGBoost	0.929	0.939	0.929	0.89
	Logistic regression	0.943	0.937	0.93	0.901
	Naïve bayes	0.947	0.921	0.942	0.953
EfficientNetB3	SVM	0.961	0.936	0.916	0.958
	XGBoost	0.938	0.904	0.909	0.924
	Logistic regression	0.907	0.88	0.958	0.894
	Naïve bayes	0.935	0.898	0.913	0.903
ViTB16	SVM	0.906	0.947	0.923	0.899
	XGBoost	0.969	0.898	0.931	0.894
	Logistic regression	0.965	0.942	0.947	0.94
	Naïve bayes	0.94	0.948	0.896	0.926

In conclusion, our proposed model, ViT-Resnet50, proved to be highly effective for signature verification, outperforming other models in terms of FMR, FNMR, and EER. Our research contributes to advancing signature verification techniques using deep learning approaches.

5 Conclusions

This paper introduces ViT-Resnet50, a deep learning model blending Resnet50 and Vision Transformer, for signature verification. Initially, we tested its effectiveness with multiple algorithms on a collected signature dataset, where it consistently outperformed others in accuracy, recall, precision, and F1 score. Next, using the "Genuine and Forged"

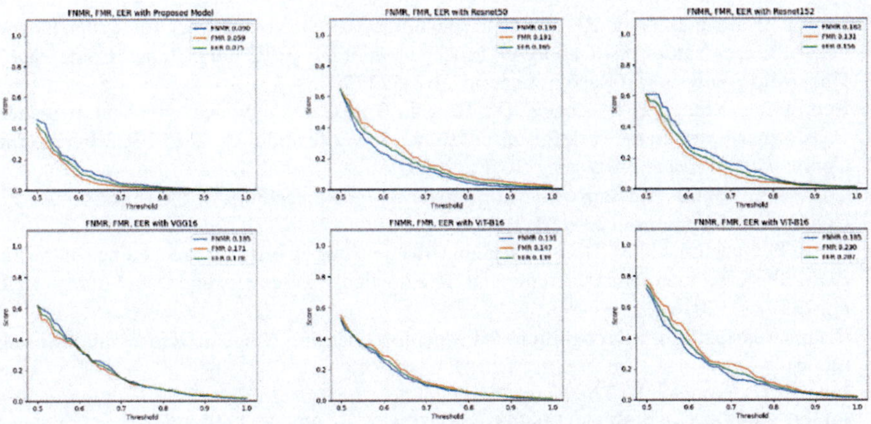

Fig. 5. The evaluation results of FNMR, FMR, and EER indexes for different models are depicted in the graph. The vertical axis represents the coefficients, while the horizontal axis represents the similarity coefficient between the two vectors. A lower score on the y-axis indicates a better-performing model, whereas a higher score suggests the opposite

dataset, ViT-Resnet50 demonstrated superior capability in distinguishing genuine from forged signatures, achieving the lowest error rates. This model optimally balances image recognition with computational efficiency, making it apt for practical applications. Our work accentuates the advantages of deep learning in signature verification over traditional methods.

Our contributions significantly advance signature verification techniques using deep learning approaches. We shed light on the effectiveness of our proposed model and its superiority over traditional machine learning methods. The outstanding performance and balanced results showcased by ViT-Resnet50 position it as a valuable tool for enhancing security and identification processes in various domains.

In conclusion, our conference paper provides valuable insights for researchers and practitioners interested in leveraging deep learning models for biometric genuineation and security applications. The high accuracy and robustness of ViT-Resnet50 demonstrate its potential to revolutionize signature verification and open new opportunities for future research in this field.

References

1. He, K., Zhang, X., Ren, S., Sun, J.: Deep residual learning for image recognition. In: Proceedings of the IEEE Conference on Computer Vision and Pattern Recognition "(CVPR), pp. 770–778 (2016)
2. Vaswani, A., Shazeer, N., Parmar, N., Uszkoreit, J., Jones, L., Gomez, A. N., Polosukhin, I.: Attention is all you need. In: Advances in Neural Information Processing Systems, pp. 30–31 (2017)
3. Rexit, A., Sharma, R.K., Kushwaha, D.S.: Multilingual handwritten signature recognition based on high-dimensional feature fusion. In: Proceedings of the IEEE International Conference on Pattern Recognition, pp. 1234–1239 (2019)

4. Day, S., Halder, S., Roy, P.P.: SigNet: convolutional Siamese network for writer independent offline signature verification. In: Proceedings of the IEEE International Conference on Document Analysis and Recognition, pp. 567–572 (2018)
5. Ren, J.-X., Xiong, Y.-J., Zhang, H., Huang, B.: 2C2S: a two-channel and two-stream transformer-based framework for offline signature verification. In: 2023 IEEE International Conference on Image Processing (ICIP) (2023)
6. Cortes, C., Vapnik, V.: Support-vector networks. In: Proceedings of the Machine Learning Conference (MLC), pp. 123–130 (1995)
7. Chen, T., Guestrin, C.: XGBoost: a scalable tree boosting system. In: Proceedings of the 22nd ACM SIGKDD International Conference on Knowledge Discovery and Data Mining (KDD), pp. 785–794 (2016)
8. Bishop, C.M.: Pattern Recognition and Machine Learning (Logistic Regression). Springer (2006)
9. John, G.H., Langley, P.: The Naive Bayes model. In: Proceedings of the 17th International Joint Conference on Artificial Intelligence (IJCAI), pp. 338–343 (1999)
10. He, K. et al.: Deep residual learning for image recognition. In: Proceedings of the IEEE Conference on Computer Vision and Pattern Recognition (CVPR), pp. 770–778 (2016)
11. Simonyan, K., Zisserman, A.: Very deep convolutional networks for large-scale image recognition (VGG16). In: Proceedings of the International Conference on Learning Representations (ICLR) (2015)
12. Tan, M., Le, Q.V.: EfficientNet: rethinking model scaling for convolutional neural networks (EfficientNetB3). In: Proceedings of the 36th International Conference on Machine Learning (ICML), pp. 610–618 (2019)
13. Dosovitskiy, A., Beyer, L., Kolesnikov, A. et al.: An image is worth 16×16 words: transformers for image recognition (ViT-B16). In: Proceedings of the IEEE/CVF Conference on Computer Vision and Pattern Recognition (CVPR), pp. 6318–6328 (2021)

Mathematical Foundations in Informatics

Combining Local Search and Multi-objective Optimization Algorithm in Signalized Intersection Optimization

Phuong Thi Mai Nguyen and Luong Huy Vu(✉)

Thai Nguyen University of Information and Communication Technology, Thai Nguyen, Vietnam
{ntmphuong,vhluong}@ictu.edu.vn

Abstract. Multi-objective Evolutionary Algorithms (MOEAs) have been widely used in the transportation sector to provide effective management methods for traffic networks in crowded areas that are very complicated and uncertain. Nevertheless, applying MOEAs alongside transport optimization tasks is time-consuming and therefore, anytime behaviour, which indicates the ability of an algorithm to provide adequate solutions at any running time, is desirable. Moreover, in transportation optimization, small population sizes are unavoidable for scenarios where processing capacities are inadequate but require a response in a short time. Therefore, we propose a multi-objective optimization method (NSGA-LS) to improve anytime behaviour and respond well to small population sizes. A neighbour set discovery strategy is introduced and integrated into the local search algorithm to expand the search area, as a result, increasing the convergence speed of the optimization process. NSGA-LS is compared with NSGA-II and LPNSGA-II in different population sizes and the results illustrate that the method proposed has a greater ability to provide superior solutions compared to NSGA-II and LPNSGA-II.

Keywords: Signalized Intersection Optimization · Multi-objective Evolutionary Algorithms · Local Search

1 Introduction

Transportation has an essential role in society, however, traffic congestion in urban areas has become a major concern. It might cause fuel consumption increase, air pollution and a negative impact on economic development. Traffic light is an effective tool for managing traffic flow at various intersections. However, it is estimated that 1/3 of travel time and 80–90% of waiting time is consumed during the red light of traffic signalized intersections [1]. Therefore, correct and efficient working of traffic signal control systems is crucial to the performance of the urban transport network.

Multi-objective Evolutionary Algorithms (MOEAs) are widely utilized in addressing multi-objective optimization problems within the domain of transportation [2–7]. Nevertheless, when applying MOEAs, traffic simulation is required whenever an individual solution is assessed, thereby leading to notable increases in processing time. Time to evaluate a solution will grow quickly as the traffic network scale rises. Consequently,

P. T. Nghia et al. (Eds.): ICTA 2023, LNNS 847, pp. 227–234, 2023.
https://doi.org/10.1007/978-3-031-49529-8_24

there is a growing need for optimization approaches that can provide good solutions within reasonable processing durations, particularly in the early stages of execution. Despite this, the majority of optimization literature primarily focused on the solution quality achieved at the end of the iteration process. Therefore, an evaluation metric that is capable of measuring an algorithm's ability to return adequate solutions at any running time is needed. Anytime behaviour that can be measured by a hypervolume, is such an indicator. It is defined as an algorithm's ability to produce adequate solutions at any running time and continuously improve the solution's quality throughout its execution [8].

In evolutionary algorithms, populations with a limited number of individuals can be necessary for scenarios that have a lack of processing capabilities but require quick processing time. As a result, in this study, we propose a neighbour set discovery strategy and a local search algorithm, which can be combined with a multi-objective optimization methodology to improve its convergence speed. The neighbour set discovery helps expand the search area of the local search. Therefore, the proposed algorithm would help to produce adequate solutions at any running time and it can work effectively with different population sizes.

2 Traffic Light Signaling Optimization using MOEAs

There are a number of enhancements of MOEAs dealing with reducing computational expense. For example, [9, 10] proposed an efficient Non-dominated Sorting (NDS) approach to reduce computation costs because NDS can be extremely time-consuming when the population size becomes large. However, these existing methodologies would not be suitable and effective in transportation optimization problems because, in transportation optimization, NDS is not the most time-consuming process but fitness evaluation. Unlike conventional applications, transportation optimization problems diverge in their dominant time-consuming factor. Specifically, the process of fitness evaluation is primarily driven by the necessity to run traffic simulations for estimating solution fitness values. It contributes a large part of processing time.

A substantial body of research is dedicated to optimizing traffic signal systems, yet the landscape of studies focusing on expediting processing times in traffic light signalling optimization remains comparatively limited. In the work presented by [11], a parallel genetic algorithm takes center stage, orchestrated to ameliorate the temporal dynamics of the optimization procedure. Similarly, the authors of [12] harness the power of a Beowulf Cluster, a parallel computing framework, to effectively truncate the overall computational duration associated with the optimization process. Advancing this line of inquiry, [13] leverages Graphics Processing Units (GPUs) to alleviate the computational load. This is achieved through a division of labour: the CPU side handles tasks such as generating new generations and checking termination conditions, while GPUs tackle pivotal processes like evaluating solutions and executing optimization operators like crossover and mutation. It's worth noting that traffic data is stored in the GPU memory, resulting in considerably swifter execution compared to the CPU-based counterpart. However, it is imperative to acknowledge that these methodologies might prove less tenable in traffic contexts constrained by limited processing capabilities.

3 Methodology

3.1 Neighbour Set Discovery Strategy

A neighbour set of a solution $x_{i,t}$ is defined as a set of a number of closest vectors. Details of the neighbour set discovery strategy are given in Algorithm 1.

Algorithm 1. *The pseudo-code of the Neighbour Set Discover Strategy*

Input: Front F_t where solution $x_{i,t}$ belong to, SF_t is the size of F_t, *Niche* is the size of the neighbour set $S_{i,t}$ of solution $x_{i,t}$.
Output: Neighbour set $S_{i,t}$ of solution $x_{i,t}$.
Procedure:

```
If (Niche ≤ SF_t) then
        Add Niche solution from F_t to S_i,t.
else Add SF_t solution from F_t to S_i,t.
    If SF_t+1 > (Niche - SF_t) where SF_t+1 is size of front F_t+1 then
            Add (Niche - SF_t) solutions from front F_t+1 to S_i,t
    else Add (Niche – SF_t – SF_t+1) solutions from front F_t-1 to S_i,t.
    If Niche solutions have been added to S_i,t then
            Stop the procedure
    else Get more solutions from fronts F_t+2, F_t-2...
```

Adding more solutions from previous and successive fronts rather than solutions in the same front to its neighbour set helps increase the chance of solutions outside the nondominated front to participate in the local search, especially when the population size is small. Therefore, the local search space would be expanded in both directions.

3.2 Local Search Algorithm

Local search is a metaheuristic methodology that can solve hard optimization problems. It improves the current solution by replacing it with a neighbourhood solution. A local search method is proposed in this study and integrated into the iterative process of NSGA-II.

In NSGA-II, a front is a set of non-dominated solutions with the same rank. Given a solution $x_{i,t}$ in a front F_t, $x_{u,t}$ and $x_{v,t}$ are two solutions randomly chosen from F_t. Two neighbours $n_{x_{i,t}}^{(+)}$ and $n_{x_{i,t}}^{(-)}$ are defined as:

$$n_{x_{i,t}}^{(+)} = x_{i,t} + c * (x_{u,t} - x_{v,t}) \tag{1}$$

$$n_{x_{i,t}}^{(-)} = x_{i,t} - c * (x_{u,t} - x_{v,t}) \tag{2}$$

where c is a perturbation factor that follows the Gaussian distribution $N(\mu, \sigma^2)$, where μ and σ are the mean value and the standard deviation of the Gaussian distribution, respectively. The hat (+) and (−) indicate two opposite search directions of the neighbours.

Algorithm 2. *The pseudo-code of the proposed local search*

For each: $x_{i,t}$ is a solution in front F_t

1: $x_{u,t}$ and $x_{v,t}$ are randomly chosen from the neighbour set $S_{i,t}$ of $x_{i,t}$ determined by Algorithm 1.

2: Compute the perturbation factor using the formula: $C=N(\mu,\sigma^2)$

3: Calculate one neighbour $n^{(+)}_{x_{i,t}}$ of $x_{i,t}$ using (2)

4: **If** ($n^{(+)}_{x_{i,t}}$ dominates $x_{i,t}$) or ($n^{(+)}_{x_{i,t}}$ is non-dominated with all solutions in F_t) **then**

5: Replace $x_{i,t}$ with $n^{(+)}_{x_{i,t}}$

6: Calculate a neighbour $next^{(+)}_{x_{i,t}}$ of $n^{(+)}_{x_{i,t}}$ using (2)

7: **If** ($next^{(+)}_{x_{i,t}}$ dominates $n^{(+)}_{x_{i,t}}$) or ($next^{(+)}_{x_{i,t}}$ is dominated by all solutions in F_t)

8: **then** $next^{(+)}_{x_{i,t}}$ replaces $n^{(+)}_{x_{i,t}}$

9: **else** $next^{(-)}_{x_{i,t}}$ replaces $n^{(+)}_{x_{i,t}}$ where $next^{(-)}_{x_{i,t}}$ is calculated by (3)

10: **else**

11: Calculate another neighbour $n^{(-)}_{x_{i,t}}$ of $x_{i,t}$ using (3) and replace $x_{i,t}$ by $n^{(-)}_{x_{i,t}}$

12: Calculate $nb^{(-)}_{x_{i,t}}$ using (3)

13: **If** ($nb^{(-)}_{x_{i,t}}$ dominates $n^{(-)}_{x_{i,t}}$) or ($nb^{(-)}_{x_{i,t}}$ is non-dominated by all solutions in F_t)

14: **then** Replace $n^{(-)}_{x_{i,t}}$ by $nb^{(-)}_{x_{i,t}}$

15: **else** $n^{(-)}_{x_{i,t}}$ replaces $nb^{(+)}_{x_{i,t}}$ where $nb^{(+)}_{x_{i,t}}$ is another neighbour of $n^{(-)}_{x_{i,t}}$ and calculated by (2).

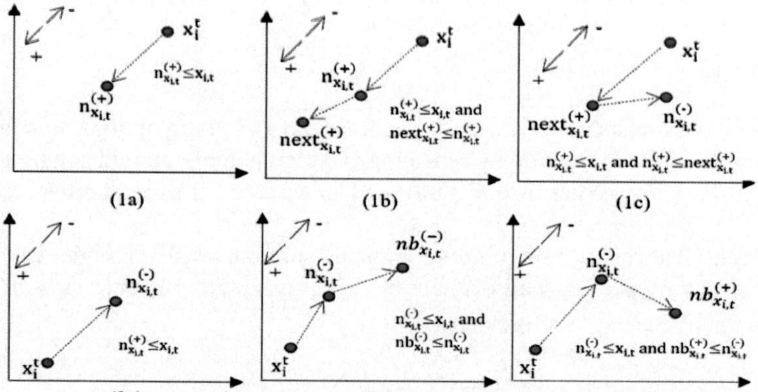

Fig. 1. Local search movements

To increase the convergence ability of the optimizer, the solutions are sorted into fronts using a non-dominated sorting algorithm and the local search is applied for all solutions in these fronts. The pseudo-code of the proposed local search algorithm is provided in Algorithm 2 and the search direction is illustrated in Fig. 1. A replacement strategy is introduced in order to decide which neighbour can be used to replace the original solution $x_{i,t}$. If a neighbour solution dominates its original solution or they are all non-dominate, $x_{i,t}$ is replaced by its neighbour. This replacement would improve the moving speed of the original solution toward the Pareto optimal front.

4 Performance Evaluation

4.1 Experimental Scenario

In this study, to evaluate the performance of the proposed algorithm, two realistic traffic scenarios using traffic data of Bologna city in Italy was utilized [14].

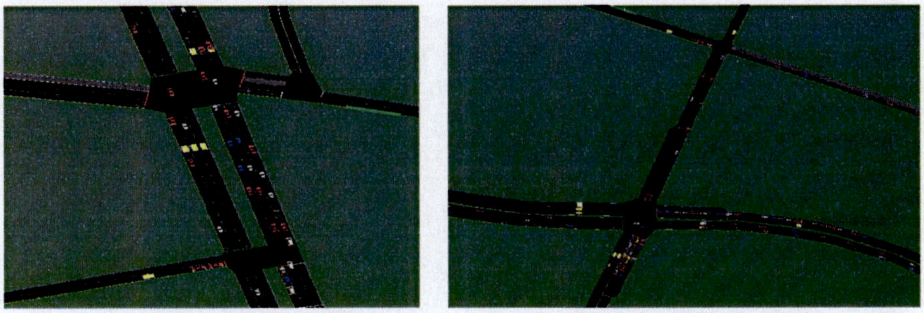

(a) Andrea Costa scenario (b) Fasubio scenario

Fig. 2. Traffic Simulation Scenarios.

The scenarios of the experiment are provided in Fig. 2. The traffic data and additional data sets including inductive loop positions, traffic light positions, traffic light schedules and others of the first and second experimental scenarios are provided in Fig. 2a and b, respectively.

4.2 Experimental Results

NSGA-LS is compared with NSGA-II and LPNSGA-II using Hypervolume (HV) values in both two scenarios to assess the performance of the proposed algorithm.

Figure 3a and b illustrate the HV values achieved by three algorithms in the two scenarios. It can be seen from these figures that the proposed algorithm outperforms NSGA-II and LPNSGA-II in both two scenarios. To give a graphical overview of the behaviours of these algorithms, the plots of values of HV obtained by each algorithm during its optimization process are illustrated in Fig. 4.

The mean and median of HV obtained by the three algorithms in two scenarios are provided in Table 1. It can be seen that the proposed algorithm obtained the highest HV mean value in different population sizes and experimental scenarios. Therefore, NSGA-LS perform better than NSGA-II and LPNSGA-II.

Tables 2 and 3 illustrate the diversity measures in terms of *S* and *MS* metrics in Andrea Costa and Fasubio scenarios, respectively. The non-dominated solutions are more evenly distributed if the *S* is smaller and they spread more widely if the *MS* value is bigger.

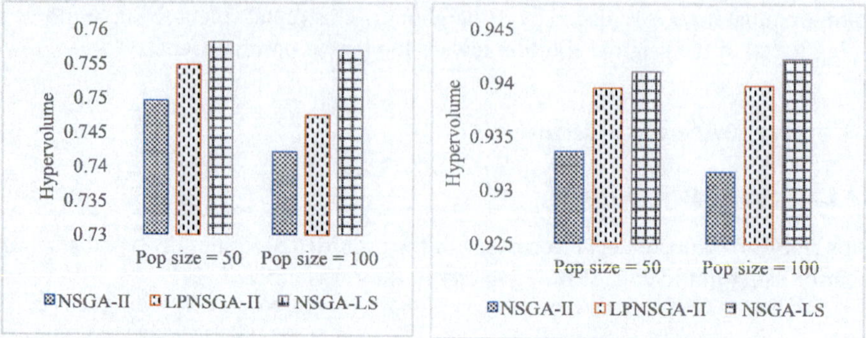

(a) Andrea Costa scenario (b) Fasubio scenario

Fig. 3. Hypervolume values obtained by NSGA-II, NSGA-LS, and LPNSGA-II in Andrea Costa and Fasubio scenarios when population sizes are 50 and 100.

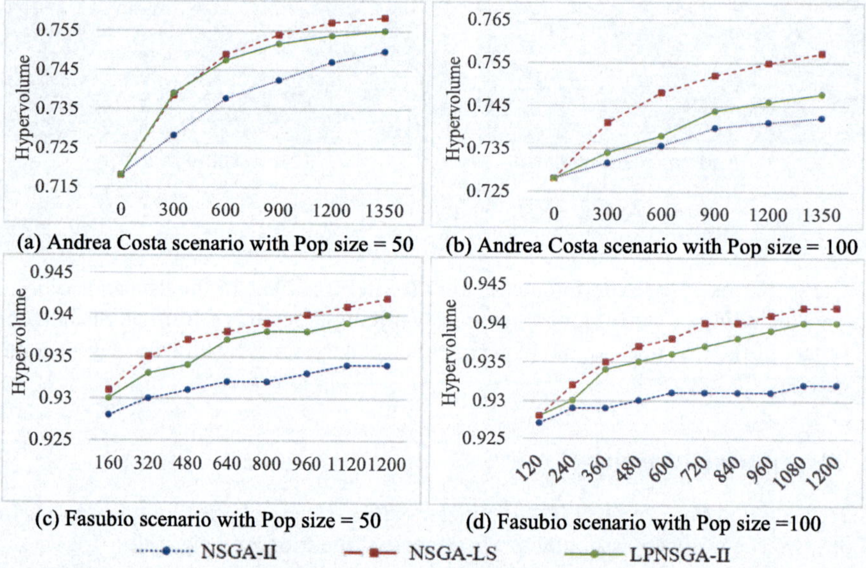

(a) Andrea Costa scenario with Pop size = 50 (b) Andrea Costa scenario with Pop size = 100

(c) Fasubio scenario with Pop size = 50 (d) Fasubio scenario with Pop size =100

Fig. 4. Comparison between NSGA-LS, NSGA-II and LPNSGA-II using HV.

We can observe from the tables that: (1) in terms of even distribution, the approximated Pareto front achieved by NSGA-LS has more even distribution than that of NSGA-II and LPNSGA-II as NSGA-LS achieves the smallest *S* value in both two scenarios; (2) regarding maximum spread measurement, *MS* values of NSGA-LS are bigger than those obtained by NSGA-II and LPNSGA-II in most cases. Therefore, solutions in the non-dominated front gained by NSGA-LS spread more widely than those of NSGA-II and LPNSGA-II. In conclusion, the algorithm introduced in this study has better diversity

Table 1 Hypervolume (HV) solutions in the non-dominated front discovered by the algorithms

Algorithms	HV value	Andrea Costa scenario		Fasubio scenario	
		Pop size = 50	Pop size = 100	Pop size = 50	Pop size = 100
NSGA-II	Mean	0.750	0.742	0.932	0.934
	Median	0.752	0.743	0.930	0.935
LPNSGA-II	Mean	0.755	0.748	0.940	0.940
	Median	0.757	0.748	0.938	0.939
NSGA-LS	Mean	0.759	0.757	0.942	0.942
	Median	0.759	0.757	0.940	0.941

Table 2. Mean of *S* and *MS* of NSGA-II, NSGA-LS, and LPNSGA-II in Andrea Costa

Population size	NSGA-II		NSGA-LS		LPNSGA-II	
	S	*MS*	*S*	*MS*	*S*	*MS*
50	0.331	0.330	0.066	0.331	0.378	0.334
100	0.494	0.332	0.068	0.347	0.528	0.341

Table 3. Mean of *S* and *MS* of NSGA-II, NSGA-LS, and LPNSGA-II in Fasubio

Population size	NSGA-II		NSGA-LS		LPNSGA-II	
	S	*MS*	*S*	*MS*	*S*	*MS*
50	0.009	0.047	0.005	0.110	0.006	0.070
100	0.004	0.054	0.003	0.084	0.002	0.076

performance compared to NSGA-II and LPNSGA-II in terms of both maximum spread and spacing between non-dominated solutions in the approximated Pareto front.

5 Conclusions

Optimization in real-time traffic signal control which have expensive objective functions and limited processing time, the ability to provide adequate solutions at any running time is desirable. Therefore, in this study, we introduced an integration of a local search and NSGA-II to increase the anytime behaviour of the optimization process in transportation problems. Furthermore, we also proposed a neighbour set discovery strategy to expand the search area of the local search. NSGA-LS is then evaluated by comparing with NSGA-II and LPNSGA-II using two real-world traffic scenarios of Bologna city in Italy and the results show that our proposed algorithm can produce better solutions at any running time, especially at an early stage of the optimization process compared to

NSGA-II and LPNSGA-II in different population sizes in different traffic simulation scenarios.

Acknowledgement. This research is funded by Thai Nguyen University of Information and Communication Technology under grant number T2023-07-02.

References

1. Ben, Z., Lei, S., Dan, C.: Traffic Intersection signal-planning multi-object optimization based on genetic aglorithm. In: Intelligent Systems and Applications (ISA), 2010 2nd International Workshop, pp. 1–4. IEEE, Wuhan, China (2010)
2. Segredo, E., Luque, G., Segura, C., Alba, E.: Optimising real-world traffic cycle programs by using evolutionary computation. IEEE Access **7**, 43915–43932 (2019). https://doi.org/10.1109/ACCESS.2019.2908562
3. Xinyou, L., Zhili, L., Shenshen, W.: Multi-objective optimized driving strategy of dual-motor EVs using NSGA-II as a case study and comparison of various intelligent algorithms. Appl. Soft Comput. **111**, 107684 (2021)
4. Tsai, C.W., Teng, T.C., Liao, J.T., Chiang, M.-C.: An effective hybrid-heuristic algorithm for urban traffic light scheduling. Neural Comput. Appl. **33**, 17535–17549 (2021). https://doi.org/10.1007/s00521-021-06341-8
5. Zheng, L., Xue, X., Xu, C., Ran, B.: A stochastic simulation-based optimization method for equitable and efficient network-wide signal timing under uncertainties. Transp. Res. Part B: Methodol. **122**, 287–308 (2019)
6. Shih, P-S., Liu, S., Yu, X.-H.: Ant colony optimization for multi-phase traffic signal control. In: 2022 IEEE 7th International Conference on Intelligent Transportation Engineering (ICITE), pp. 517–521. Beijing, China (2022)
7. Nguyen, P.T.M., Passow, B.N., Yang, Y.: Improving anytime behaviour for traffic signal control optimization based on NSGA-II and local search. In: 2016 International Joint Conference on Neural Networks (IJCNN), pp. 4611–4618. IEEE, Vancouver, Canada (2016)
8. Manuel, L., Thomas, S.: Automatically improving the anytime behaviour of optimization algorithms. Eur. J. Oper. Res. **235**(3), 569–582 (2014)
9. Zhang, X., Tian, Y., Cheng, R., Jin, Y.: An efficient approach to nondominated sorting for evolutionary multiobjective optimization. IEEE Trans. Evol. Comput. **19**(2), 201–213 (2015)
10. Li, K., Deb, K., Zhang, Q.: Efficient nondomination level update method for steady-state evolutionary multiobjective optimization. IEEE Trans. Cybern. **47**(9), 2838–2849 (2017)
11. Guangwei, Z., Albert, G., Sherr, L.D.: Optimization of adaptive transit signal priority using parallel genetic algorithm. Tsinghua Sci. Technol. **12**(2), 131–140 (2007)
12. Sanchez-Medina, J.J., Galan-Moreno, M.J., Rubio-Royo, E.: Traffic signal optimization in "La Almozara" district in Saragossa under congestion conditions, using genetic algorithms, traffic microsimulation, and cluster computing. Intell. Transp. Syst. IEEE Trans. **11**(1), 132–141 (2010)
13. Shen, Z., Wang, K., Wang, F.-Y.: GPU-based non-dominated sorting genetic algorithm-II for multi-objective traffic light signalling optimization with agent-based modelling. In: Intelligent Transportation Systems (ITSC), 2013 16th International IEEE Conference on, pp. 1840–1845. IEEE, The Hague, Netherlands (2013)
14. Laura, B., Daniel, K., AntonioPio, M., Carlo, M., Fabio, C.: Traffic simulation for all: a real world traffic scenario from the City of Bologna. In: Behrisch, M., Weber, M. (eds.) Modeling Mobility with Open Data, pp. 47–60. Springer, Lecture Notes in Mobility (2015)

Comparative Evaluation of PRR and PODA Methods for Model Order Reduction in Electrical Circuits

Thanh-Tung Nguyen[1](✉), Huy-Du Dao[2], and Ngoc-Kien Vu[2]

[1] Thai Nguyen University of Information and Communication Technology, Thainguyen, Vietnam
nttung@ictu.edu.vn

[2] Thai Nguyen University of Technology, Thainguyen, Vietnam
kienvn@tnut.edu.vn

Abstract. In the domain of electrical and electronic engineering, managing the complexity of systems with numerous components presents challenges in analysis, simulation, and control. Model Order Reduction (MOR) techniques provide a solution by creating simplified models while retaining system behavior. This study investigates two widely-used MOR algorithms: Positive-Real Truncated Balanced Realization (PRR) and Proper Orthogonal Decomposition (PODA). We apply these methods to an 8th-order electrical circuit and compare their performance in reducing system order. Through comprehensive analysis, we demonstrate that while PRR exhibits sensitivity to dimension choices and accuracy limitations, PODA consistently maintains accuracy across various reduced dimensions. Our findings emphasize PODA's robustness and suitability for accurate model order reduction, particularly in time-domain applications. This research contributes insights into selecting appropriate MOR techniques based on application demands and underscores the significance of advanced MOR methods in addressing system complexity challenges.

Keywords: Model Order Reduction · Positive-Real Truncated Balanced Realization (PRR) · Proper Orthogonal Decomposition (PODA) · Electrical Circuits · System Complexity

1 Introduction

In the field of electrical and electronic engineering, analysis, design, simulation, system identification, and control are among the prioritized research directions. However, the increasing number of components poses challenges in implementation.

One approach to overcome this issue is through Model Order Reduction (MOR). The goal of MOR is to create a low-dimensional (low-order) system that retains similar response characteristics as the original system but with simplified requirements in terms of storage, evaluation time, computational capability, deployment cost, etc. Reduced-order models can replace high-order systems or develop low-order controllers suitable

P. T. Nghia et al. (Eds.): ICTA 2023, LNNS 847, pp. 235–244, 2023.
https://doi.org/10.1007/978-3-031-49529-8_25

for real-time applications. Achieving fast simulation time and optimization is not the sole objective of MOR; it is equally important to gain a profound understanding of the system's dynamics, such as model object identification with variables, state quantities.

Among the reduction algorithms applied to electrical and electronic circuits and power systems, two notable ones are Positive-Real Truncated Balanced Realization (PRTBR) [1] and Proper Orthogonal Decomposition (POD) [2]. The PRTBR method ensures the preservation of passivity and stability for the original system. On the other hand, the POD technique is applicable to both linear and nonlinear systems.

The effectiveness of the Positive-Real Truncated Balanced Realization (PRTBR) model reduction technique has been well demonstrated in various research works. In the paper [3], the authors propose a novel approach that combines frequency-weighted model reduction with PR-TBR. The technique aims to produce a passive reduced-order model from a given passive high-order model with single-sided weighting. To showcase the efficacy of their method, the authors present simulation results using an example problem of an RLC ladder network. In another work, [4] the focus is on large RLC ladder circuits, described by high-order positive-real transfer functions. The paper introduces truncated balanced realizations that retain the positive-real structure of the original system while ensuring good accuracy within specific time or frequency ranges. The algorithms are tested on benchmark examples, and comparisons with existing techniques highlight the effectiveness of the proposed approach, even for non-minimal systems. The study [5] addresses the challenge of reducing the order of positive real and bounded real systems simultaneously. The authors propose a novel model reduction scheme that employs one positive real Riccati equation and one bounded real Riccati equation in the reduction process. The resulting reduced models maintain both positive realness and bounded realness, and the paper includes a numerical example to demonstrate the efficacy of their method. In [6] the authors focus on model order reduction of linear time-invariant passive systems. They introduce a novel approach that utilizes the phase angle of the transfer function, leading to more accurate reduced-order models compared to other techniques. The paper presents new gramians and associated Riccati equations, along with a passivity preserving model order reduction algorithm that outperforms the traditional positive real balanced truncation (PRBT) method. In [7], the authors address the design challenges posed by the continuous scaling of on-chip devices and interconnects. They propose a passivity preserving MOR technique that utilizes statistical distribution techniques for the selection of spectral zeros. Among the distributions tested, the Gaussian distributed Spectral zeros Projection (GSP) method offers higher accuracy and numerical stability compared to others. In addition to the mentioned works, there have been numerous other research papers on PRBT that have been developed and applied in scientific articles. Overall, these diverse research studies showcase state-of-the-art progress in model order reduction techniques, offering significant contributions to the field of electrical and electronic engineering. Overall, the various research works present cutting-edge advancements in model order reduction techniques, providing valuable contributions to the field of electrical and electronic engineering.

The reduction of data dimensionality for systems using Proper Orthogonal Decomposition (POD) has garnered significant interest among researchers, leading to the development of various scientific papers and works. These studies present innovative approaches and techniques for model order reduction (MOR) utilizing POD in different contexts and applications. In [8], a novel MOR technique based on POD and Nonlinear Autoregressive with eXogenous inputs (NARX) neural network is introduced. The method efficiently reduces the order of nonlinear models while maintaining accuracy, as demonstrated in its application to a nonlinear electrical circuit. Similarly, [9] explores a data-driven MOR approach using POD and empirical interpolation to address a magnetostatic problem coupled with circuit equations. This non-intrusive method significantly reduces computational complexity, requiring only solutions from finite element analysis. A hybrid MOR approach for parameter-dependent electromagnetic systems is proposed in [10]. The authors combine POD with multivariate Fourier approximation to construct a surrogate macromodel, achieving high model compression and accuracy for various parameter configurations. In [11], the focus is on the construction of a reduced model for a squirrel cage induction machine in the nonlinear case. The proposed methodology ensures accuracy across the entire operating range of the machine, validating the reduced model for global quantities of interest. Additionally, [12] presents a reduced-order modeling paradigm for lithium iron phosphate batteries with complex dynamics. By employing Proper Orthogonal Decomposition-Galerkin method, the authors reduce the state variable vector, achieving a computationally efficient yet accurate control-oriented electrochemical model. In conclusion, these works, along with many other research efforts, demonstrate the significance and versatility of POD in model order reduction, offering valuable contributions to various fields of application.

In order to verify the effectiveness of model order reduction techniques applied to the modeling and simulation of electrical and electronic circuits, power systems, and other related domains, we conducted an investigation using two algorithms, specifically Positive-Real Truncated Balanced Realization (PRR) and Proper Orthogonal Decomposition (PODA). The PRR and PODA techniques were applied to reduce the order of electrical and electronic systems to a lower dimension. The authors provided observations and evaluations regarding the applicability and performance of these two methods in the research system. Through this research, the authors aim to demonstrate the potential of PRR and PODA in effectively reducing the order of complex electrical and power systems, facilitating more efficient and manageable modeling and simulation processes. The findings and assessments obtained from this investigation contribute valuable insights to the application of these model order reduction techniques in practical scenarios.

2 Positive-Real Truncated Balanced Realization Algorithm

The Positive-Real Truncated Balanced Realization (PRR) algorithm is a powerful and widely used technique in the field of model order reduction. This algorithm has gained significant attention and interest due to its effectiveness in reducing the complexity of large-scale systems while preserving key system properties. The PRTBR algorithm combines concepts from balanced truncation and positive realness to achieve its objectives. By leveraging balanced truncation, the algorithm reduces the order of the system while maintaining its stability and dynamics. Additionally, the PRTBR approach further guarantees that the resulting reduced model retains the positive-real property, which is essential for stability and energy conservation [1, 13]. An essential aspect in dynamic systems is the consideration of the positive (passive) real system class. In a physical context, positive realism implies that the system's energy output cannot exceed its energy input. The circuit under study is a component of a positive real dynamics system. For a given asymptotically stable system (1), it can be deemed positive real if the following conditions hold: $m = p, D^{T} + D \geq 0; G^{\sim}(iw) + G(iw) \geq 0 (\forall w \in R)$. . The Positive-Real Balanced Truncation (PRTBR) algorithm is outlined as follows:

Input: An asymptotically stable system and a balanced minimum perception (A, B, C, D) of order n. Specify the desired reduced order, r ($r < n$).

- Step 1: Compute the control Gramian $\mathbf{X_p}$ and the observability Gramian $\mathbf{Y_p}$ using two Riccati Eqs. (1) and (2)

$$\mathbf{A^{T}X_p + X_pA + \left(X_pB - C^{T}\right)\left(D + D^{T}\right)^{-1}\left(B^{T}X_p - C\right) = 0} \tag{1}$$

$$\mathbf{AY_p + Y_pA^{T} + \left(Y_pC^{T} - B\right)\left(D + D^{T}\right)^{-1}\left(CY_p - B^{T}\right) = 0} \tag{2}$$

- Step 2: Perform Cholesky factorization on $\mathbf{X_p}$ and $\mathbf{Y_p}$ to obtain triangular matrices.

$$\mathbf{X_p = R_pR_p^{T}} \tag{3}$$

$$\mathbf{Y_p = L_pL_p^{T}} \tag{4}$$

- Step 3: Conduct Singular Value Decomposition (SVD).

$$\mathbf{L_p^{T}R_p = U_p\Sigma_pV_p^{T}} \tag{5}$$

- Step 4: Calculate the transformation matrices.

$$\mathbf{T_p = R_pV_p\Sigma_p^{-\frac{1}{2}}; T_p^{-1} = \Sigma_p^{-\frac{1}{2}}U_p^{T}L_p^{T}} \tag{6}$$

- Step 5: Apply the balancing transformation.

$$\left[\mathbf{T_p^{-1}AT, T_p^{-1}B, CT_p, D}\right] = \left[\begin{bmatrix}\mathbf{A_{11}} & \mathbf{A_{12}} \\ \mathbf{A_{21}} & \mathbf{A_{22}}\end{bmatrix}, \begin{bmatrix}\mathbf{B_1} \\ \mathbf{B_2}\end{bmatrix}, [\mathbf{C_1} \; \mathbf{C_2}], \mathbf{D}\right] \tag{7}$$

Output: Obtain the reduced asymptotically stable system of order r as: $(\mathbf{A_r}, \mathbf{B_r}, \mathbf{C_r}, \mathbf{D_r}) = (\mathbf{A_{11}}, \mathbf{B_1}, \mathbf{C_1}, \mathbf{D})$ where $\mathbf{A}_r \in R^{r \times r}, \mathbf{B}_r \in R^{r \times m}, \mathbf{C}_r \in R^{p \times r}, \mathbf{D}_r \in R^{p \times m} \mathbf{A}_r \in R^{r \times r}, \mathbf{B}_r \in R^{r \times m}, \mathbf{C}_r \in R^{p \times r}, \mathbf{D}_r \in R^{p \times m}$.

3 Proper Orthogonal Decomposition Algorithm

The Proper Orthogonal Decomposition Algorithm (PODA) is a powerful data-driven technique widely used in various scientific and engineering disciplines. It is a dimensionality reduction method that enables the extraction of dominant spatial and temporal structures from high-dimensional data sets. The main objective of the PODA algorithm is to capture the most significant modes of variation present in the data. It achieves this by performing a singular value decomposition (SVD) on the data matrix, which decomposes the data into a set of orthogonal spatial modes, temporal coefficients, and temporal modes [2, 14]. The spatial modes represent the spatial patterns of variation, while the temporal coefficients and modes capture the temporal evolution of these patterns. In practical applications, the PODA is commonly used for model order reduction, data compression, and system identification. In model order reduction, it enables the construction of reduced-order models that capture the essential dynamics of a system while significantly reducing computational complexity. In data compression, it allows for the compact representation of data, saving storage space and facilitating efficient data processing. In system identification, it aids in identifying dominant system behavior from measured data, leading to improved understanding and control of complex systems. The basic PODA is described as follows:

Input: Snapshots matrix **X** containing $\mathbf{N_t}$ snapshots of the state vector $\mathbf{X}_\mu$, where N_t is the number of time instances, and $\mathbf{N_{state}}$ is the dimension of the state vector.

1 Step 1: Compute the singular value decomposition (SVD) of the snapshots matrix:

$$\mathbf{X} = \mathbf{U}_\mu \Sigma_\mu \mathbf{V}_\mu^T \tag{8}$$

where $\mathbf{U}_\mu$ is the matrix of singular vectors with dimensions
$\mathbf{N_{state}} \times \mathbf{N_t}$, Σ_μ is the diagonal matrix containing singular values.
$\lambda_1 \geq \lambda_2 \geq \ldots \geq \lambda_{N_{\text{state}}} \geq 0$, and $\mathbf{V}_\mu^T$ is the transpose of $\mathbf{V}_\mu$.

1 Step 2: Define $\mathbf{I}(m)$ as the cumulative energy captured by the first m singular values:

$$\mathbf{I}(m) = \sum_{i=1}^{m} \lambda_i \tag{9}$$

- Step 3: Choose the dimension of the PODA basis, k, as the minimum value of m satisfying $I(m) \geq \gamma$, where $0 \leq \gamma \leq 1$ is the specified percentage of total information captured by the reduced space:

$$\mathbf{U}_\mu = \text{span}\left\{u_\mu^1, u_\mu^2, \ldots, u_\mu^k\right\} \tag{10}$$

- Step 4: Calculate the balancing transformation matrix $\mathbf{T_b}$ as shown in (17).

$$\mathbf{T_b} = \mathbf{R_b V_b \Sigma_b^{-\frac{1}{2}}};\ \mathbf{T_b^{-1}} = \mathbf{\Sigma_b^{-\frac{1}{2}} U_b^T L_b^T} \tag{11}$$

- Step 5: Transform the system into a positive real balanced using the expressions in (18)

$$\mathbf{T_b^{-1} A T} = \mathbf{A_b};\ \mathbf{T_b^{-1} B} = \mathbf{B_b};\ \mathbf{C T_b} = \mathbf{C_b} \tag{12}$$

Output: PODA basis matrix $\mathbf{U}_\mu$ containing the dominant spatial modes. The resulting PODA basis matrix $\mathbf{U}_\mu$ contains the dominant spatial modes, enabling efficient dimensionality reduction and data representation for various applications, such as model order reduction and data compression.

4 Comparative Analysis of PRR and PODA Methods for Model Order Reduction

Considering an 8th-order electrical circuit, described by the transfer function:

$$G(s) = \frac{0.01s^8 + 2.2s^7 + 33.32s^6 + 143.9s^5 + 246.6s^4 + 287.8s^3 + 133.3s^2 + 17.6s + 0.16}{s^8 + 20s^7 + 132s^6 + 392s^5 + 664s^4 + 784s^3 + 528s^2 + 160s + 16}$$

We implement two algorithms, PRR and PODA, in MATLAB. Subsequently, we apply each method to reduce the order of the original system to 4th order, we obtain the reduced-order transfer functions corresponding to PRR as GrPP(s) and to PODA as GrPODA(s), as follows:

$$GrPRR(s) = \frac{0.1414s^4 - 0.3613s^3 - 17.87s^2 - 86.65s + 73.46}{s^4 - 14.78s^3 + 1.801s^2 + 36.15s + 161.2}$$

$$GrPODA(s) = \frac{0.1414s^4 + 2.806s^3 + 4.563s^2 + 5.578s + 0.5576}{s^4 + 5.701s^3 + 8.334s^2 + 11.37s + 3.937}$$

The resulting step responses of the reduced systems are plotted against the original system, as shown in Fig. 1.

From Fig. 1, it is evident that reducing the system to 4th order using the PRR method results in significant discrepancies in the step response compared to the original system. Conversely, applying the PODA method to reduce the original 8th-order system to 4th order yields a step response that closely matches the original system's response throughout the time domain. Hence, the PODA technique can effectively replace the original 8th-order system with the reduced 4th-order system for applications within the time domain.

The comparative analysis indicates that while PRR can lead to substantial deviations in the reduced system's step response, PODA maintains the accuracy of the response across the time domain when reducing the system order. This underscores the suitability of the PODA technique for order reduction tasks in time-domain applications.

Table 1 presents a comprehensive comparison between the PRR and PODA methods in terms of their norm error values at different reduced dimensions. The norm errors are calculated to evaluate the accuracy of the reduced models obtained through these methods. The reduced dimension (r) indicates the number of retained modes or components in the reduced models. The norm errors demonstrate how closely the reduced models capture the dynamics of the original 8th-order system. The norm errors for both methods decrease as the reduced dimension increases. Notably, the PODA method consistently exhibits lower norm errors across a range of reduced dimensions, indicating its superior accuracy in capturing essential system behavior during model order reduction.

In this comparison between the PRR and PODA methods for model order reduction, several key observations can be made based on the calculated errors for different reduced dimensions (r):

- Effectiveness of Model Reduction: When analyzing the performance of both PRR and PODA, it's evident that both methods successfully reduce the original 8th-order system to lower dimensions. The reduction is achieved by retaining only a subset of the most significant modes of the system.

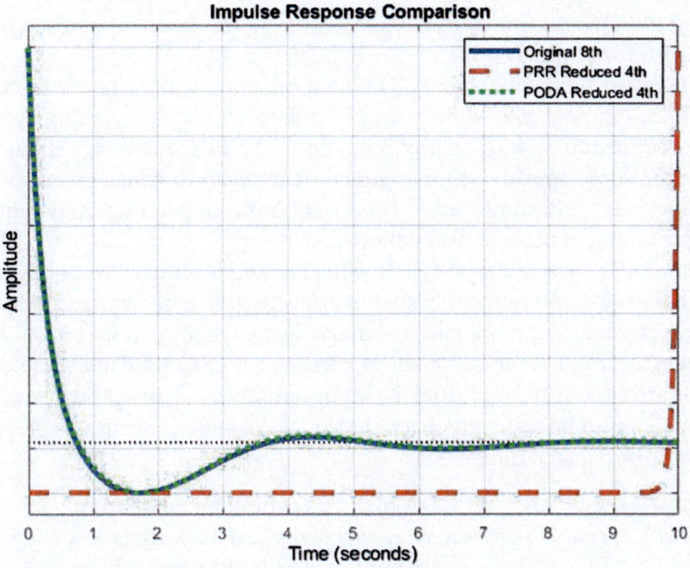

Fig. 1. Impulse plot between base system and reduction order using PRR and PODA

Table 1 Comparison of norm errors for PRR and PODA

Reduced dimension (r)	PRR norm error	PODA norm error
1	0.6836	0.6383
2	0.8011	0.2154
3	1.0635	0.1802
4	1.0725	0.1389
5	0.5466	0.1343
6	0.8758	0.1316
7	0.4220	0.1316

- Norm Errors: The norm errors, as quantified by the infinity norm, are used to measure the discrepancy between the original system and the reduced models obtained from both PRR and PODA methods. These errors reveal the extent to which each reduced model deviates from the original dynamics. It's notable that for most values of r, the POD method tends to yield lower norm errors compared to the PRRmethod.
- Comparison of Norm Errors: For $r = 1$ to 3: Both PRR and PODA exhibit increasing norm errors as the reduced dimension (r) grows. This could indicate that reducing the model too aggressively leads to a loss of accuracy in capturing the original system's dynamics. For $r = 4$ to 7: Both PRR and PODA show some fluctuations in the norm errors, but overall, PODA maintains lower errors compared to PRR. This could

suggest that PODA is more effective in capturing the essential characteristics of the system with fewer dimensions.
- Robustness and Accuracy: The PODA method demonstrates better accuracy across most reduced dimensions, showcasing its robustness in maintaining system behavior while achieving dimensionality reduction. The PRR method, although effective, seems to be more sensitive to the choice of reduced dimensions, as evidenced by higher norm errors in some cases. This could indicate that PRR may require careful parameter tuning to achieve optimal results.
- Advantages and Disadvantages: PRR: This method leverages the concepts of proper orthogonal modes and rational basis functions to derive the reduced model. While it is capable of capturing dominant dynamics, it may require more careful handling of parameters to ensure accurate results. PODA: As a widely-used method, PODA's basis functions are directly derived from the input-output data. It has shown its effectiveness in achieving accurate results with a reduced model and demonstrates greater stability across various reduced dimensions.

In summary, both PRR and PODA methods exhibit potential for model order reduction, but PODA appears to provide better accuracy and stability over a range of reduced dimensions. The choice between these methods should consider factors such as accuracy requirements, computational efficiency, and robustness in handling different system dynamics.

5 Conclusion

In the realm of electrical and electronic engineering, the challenges posed by complex systems with a high number of components necessitate efficient solutions for analysis, design, and simulation. Model Order Reduction (MOR) emerges as a powerful technique to address these challenges by creating lower-dimensional models that retain essential system behavior. This study delved into two prominent MOR algorithms, Positive-Real Truncated Balanced Realization (PRR) and Proper Orthogonal Decomposition (PODA), and examined their applicability in the context of an 8th-order electrical circuit.

The comparative assessment revealed distinct traits of PRR and PODA when applied to model order reduction. PRR, founded on the principles of balanced truncation and positive realness, exhibited sensitivity to the choice of reduced dimensions, resulting in varying norm errors. On the other hand, PODA, rooted in data-driven methods, consistently showcased superior accuracy across different reduced dimensions. The reduction of the 8th-order system to a 4th-order system through PRR led to significant discrepancies in the step response compared to the original system. In contrast, PODA exhibited the capability to closely match the original system's response, highlighting its efficacy in maintaining accuracy during order reduction.

The findings underscore the robustness and versatility of PODA in handling complex systems across a spectrum of reduced dimensions. This advantage positions PODA as a dependable method for achieving accurate model order reduction in diverse applications, particularly within the time-domain domain. While both PRR and PODA demonstrated potential for model order reduction, the results demonstrate the suitability of PODA for tasks that demand accuracy and stability across various system dynamics.

In conclusion, this research contributes valuable insights into the effectiveness of PRR and PODA in model order reduction for electrical and electronic circuits. It presents a comprehensive comparative analysis that aids practitioners in selecting the appropriate method based on specific application requirements, computational efficiency, and accuracy considerations. The study reinforces the importance of advanced MOR techniques in addressing complexity challenges and paves the way for further advancements in the field of model order reduction.

Acknowledgments. This research result is the product of a university-level scientific research project with code ĐH2023-TN07-01, funded by the Thai Nguyen University of Information and Communication Technology (ICTU).

References

1. Salehi, Z., Karimaghaee, P., Khooban, M.H.: Model order reduction of positive real systems based on mixed Gramian balanced truncation with error bounds. Circ. Syst. Sig. Process. **40**(12), 5309–5327 (2021)
2. Block, B., Chen, X., Stockar, S.: Stabilization of a POD/Galerkin reduced order Payne-Whitham traffic model. In: 2023 American Control Conference (ACC), San Diego, CA, USA, pp. 4443–4448 (2023)
3. Kumar, D., et al.: Positive-real truncated balanced realization based frequency-weighted model reduction. In: 2019 Australian & New Zealand Control Conference (ANZCC), Auckland, New Zealand, pp. 145–147 (2019)
4. Zulfiqar, U., Imran, M., Ghafoor, A., Liaqat, M.: Time/frequency-limited positive-real truncated balanced realizations. IMA J. Math. Control Inform. **37**(1), 64–81 (2020)
5. Salehi, Z., Karimaghaee, P., Khooban, M.-H.: Mixed positive-bounded balanced truncation. IEEE Trans. Circ. Syst. II: Express Briefs **68**(7), 2488–2492 (2021)
6. Salehi, Z., Karimaghaee, P., Khooban, M.-H.: A new passivity preserving model order reduction method: conic positive real balanced truncation method. IEEE Trans. Syst. Man, Cybern. Syst. **52**(5), 2945–2953 (2022)
7. Akram, N., Alam, M., Hussain, R., Massoud, Y.: Statistically inspired passivity preserving model order reduction. IEEE Access **11**, 52226–52235 (2023)
8. Nagaraj, S., Seshachalam, D., Hucharaddi, S.: Model order reduction of nonlinear circuit using proper orthogonal decomposition and nonlinear autoregressive with exogenous input (NARX) neural network. In: 2018 16th ACM/IEEE International Conference on Formal Methods and Models for System Design (MEMOCODE), Beijing, China, pp. 1–4 (2018)
9. Pierquin, A., Henneron, T.: Nonlinear data-driven model order reduction applied to circuit-field magnetic problem. IEEE Trans. Magn. **57**(11), 1–9 (2021)
10. Torchio, R., Zanco, A., Lucchini, F., Alotto, P., Grivet-Talocia, S.: Mixed proper orthogonal decomposition with harmonic approximation for parameterized order reduction of electromagnetic models. In: 2022 International Symposium on Electromagnetic Compatibility—EMC Europe, pp. 349–354. Gothenburg, Sweden (2022)
11. Delagnes, T., Henneron, T., Clenet, S., Fratila, M.: Development of a FE reduced model on a large operating range for a squirrel cage induction machine in non-linear case. In: 2022 IEEE 20th Biennial Conference on Electromagnetic Field Computation (CEFC), Denver, CO, USA, pp. 1–2 (2022)

12. Fasolato, S., Allam, A., Li, X., Lee, D., Ko, J., Onori, S.: Reduced-order model of lithium-iron phosphate battery dynamics: a POD-Galerkin approach. IEEE Control Syst. Lett. **7**, 1117–1122 (2023)
13. Guiver, C., Opmeer, M.R.: A counterexample to "positive realness preserving model reduction with $\mathscr{H}_\infty$ norm error bounds." IEEE Trans. Circ. Syst. I: Reg. Papers **58**(6), 1410–1411 (2011)
14. Yang, Y., Shang, W., Yang, N., Wang, E., Xu, X.: Surrogate model based on POD-kriging for fast analysis of electrical devices. In: 2023 IEEE 6th International Electrical and Energy Conference (CIEEC), Hefei, China, pp. 3190–3194 (2023)

Data-Driven Narratives: Unleashing the Potential of R for Journalistic Storytelling

Tran Quang Quy[1], Nguyen Duc-Binh[1], and Pham Chien Thang[2](✉)

[1] Thai Nguyen University of Information and Communication Technology, Thai Nguyen 24000, Vietnam
{tqquy,ndbinh}@ictu.edu.vn
[2] TNU-University of Sciences, Thai Nguyen 24000, Vietnam
thangpc@tnus.edu.vn

Abstract. The digital age has seen a significant rise in data journalism and the importance of data-driven narratives. However, the creation of these narratives is often constrained by insufficient data models and the need for efficient tools to facilitate the process. This paper explores the potential of R, a programming language and free software environment for statistical computing and graphics, in enhancing journalistic storytelling. R offers a wide range of packages that can be particularly useful for journalists, including ggplot2 for data visualization, dplyr for data manipulation, tidytext for text analysis, and sf for spatial data analysis. These packages facilitate data cleaning, analysis, and visualization, crucial for data-driven storytelling. However, using R in journalistic contexts is challenging, underscoring the need for further research and development. This paper concludes that R holds significant potential for enhancing journalistic storytelling, but more work needs to be done to realize this potential and overcome the existing challenges fully.

Keywords: Data Journalism · R Programming · Computational Journalism

1 Introduction

With the dawn of the digital era, journalism has undergone a profound metamorphosis, leading to the emergence of data journalism. This form of journalism leverages the wealth of data available in the digital world to extract knowledge or useful insights for smart decision-making in various application domains, including journalistic storytelling [1]. Data journalism distinguishes itself by applying data science and sophisticated analytical techniques, such as machine learning models, to extract actionable insights or a more profound understanding of data. This approach automates the computational process and enhances its intelligence, thereby augmenting the capabilities of journalistic applications through informed decision-making in diverse situations [1].

Based on data analysis and interpretation, these narratives provide a richer and more complex view of events, extending the realm of automated journalism from simple

Nguyen Duc-Binh these authors contributed equally to this work as the first author.

P. T. Nghia et al. (Eds.): ICTA 2023, LNNS 847, pp. 245–252, 2023.
https://doi.org/10.1007/978-3-031-49529-8_26

descriptions to event-driven narratives [2]. A significant limitation hampering the potential for journalistic writing automation is the lack of data models that can adequately encapsulate the knowledge required for automatically crafting event-based narratives. Mitigating this limitation necessitates depicting journalistic knowledge as structured events and narrative data [2].

In the current age of the Fourth Industrial Revolution (Industry 4.0 or 4IR), more and more industries are capitalizing on digital innovation for sustainable business growth. This includes the journalism industry, which is increasingly using analytics-based data-driven innovation (DDI) for product conceptualization, data acquisition, data refinement, data storage and retrieval, distribution, presentation, and market feedback [3].

In summary, the rise of data journalism and the importance of data-driven narratives in today's digital age are closely tied to the broader trends of digital transformation and the increasing use of data science and advanced analytics in various fields.

Creating data-driven narratives in journalism is not without its challenges. A noteworthy impediment in automated journalism is the deficiency of data models that can adequately encode the necessary journalistic knowledge to generate event-driven stories automatically [2]. This highlights the need for efficient tools to facilitate the creation of data-driven narratives.

Another challenge is the variability of publicly available data pools. The volume, velocity, and variety of data in the digital age present significant computational challenges, making it difficult to generate new predictive models for data-driven narratives [4]. This underscores the need for advanced artificial intelligence approaches that can handle the complexity of big data and generate reliable and accurate narratives.

Furthermore, automated journalism remains somewhat rare despite the escalating volumes of data and continuous advancements in natural language generation techniques. Developing a journalistic natural language generation system poses several challenges, including transparency, accuracy, modifiability, and transferability [5]. In light of these challenges, there is a pressing need for efficient tools and methodologies to facilitate the creation of data-driven narratives in journalism. Developing such tools would address the current challenges and significantly enhance the potential of data journalism.

This study investigates the capability of R—a freely available software environment and programming language tailored for statistical computing and graphic representation, to augment journalistic storytelling. This exploration is motivated by the increasing prominence of data storytelling in digital journalism and the need for more efficient tools to facilitate the creation of data-driven narratives [6]. Data journalism has become an integral facet of digital journalism, leveraging data science and avant-garde analytical methods to yield actionable insights or deepen understanding of data. However, despite its potential to enrich debates in the public sphere and hold data organizations accountable, its adoption outside large news organizations remains low [7].

Moreover, despite excellent data storytelling practices, a methodical approach is needed to fully understand the traits of successful data stories and their creation process [8]. This study aspires to contribute to this initiative by delving into the potential of R in bolstering journalistic storytelling. In journalism austerity, it is also important to consider the sustainability of data journalism. The future of data journalism could take several paths, including becoming an essential skillset for every professional journalist, remaining a niche storytelling format, or being abandoned by the mainstream media due to financial and personnel limitations [9]. This paper explores how the use of R can contribute to the sustainability of data journalism.

2 Literature Review

2.1 Data Journalism: Review Existing Literature on Data Journalism and Data-Driven Narratives

Data journalism, distinguished by its utilization of data science and cutting-edge analytics to yield actionable insights and deepen data understanding, has emerged as a defining feature of digital journalism. Yet, despite its potential to invigorate public discourse and hold data-centric organizations accountable, its uptake outside of major news corporations remains limited [2].

Automated journalism, a facet of data journalism, is marked by a significant limitation—the lack of adequate data models to encapsulate the journalistic knowledge required for the automated creation of event-driven narratives. This underlines the need to express journalistic knowledge through structured events and narrative data [2]. Although data availability and the sophistication of natural language generation techniques are rising, advancements in automated journalism are still relatively few. Establishing a journalistic natural language generation system presents multiple challenges, encompassing transparency, accuracy, modifiability, and transferability [5].

In underserved communities, data journalism has served to amplify their voices. Despite the inherent obstacles, four key factors typically support these organizations in creating data stories: citizen involvement, activism, collaboration, and the humanization of data [10]. Lastly, deep journalism, a data-oriented, deep learning-based methodology, has been proposed to uncover and analyze cross-sectional, multi-perspective information. This approach seeks to facilitate superior decision-making and forge enhanced tools for academic, corporate, national, and international governance [11].

2.2 R in Journalism: Discuss Previous Studies That Have Used R in Journalistic Contexts

While the use of R in journalistic contexts is not explicitly discussed in the literature, the application of computational methods and data-driven approaches, which R is well-suited for, is a recurring theme in studies on data journalism. For instance, API-based research, a computational social science and digital sociology methodology reliant on retrieving records from datasets offered by online platforms via their application programming interfaces, has been recognized as a crucial approach in digital journalism [12]. With its extensive package ecosystem, R is particularly well-suited for API-based research.

In the context of immersive storytelling techniques, such as Virtual Reality (VR), the use of computational methods to examine these technologies has been proposed. Given R's statistical analysis and data visualization capabilities, it can be valuable in such examinations [13].

Additionally, the routine practice of data journalism has been observed to chiefly involve relatively straightforward data visualizations exhibiting minimal interactivity, predominantly depending on institutional sources and providing little original data gathering [14]. R, with its powerful data manipulation and visualization capabilities, can be used to enhance these aspects of data journalism.

Lastly, the roles and practices of ancillary participants in data journalism, including non-profit organizations and civic technologists, have been scrutinized in both European and African contexts. These actors often engage in data-driven practices, for which R can be useful [15].

3 The Potential of R for Data-Driven Journalism

3.1 R and Data Analysis: Discuss How R Can Be Used for Data Cleaning, Analysis, and Visualization, Which Are Crucial for Data-Driven Storytelling

R, a complimentary software environment and programming language developed for statistical computation and graphic representation is well-adapted for data cleaning, analysis, and visualization—all pivotal for data-oriented storytelling. For instance, in the context of omics data, R has been used to extract meaningful relationships across different omics layers. The xMWAS software, crafted in R, enables data integration, network visualization, clustering, and differential network analysis of data derived from biochemical and phenotypic assays and two or more omics platforms [16].

R has also been used to reduce the complexity of data provenance graphs using hierarchical and motif-based aggregation. This adaptive visualization approach, tailored to user interest, permits users to examine and articulate intricate multi-step analyses, potentially based on hundreds of files processed via numerous workflows [17].

R offers a wide range of packages that can be particularly useful for journalists, including ggplot2 for data visualization, dplyr for data manipulation, tidytext for text analysis, and sf for spatial data analysis. These packages facilitate data cleaning, analysis, and visualization, which are crucial for data-driven storytelling [18, 19].

3.2 Case Study

The dataset used in this article is the coronavirus dataset in the R package [20]. This open-source dataset was collected and updated by Rami Krispin from the John Hopkins University data source. This dataset provides information on the number of cases, deaths, and recoveries related to the COVID-19 pandemic in countries and regions worldwide. The coronavirus package provides a tidy format for the COVID-19 dataset amassed by the Center for Systems Science and Engineering (CSSE) at Johns Hopkins University. This dataset incorporates daily new cases, deaths from January 2020 to January 2023, and recovery cases until August 2022. The dataset is very useful for people in the medical field, people interested in the worldwide Covid-19 situation, and people working in data analysis and forecasting.

Before conducting time series modeling and predicting the number of Covid-19 cases in Vietnam, we need to collect data to get an overview of the epidemic. This helps us better understand the scale and development trends of the pandemic. This information is an important basis for us to continue to monitor, evaluate and predict the rise of Covid-19. Understanding the trends and fluctuations of the epidemic will help us develop more effective prevention and management measures in epidemic control and public health protection. After we have collected and preprocessed the data, we can visualize the distribution of the number of Covid-19 cases globally. First, we can use a line chart to visualize the change in the number of Covid-19 cases over time using the Plotly library. The line chart helps us to see the increasing and decreasing trend of the number of cases in each period and get an overview of the epidemic situation (see Fig. 1).

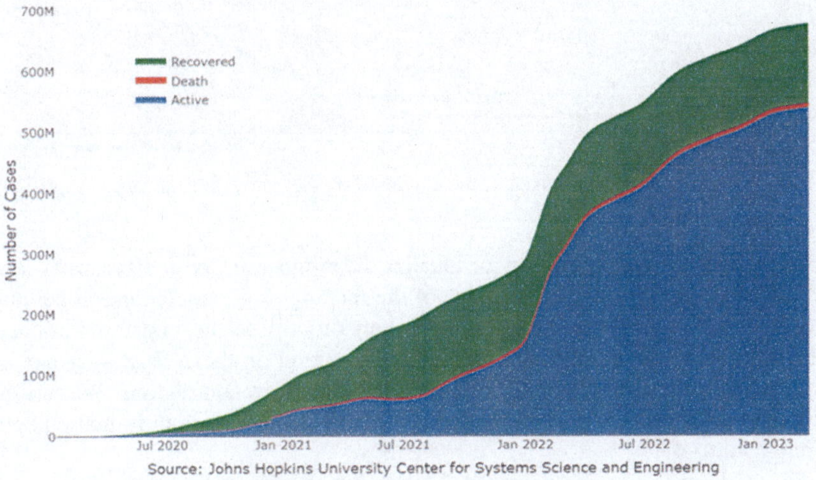

Fig. 1. Cumulative chart from June 2020 to January 2023.

To visually describe the covid case in Vietnam, in this article we use Covid 19 data for Vietnam in the period from January 23 to April 4, 2020, this is the period The most typical example of the Covid epidemic in Vietnam. In Vietnam, as of the end of March, the epidemic was divided into three stages. **Phase 1** from January 23 to February 13, 16 cases of infection were recorded and completely cured. **Phase 2** from March 6 to March 21, with 78 cases testing positive for the virus and the number of close contacts of sick people who had to be closely monitored reached hundreds of people. **Phase 3** began after the number of infected cases exceeded 100 on March 22. The number of 100 cases is an important number, because after this period, controlling the epidemic becomes more difficult. To perform data visualization, we first preprocessed COVID 19 data from the John Hopkins University source, next the data was extracted and filtered by the keyword "Vietnam". Using data dimensional transformation and ggplot2 data visualization library, we obtain data visualization as shown in Fig. 2.

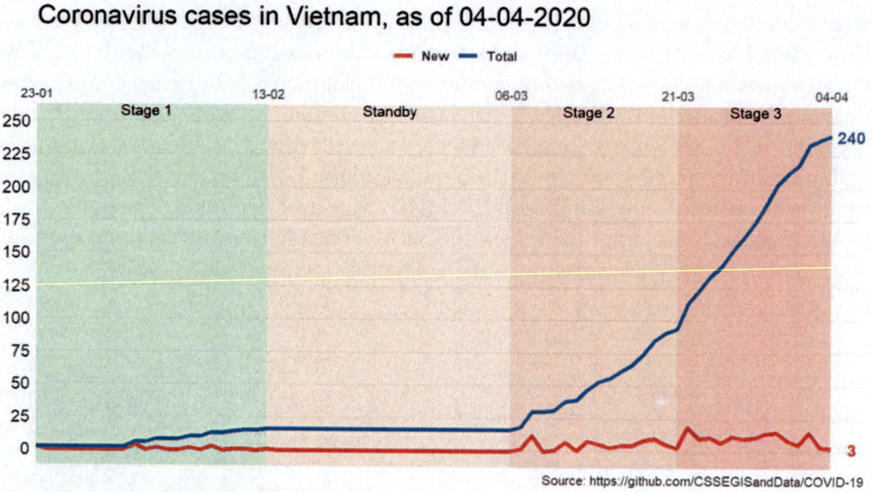

Fig. 2. Chart depicting the stages of the Covid 19 in Vietnam from January to April 2020

Next, we have a Treemap, used to visualize the number of Covid-19 cases by country. The Treemap chart will display the size of the rectangles corresponding to the number of cases in each country (see Fig. 3). This helps us to see the impact of disease and the distribution of the number of cases globally. Based on these charts, we can assess the spread and impact of Covid-19 on individual countries and regions. We can see the countries in the most severe epidemic situation and also compare the situation between different countries.

Fig. 3. Treemap chart by countries

4 Conclusion

The rise of data journalism and the increasing importance of data-driven narratives in today's digital age has necessitated the development of efficient tools to facilitate the creation of these narratives. This paper has explored the potential of R, a programming language and free software environment for statistical computing and graphics, in enhancing journalistic storytelling.

Nevertheless, applying R in journalistic settings does not come without its difficulties. For example, automated journalism, a branch of data journalism, is hindered by the insufficiency of data models capable of automatically encoding the necessary journalistic knowledge to generate event-driven narratives. This fact accentuates the demand for additional research and development in this field.

In conclusion, R holds significant potential for enhancing journalistic storytelling. By leveraging the capabilities of R and its extensive package ecosystem, journalists can create more compelling, insightful, and interactive data-driven narratives. However, more work must be done to realize this potential and fully overcome the existing challenges.

References

1. Sarker, I.H.: Data science and analytics: an overview from data-driven smart computing, decision-making and applications perspective. SN Comput. Sci. **2**(5), 377 (2021)
2. Caswell, D., Dörr, K.: Automated journalism 2.0: event-driven narratives: from simple descriptions to real stories. Journal. Pract. **12**(4), 477–496 (2018)
3. Sultana, S., Akter, S., Kyriazis, E., Wamba, S.F.: Architecting and developing big data-driven innovation (DDI) in the digital economy. J Glob. Inform. Manage. (JGIM) **29**(3), 165–187 (2021)
4. Ciallella, H.L., Zhu, H.: Advancing computational toxicology in the big data era by artificial intelligence: data-driven and mechanism-driven modeling for chemical toxicity. Chem. Res. Toxicol. **32**(4), 536–547 (2019)
5. Leppänen, L., Munezero, M., Granroth-Wilding, M., Toivonen, H.: Data-driven news generation for automated journalism. In: Proceedings of the 10th International Conference on Natural Language Generation (pp. 188–197) (2017)
6. Borges-Rey, E.: Unravelling data journalism: a study of data journalism practice in British newsrooms. Journal. Pract. **10**(7), 833–843 (2016)
7. Cushion, S., Lewis, J., Callaghan, R.: Data journalism, impartiality and statistical claims: towards more independent scrutiny in news reporting. Journal. Pract. **11**(10), 1198–1215 (2017)
8. Ojo, A., Heravi, B.: Patterns in award winning data storytelling: story types, enabling tools and competences. Digit. Journal. **6**(6), 693–718 (2018)
9. Stalph, F., Borges-Rey, E.: Data journalism sustainability: an outlook on the future of data-driven reporting. Digit. Journal. **6**(8), 1078–1089 (2018)
10. de-Lima-Santos, M.F., Mesquita, L.: Data journalism in favela: made by, for, and about forgotten and marginalized communities. Journal. Pract. **17**(1), 108–126 (2023)
11. Ahmad, I., Alqurashi, F., Abozinadah, E., Mehmood, R.: Deep journalism and DeepJournal V1. 0: a data-driven deep learning approach to discover parameters for transportation. Sustainability **14**(9), 5711 (2022)
12. Venturini, T., Rogers, R.: "API-based research" or how can digital sociology and journalism studies learn from the Facebook and Cambridge Analytica data breach. Digit. Journal. **7**(4), 532–540 (2019)

13. Mabrook, R., Singer, J.B.: Virtual reality, 360 video, and journalism studies: conceptual approaches to immersive technologies. Journal. Stud. **20**(14), 2096–2112 (2019)
14. Zamith, R.: Transparency, interactivity, diversity, and information provenance in everyday data journalism. Digit. Journal. **7**(4), 470–489 (2019)
15. Cheruiyot, D., Baack, S., Ferrer-Conill, R.: Data journalism beyond legacy media: the case of African and European civic technology organizations. Digit. Journal. **7**(9), 1215–1229 (2019)
16. Uppal, K., Ma, C., Go, Y.M., Jones, D.P.: XMWAS: a data-driven integration and differential network analysis tool. Bioinformatics **34**(4), 701–702 (2018)
17. Gratzl, S., Lex, A., Gehlenborg, N., Cosgrove, N., Streit, M.: From visual exploration to storytelling and back again. In: Computer Graphics Forum, Vol. 35, No. 3, pp. 491–500 (2016)
18. Pebesma, E.J.: Simple features for R: standardized support for spatial vector data. R J. **10**(1), 439 (2018)
19. Mangiola, S., Doyle, M.A., Papenfuss, A.T.: Interfacing Seurat with the R tidy universe. Bioinformatics **37**(22), 4100–4107 (2021)
20. https://github.com/CSSEGISandData/COVID-19

Dynamic Analysis of Capsubot Model in Liquid Environment by Numerical Method

Quoc-Huy Ngo[1](✉), Ky-Thanh Ho[1], and Van-Dam Vu[2]

[1] Thai Nguyen University of Technology, Thai Nguyen, Vietnam
ngoquochuy24@gmail.com

[2] University of Information and Communication Technology-ICTU, Thai Nguyen, Vietnam

Abstract. Vibration-driven locomotion system that exploit the vibration-impact effect has strong nonlinearity, which should be evaluated for behavior in fluid environments. In this paper, the dynamic response of new physical and mathematical model of vibro-impact capsubot in liquids was investigated by numerical simulation tool of the XPP Auto software. The results show that square-pulse excitation signal form can make the capsubot moving faster than the full sine and half sine excitations. Using square-pulse excitation for analyzing dynamic response of this system, the results also show that in the liquid environment, the vibro-impact capsubot is completely capable of moving forward and backward by changing frequency and intensity of excitation current. These results are able to use for the design, building a verified experimental capsubot. This result can also be used to analyze the stability of the system as well as further analysis such as phase portrait, Poincaré map, and bifurcation diagram.

Keywords: Capsule robot · Vibro-impact · Excitation · Numerical

1 Introduction

Using the capsule endoscope, also called capsule robot or capsubot, in medical applications is limited by its passive movement resulting from the natural peristalsis of the gastrointestinal tract [1, 2]. Therefore, about the last 2 decades, many scholars have proposed different driving mechanisms to overcome capsubot passivity, for example: using an external magnetic field [2–4], using an external drive system (legs, grips, wheels, propeller blades) [2, 4]. Although currently being used quite commonly in practice, the use of an external magnetic field causes significant harm to the patient's health [5, 6]. While the external drive system can damages the patient's gastrointestinal tract [7]. Some scholars proposed vibro-impact mechanism to create force interaction between the capsubot body and the environment [8–12].

The autonomous systems based on vibro-impact mechanism have being studied in many different aspects, such as modeling, dynamic analysis and control, experimental design and verification. However, those systems have some disadvantages, such as: strong nonlinear behavior, very sensitive to structural and operating parameters and

P. T. Nghia et al. (Eds.): ICTA 2023, LNNS 847, pp. 253–260, 2023.
https://doi.org/10.1007/978-3-031-49529-8_27

noise. So those need to be designed and selected very carefully, based on results of dynamic analysis of the system [9, 10, 13].

In the vibro-impact capsubot, the resistance plays an important role; the dynamic behavior of the system depends strongly on the friction model and on the ratio of the excitation force to the friction force. Many friction models have been mentioned by previous authors, such as Coulomb, Coulomb-Stribeck, viscous... [8, 9, 13, 14]. Most of the previous models have not yet reflected the reality of the influence of viscous drag force on capsubot in liquid environments, including the shape size, velocity of the capsubot and viscosity of fluids.

This paper presents a new mathematical model that adds the influencing factor of the viscous drag force acting on the capsubot when moving in liquid environments. The numerical method is used to determine solutions and analysis the affection of parameters for design and control of an active capsubot.

2 Modelling of Capsule Robot in the Liquid Environment

2.1 Physical and Mathematical Models

Vibro-impact capsubot moving in fluid is described as a two degrees-of-freedom, as shown in Fig. 1a. The difference in this new model is subjected to the F_D viscous drag component when the capsubot moves in a liquid environment. The inertial mass m_1, excited by the electromagnetic force F_m, moves back-forth and impact with the capsubot body m_2. When the impact force is large enough to overcome the friction force F_f and the drag force F_D, the capsubot will move. A linear spring k and a viscous damper c are used to connect between two masses m_1 and m_2, while the friction between them is assumed to be zero. X_1 and X_2 represent the absolute displacements of the internal mass and the capsule, respectively. In the initial position when m_1 is not excited, the distance between m_1 and m_2 is denoted G. The internal mass m_1 will impact the capsubot body when the relative displacement X_1-X_2 is larger or equals the gap G. The stiffness at the collision site between two masses m_1 and m_2 is characterized by a linear spring k_0.

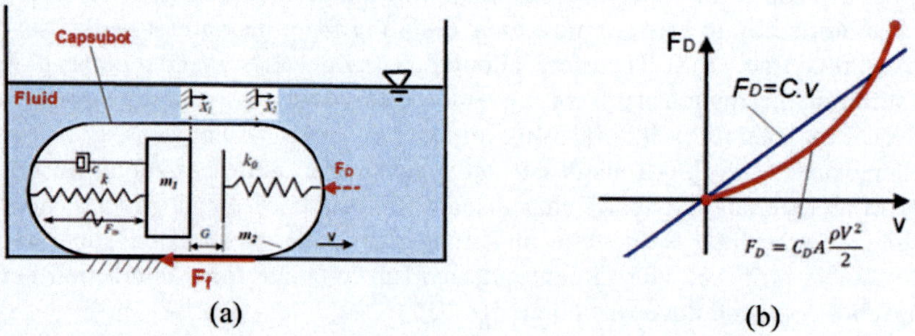

Fig. 1. (a) The vibro-impact capsubot model moving in fluid developed from [9, 13]; (b) model of drag resistance in fluid

The mathematical model of capsule robot in fluid, as illustrated in Fig. 1a, is represented in the system of Eq. (1) below.

$$\begin{cases} m_1 \dfrac{d^2X_1}{dt^2} = F_m - k(X_1 - X_2) - c\left(\dfrac{dX_1}{dt} - \dfrac{dX_2}{dt}\right) - H\,[k_0(X_1 - X_2 - G)] \\ m_2 \dfrac{d^2X_2}{dt^2} = -F_m + k(X_1 - X_2) + c\left(\dfrac{dX_1}{dt} - \dfrac{dX_2}{dt}\right) + H\,[k_0(X_1 - X_2 - G)] - F_f - F_D \end{cases} \quad (1)$$

The H is Heaviside function defined as:

$$H = \begin{cases} 1, & \text{if}(X_1 - X_2 - G) > 0 \\ 0, & \text{if}\,(X_1 - X_2 - G) \leq 0 \end{cases} \quad (2)$$

The frictional force F_f, based on Coulomb friction model, describes in Eq. (3):

$$F_f = \mu.g.(m_1 + m_2).sgn(dX_2/dt) \quad (3)$$

The excitation force F_m depended on the exciting signals includes full sine function, square-pulse function and half sine function, as shown in Eqs. (4a), (4b) and (4c), respectively.

$$F_m = B.L.i.sin(2\pi ft) \quad (4a)$$

$$F_m = B.L.i.\,heav\,(sin\,(2\pi ft)) \quad (4b)$$

$$F_m = -B.L.i.heav\,(sin\,(2\pi ft)).\,sin(2\pi ft) \quad (4c)$$

where: g is the gravitational acceleration (m/s^2); t is time (s); other parameters are described in Table 1.

The drag force F_D will be assessed to affect the ability to move when changing the shape, size, viscosity of the fluid and the movement velocity of the capsubot V, as depicted in Fig. 1b and Eq. (5):

$$F_D = C_D.A.\frac{\rho V^2}{2} \quad (5)$$

2.2 Determine Numerical Solutions and Dynamic Analysis

In this paper, the interaction dynamic responses of the capsule models in liquids are first analysed to reveal the effect of parameters by numerical method with XPPAuto software. Firstly, the initial parameters were selected, as depicted in Table 1. Secondly, the programmed algorithm assigned initial parameters and performed numerical solutions for the variables of displacement and velocity of the capsubot. Thirdly, the returned result data were the actual numeric values. Finally, Origin Lab software was used for presenting the results. The dynamic response was analyzed through time plot when changing structural and operating parameters. This is an important basis for analyzing phase portrait, Poincaré map, bifurcation diagram and the implementation of future validation experiments.

Table 1. Initial settings for mathematical equations

Parameters	Notation	Unit	Value	Note
Internal mass	m_1	*kg*	0.023	Structure
Capsubot body mass	m_2	*kg*	0.043	Structure
Spring stiffness	k	*Nm*	150	Structure
Impact stiffness	k_0	*Nm*	1.24E6	Structure
Exciting force coefficient	BL	–	3	Structure
Damping coefficient	c	*Ns/m*	15	Structure
Exciting current	i	*A*	0.1−0.8	Operation
Impact gap	G	*mm*	0−3	Structure
Coulomb coefficient of friction	μ	–	0.01−0.4	Operation
Viscous drag coefficient	C_D	–	0.2−30	Operation
Excitation frequency	f	*Hz*	10−40	Operation
Radius of capsubot body	R	*m*	0.005–0.03	Structure

3 Results and Discussion

3.1 Effect of Excitation Signal Form and Frequency

Investigating the influence of signal form of excitation force, numerical simulation was used with unchanged parameters: $i = 0.35$A; $C_D = 5$; $\mu = 0.05$; $R = 0.01$m; $G =$ 0.1mm. By changing the excitation frequency $f = [10-40]$ Hz, the scanning function of XPPAuto software was performed to evaluate the displacement trend of the capsubot after 5s with corresponding excitation force forms: full sine, half sine and square-pulse, as illustrated in Fig. 2a. While the displacement of the capsubot caused by the three exciting forms at 20 Hz after 5s is shown Fig. 2b. The results show that, the excitation force F_m as square-pulse will help the capsubot move at the fastest speed or travel distance is longer than the sine and half sine excitations. This is the basis of the premise of choosing the square-pulse excitation form for the robot model. Square-pulse exciting form also was used for the next analysis.

The value of the F_D resistance force component acting on the capsubot during each excitation cycle depends on of excitation signal form illustrated in Fig. 3. With the square pulse excitation signal, the F_D resistance force has a damped form. With the sine excitation signal form, the resistance value F_D is larger than the other cases. This is the reason why the displacement of the capsubot after 5s will be smaller than in the case of square and half-sine pulse excitation as shown in Fig. 2b.

3.2 Effect of the Coefficient of Friction

It is noted that the main work environment of capsubot is the gastric liquid and small intestine. According to many studies [15, 16], the values of the coefficient friction μ

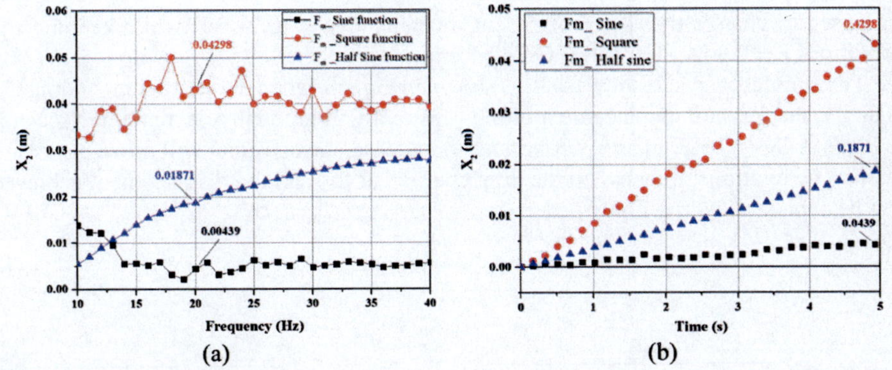

Fig. 2. Displacement of capsubot: (a) excitation frequency f = [10−40] Hz and (b) excitation frequency f = 20 Hz after 5s

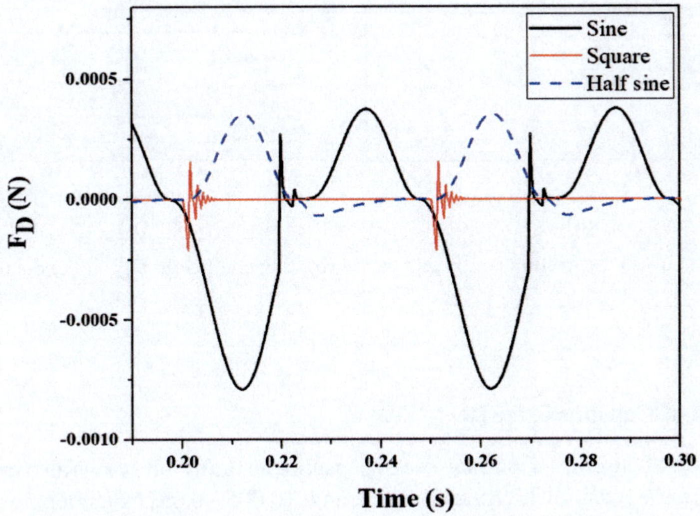

Fig. 3. The value of drag force F_D with the excitation signal forms

(muy) are changed in the range of [0.01−0.4]. This parameter was used to evaluate the displacement of capsubot X_2 with initial conditions: $i = 0.35$A; $f = 20$Hz; $G = 0.1$mm; $C_D = 5$. Simulation result is shown in Fig. 4a. The result shows that, the higher the friction coefficient μ (or frictional force F_f) is, the faster the capsubot moves.

3.3 Effect of Fluid Viscous Resistance

The drag coefficient C_D depends on the type of fluids, the area of resistance (or R parameter in Table 1) and the velocity of movement. In fact, the coefficient C_D will have different values in different environments. In this paper, numerical simulation tool was used to change the drag coefficient intensity $C_D = [0.2–30]$ while keeping the conditions $i = 0.35A$; $f = 20Hz$; $G = 0.1$ mm.

The simulation result shows that, as shown in Fig. 4b, when the coefficient of drag C_D is larger, the forward displacement of the capsubot will gradually decrease and reaches to 0 m/s at $C_D \approx 16$. And as C_D continues to increase, the capsubot will move backward faster. The relationship between the drag coefficient C_D and the displacement achieved can be expressed by the Eq. (6).

$$X_2 = 0.05729–0.00344 * C_D \tag{6}$$

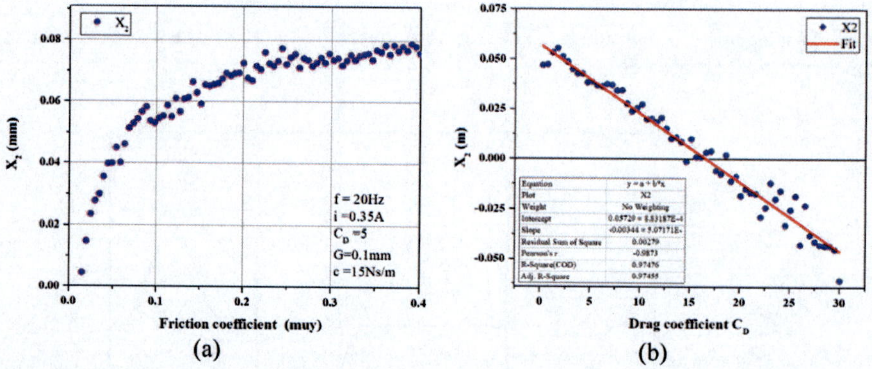

Fig. 4. Effect of (a) the friction coefficient μ and (b) drag coefficient C_D on the displacement of capsubot

3.4 Effect of Capsubot Size (R)

The influence of the cross-sectional size of capsubot on the mobility was also investigated through the parameter R, which varies in the range [0.005–0.03] m. Under the conditions ($i = 0.35$A; $f = 20$ Hz; $G = 0.1$mm; $\mu = 0.05$), the displacement of the capsubot is shown in Fig. 5. The results show that, in a liquid medium with low C_D intensity, cross-sectional size does not affect the direction of motion of the capsubot. However, when the liquid medium has a larger C_D, the influence of cross-sectional size on the direction of motion is very obvious. The larger cross-sectional size, the tendency is to move in the backward direction (that is, the direction of motion is opposite to the action of the exciting force). For endoscopy-oriented applications, the size of the capsubot needs to be as small as a capsule that can be swallowed by a human ($D < 13$ mm). So it can be seen, the capsubot's ability to move is less affected by the viscosity factor C_D.

3.5 Effect of the Impact Gap (G) and Current (I)

The results of the simultaneous investigation of the influence of two parameters impact gap ($G = [0–3]$ mm) and current ($i = [0.1–0.8]$ A) on capsubot's displacement X_2 are

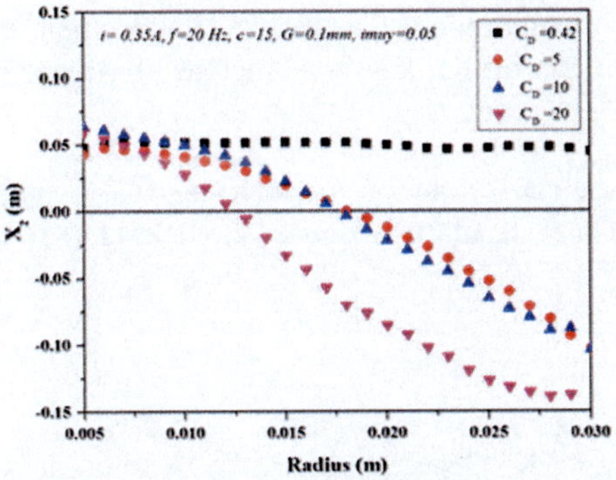

Fig. 5. Effect of the cross-sectional size *R* on movement ability of capsubot

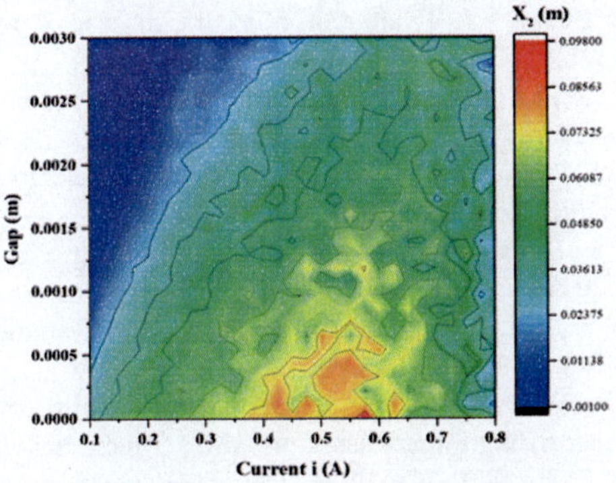

Fig. 6. Effect of the impact gap *G* and current *i* on displacement of capsubot

shown in Fig. 6 with the conditions: $i = 0.35\text{A}, f = 20$ Hz, $\mu = 0.05$, $C_D = 5$. It is easy to see that the larger the impact gap, the smaller the displacement of the capsubot. Besides, the larger or the smaller the excitation current is, the smaller the displacement of the capsubot get.

4 Conclusions

The new physical and mathematical model of vibro-impact capsubot in liquid environment was proposed. Dynamic response of this system was investigated by numerical simulation tool of the XPP Auto software. The results show that square-pulse excitation signal form can make the capsubot model moving faster than the full sine and half sine excitations. The results also show that in the liquid environment, the vibro-impact capsubot is completely capable of moving forward and backward by changing control parameters, including frequency and intensity of excitation current. These results are the premise to support the design of the model for capsubot, and the basis for building a future verified experimental model. The results can also be used to analyze the stability of the system as well as further analysis, such as phase portrait, Poincaré map, and bifurcation diagram.

Acknowledgements. [Ngo Quoc Huy] was funded by the Master, PhD Scholarship Programme of Vingroup Innovation Foundation (VINIF), code [VINIF.2022.TS161].

References

1. Hosoe, N., Takabayashi, K., Ogata, H., Kanai, T.: Capsule endoscopy for small-intestinal disorders: current status. Dig. Endosc. **31**(5), 498–507 (2019)
2. Kim, S.H., Chun, H.J.: Capsule endoscopy: pitfalls and approaches to overcome. Diagn. (Basel). **11**(10) (2021)
3. Nam, S.-J., Lee, H.S., Lim, Y.J.: Evaluation of gastric disease with capsule endoscopy. Clin Endosc **51**(4), 323–328 (2018)
4. Shamsudhin, Zverev, V.I., Keller, H., Pane, S.: Magnetically guided capsule endoscopy. Med. Phys. **44**(8), 91–111 (2017)
5. Ongel, K., Gumra, N., Ozguner, F.: The potential effects of electromagnetic field: a review. Cell Membr. Free. Radic. Res. **1**(3) (2009)
6. Norton J.C., Sarker, S., Mohammed, S.: Fundamentals of the gut for capsule engineers. Prog. Biomed. Eng. **2**(4) (2020)
7. Liu, L., Towfighian, S., Hila, A.: A review of locomotion systems for capsule endoscopy. IEEE Rev. Biomed. Eng. **8**, 138–151 (2015)
8. The-Hung, D., Van-Du, N., Ngoc-Tuan, L.: A new design for bidirectional autogenous mobile systems with two-side drifting impact oscillator. Int. J. Mech. Sci. **140**, 325–338 (2018)
9. Nguyen, K.T., La, N.T., Ho, K.T., Ngo, Q.H., Chu, N.H., Nguyen, V.D.: The effect of friction on the vibro-impact locomotion system: modeling and dynamic response. Meccanica. **56**, 121–2137 (2021)
10. Nguyen, V.-D., La, N.-T.: An improvement of vibration-driven locomotion module for capsule robots. Mech. Based Des. Struct. Mach. **50**(5), 1658–1672 (2022)
11. Liu, Y., Pavlovskaia, E., Wiercigroch, M.: Experimental verification of the vibro-impact capsule model. Nonlinear Dyn. **83**(1–2), 1029–1041 (2015)
12. Bingyong, G., Yang, L., Rauf, B., Shyam, P.: Self-propelled capsule endoscopy for small-bowel examination: proof-of-concept and model verification. Int. J. Mech. Sci. **174**, 105506 (2020)
13. Duong, T.H., Van, C.N., Ho, K.T., La, N.T., Ngo, Q.H., Nguyen, K.T., Nguyen, V.D.: Dynamic response of vibro-impact capsule moving on the inclined track and stochastic slope. Meccanica 58(2), 421–439 (2022)
14. Liu, Y., Pavlovskaia, E., Hendry, D., Wiercigroch, M.: Vibro-impact responses of capsule system with various friction models. Int. J. Mech. Sci. **72**, 39–54 (2013)
15. Tian, J., Liu, Y., Chen, J., Guo, B., Prasad, S.: Finite element analysis of a self-propelled capsule robot moving in the small intestine. Int. J. Mech. Sci. **206**, 106621 (2021)
16. Yan, Y., Liu, Y., Manfredi, L., Prasad, S.: Modelling of a vibro-impact self-propelled capsule in the small intestine. Nonlinear Dyn. **96**(1), 123–144 (2019)

Genetic Programming–A Preliminary Study of Knowledge Transfer in Mutation

Pham Thi Thuong(✉), Hoang Thi Canh, Nguyen Thi Dung, Nguyen Thu Phuong, and Nguyen Lan Oanh

Thai Nguyen University of Information and Communication Technology, Thai Nguyen, Vietnam
{ptthuong,htcanh,ntdung,ntphuong,nloanh}@ictu.edu.vn

Abstract. In this research, we introduce two new variations of Genetic Programming (GP) that reusing transferred knowledge in the mutation operator. They are called *FullTree.CM_1* and *FullTree.CM_2*. Chromosomes of the best individuals learned from previous source problems will be reused as sub-trees of children generated by mutation while training GP for the target problem. *FullTree.CM_1* takes k% good individuals in the last generation of a previous learned sub-task (source problem) into a POOL for reusing in the mutating process while *FullTree.CM_2* re-uses k% good solutions from all previous learned sub-tasks. In order to analyzing the effectiveness of these schemes, we use three evaluation criteria including errors on training data, errors on testing data, and size of the learned solutions. Experimental results show that our schemes perform better than other previous transfer GP on the most of tested problem families with all of these three evaluation criteria. This is very promising and motivating for further research to improve GP based on transfer mutation.

Keywords: Symbolic regression · Genetic programming · Transfer learning

1 Introduction

Transfer learning is one of the active research topic in machine learning [1, 2]. Transfer learning reflects the process that same as the learning process by human [3]. Naturally, human usually learns to solve new tasks using their accumulated knowledge from solving the similar previous ones.

Recently, transfer learning in GP has been received the attention of researchers [4–8]. Many researches confirmed that appling transfer learning to GP can enhance the ability of GP in solving problems. In [4], authors gave a clearly definition about transfer leaning and emphasized three important research issues that need to solve in transfer learning such as *(1) What should we tranfer from the source to the target problem? (2) How can we transfer the knowledge from the source to the target problem? And (3) How much should knowledge should we transfer?* However, they only focus on solving two first issues by proposing three transfer GP schemes: *FullTree, SubTree,* and *BesGen. FullTree* solves the 1st and 2nd issues by copying good individuals in the final generation of the previous learned sub-task to the initial population of the current learned sub-task. *SubTree* chooses

P. T. Nghia et al. (Eds.): ICTA 2023, LNNS 847, pp. 261–268, 2023.
https://doi.org/10.1007/978-3-031-49529-8_28

a subtree of each solution in the last generation of the source problemhas and re-uses them in the initial population of the target problem. *BestGen* sampled the good solutions appearing in certain generations during the evolutionary process. Then, these solutions were transferred for the target problem. Authors confirmed benefits of these schemes in obtaining smaller error on the training set, on the testing data, and in obtaining learned solutions with smaller sizes.

By realizing the important role of mutation in evolutionary theory: *"a tiny change can have a huge impact on the effectiveness of the solution"*, in this work, we propose two variations of *FullTree*, called *FullTree.CM_1* and *FullTree.CM_2* that using transfered knowledge in mutation process. Experimental results confirm these variations's advantages.

Section 2 presents briefly about Transfer Learning, and Genetic Programming combined with knowledge transfer. Next, Sect. 3 gives our schemes of applying transfer in GP. Section 4 shows the experimental settings. Section 5 presents experimental results and discussions of the effectiveness of tested techniques. Finally, Sect. 6 shows conclusions and future works.

2 Background

2.1 Transfer Learning

The transfer learning can be seen as a process that a system can recognize and use learned knowledge, skills in solving previous tasks to solve new tasks that almost the same. Transfer learning aims to achieve real life's knowledge transfer capabilities in machine learning. Tasks in transfer learning can be divided into two types such as: (1) the source task, and (2) the target task. The first task is often simpler. It usually had been learned previously. The second task, that we are aiming to address, is usually more complex. Actually, we can solve second task independently with the first task. However, due to it has a relationship with the source task, so we often solve this task by reusing the gained knowledge from the source task. This will give us the certainly benefits.

2.2 Genetic Programming (GP)

In 1992, Koza introduced GP based on an abstraction of Darwin's natural selection mechanisms to training the solutions of problems by evolving them [9]. GP is one of the machine learning techniques learning how to optimize a population of solutions of given computational tasks.

Researches about the re-use ability of GP can roughly be categorized into two classes based on their objectives. The 1st class aims to re-use some good sub-solutions (sub-structures) of one solution in other solutions [9]. This approach greatly improved the performance of GP in solving problems. The 2nd class attempts to re-use some good solutions of simpler problems in solving related problems that more difficult than previous ones. In the 2nd class, Hao et al. proposed a model for re-using solutions in Genetic Programming based TAG [6], Uy et al., Hoai et al. and Murphy et al., investigated the application of GP on dynamic problems [5]. Recently, transfer learning in Genetic Programming has been received a lot of attention from researchers. However, the results are limited, and many open issues still need to be studied thoroughly.

3 Our Methods

In this research, we use transferred knowledge in both of genetic operators: crossover (C) and mutator (M) of GP. We propose two variations of *FullTree* [4], called *FullTree.CM_1* and *FullTree.CM_2*. Our schemes are compared to other existing GP schemes such as (1) *Standard GP* (no knowledge transformation); and (2) The *FullTree* (original) scheme [4]. The amount of transferred knowledge is fixed at k = 25% for all experimental configurations re-using transfer knowledge.

3.1 FullTree.CM_1

When learning a new task (target problem), *FullTree.CM_1* takes k% good individuals in the last generation of previous learned sub-task (source problem) into the initial population of target problem. Beside to, these individuals also be pushed into a POOL for reusing in the mutating process. The mutation is performed by removing a random subtree of parent and replaced it by chromosome of random individual in the POOL to generate children for a next generation. The POOL is cleaned after the training process of a sub-task done. So, the POOL only contains k% good solutions from only one previous sub-task. Our improvements to this configuration are indicated by the grayed text box in the Fig. 1(a).

3.2 FullTree.CM_2

FullTree.CM_2 is similar to *FullTree.CM*_1. However, it re-uses k% good solutions from all previous learned sub-tasks for learning the next sub-task while mutating rather than only from one sub-task as in *FullTree.CM*_1. All transfer knowledge is contained in the POOL_M. Figure 1(b) shows these changes in the grayed text box.

4 Experimental Settings

4.1 Dynamic Problems

Six families of Dynamic Symbolic Regression problems used in experiments are presented in Table 1. In which, each problem family includes five sub-tasks that varied from simple to complex.

4.2 GP Setting

GP's parameters used in our experiments are showed in Table 2. Because we have five sub-tasks in each family, so the fitness functions also be changed five times. One hundred of generations are evolved for each task. So the total number of all generations is five hundreds. We also evaluate the generalization ability of experimental schemes by choosing the best individual in each generation and test it on an independent data set, called test set.

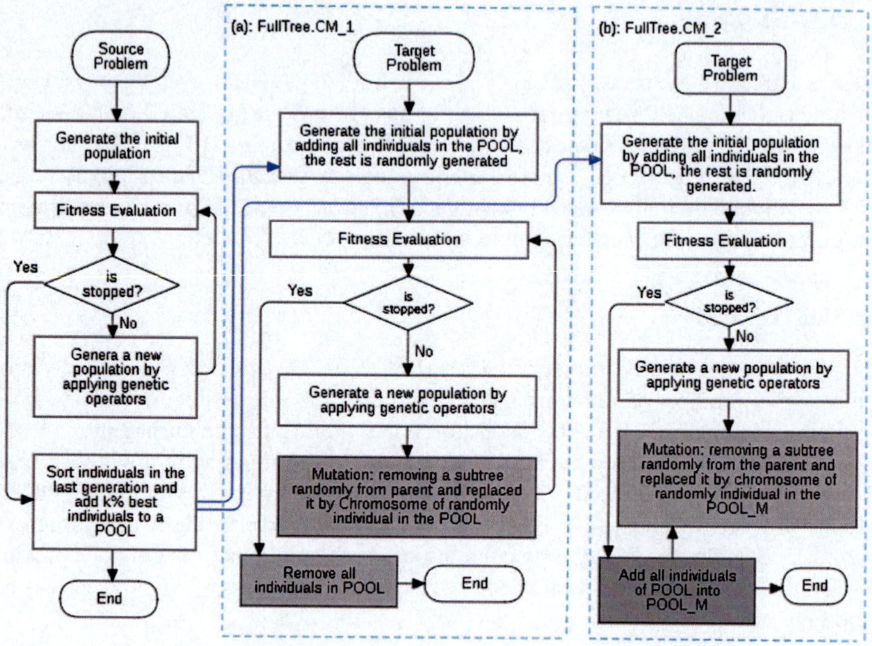

Fig.1. (a) *FullTree.CM*_1 scheme; (b) *FullTree.CM*_2 scheme.

5 Experimental Results and Discussion

In order to analyzing the effectiveness of experimental schemes, we use three evaluation criteria: the 1st metric is the average of best fitness on the training data, called the fitness. The 2nd is average of fitness on the unseen data, called generalization ability. The 3th metric is the effort, or computational complexity.

5.1 Training Error Analysis

Figure 2 shows the best fitness average of 30 runs on the training set. For each family of problems, there are four graphs corresponding to four configuarations of GP such as: (1) *GP*, (2) *FullTree,* (3) *FullTree.CM_1*, and (4) *FullTree.CM_2.*

Considering to the 1st metric, we see that transfer learning schemes achieving smaller training errors on most of problem families. In other words, with most of problem families, three transfer learning techniques have the traning errors smaller than that of standard GP. In transfer learning techniques, *FullTree.CM_1* and *FullTree.CM_2* are usually better than *FullTree* on sub-tasks such as *F_1.2, F_3.(all), F_2.2, F_2.3, F_6.2, F_6.3, F_6.4,* and *F_6.5*. This shows us that using the mutation operator with transfer knowledge obtains the positive effects on these two proposed configurations.

Table 1. Some families of dynamic symbolic regression problems.

FP	Dynamic symbolic regression subtasks
FP_1	$x^2 + x$ (F_1.1) $\Rightarrow x^3 + x^2 + x$ (F_1.2) $\Rightarrow x^4 + x^3 + x^2 + x$ (F_1.3) $\Rightarrow x^5 + x^4 + x^3 + x^2 + x$ (F_1.4) $\Rightarrow x^6 + x^5 + x^4 + x^3 + x^2 + x$ (F_1.5)
FP_2	$\cos(x) + \sin(x)$ (F_2.1) $\Rightarrow \cos(x) + \sin(x) + \sin^2(x)$ (F_2.2) $\Rightarrow \cos(x) + \sin(x) + \sin^2(x) + \sin^3(x)$ (F_2.3) $\Rightarrow \cos(x) + \sin(x) + \sin^2(x) + \sin^3(x) + \sin^4(x)$ (F_2.4) $\Rightarrow \cos(x) + \sin(x) + \sin^2(x) + \sin^3(x) + \sin^4(x) + \sin^5(x)$ (F_2.5)
FP_3	$\cos(x) + \sin(x)$ (F_3.1) $\Rightarrow \cos(x) + \sin(x) + \sin(2x)$ (F_3.2) $\Rightarrow \cos(x) + \sin(x) + \sin(2x) + \sin(3x)$ (F_3.3) $\Rightarrow \cos(x) + \sin(x) + \sin(2x) + \sin(3x) + \sin(4x)$ (F_3.4) $\Rightarrow \cos(x) + \sin(x) + \sin(2x) + \sin(3x) + \sin(4x) + \sin(5x)$ (F_3.5)
FP_4	$\log(\sin(x^2)) + \log(\sin(x^4))$ (F_4.1) $\Rightarrow \log(\sin(x^2)) + \log(\sin(x^4)) + \log(\sin(x^6))$ (F_4.2) $\Rightarrow \log(\sin(x^2)) + \log(\sin(x^4)) + \log(\sin(x^6)) + \log(\sin(x^8))$ (F_4.3) $\Rightarrow \log(\sin(x^2)) + \log(\sin(x^4)) + \log(\sin(x^6)) + \log(\sin(x^8)) + \log(\sin(x^{10}))$ (F_4.4) $\Rightarrow \log(\sin(x^2)) + \log(\sin(x^4)) + \log(\sin(x^6)) + \log(\sin(x^8)) + \log(\sin(x^{10})) + \log(\sin(x^{12}))$ (F_4.5)
FP_5	$\sin(x) - x{*}\cos(x)$ (F_5.1) $\Rightarrow \sin(x) - x{*}\cos(x) + x^2{*}\cos(x2)$ (F_5.2) $\Rightarrow \sin(x) - x{*}\cos(x) + x^2{*}\cos(x^2) - x^3{*}\cos(x^3)$ (F_5.3) $\Rightarrow \sin(x) - x{*}\cos(x) + x^2{*}\cos(x^2) - x^3{*}\cos(x^3) + x^4{*}\cos(x^4)$ (F_5.4) $\Rightarrow \sin(x) - x{*}\cos(x) + x^2{*}\cos(x^2) - x^3{*}\cos(x^3) + x^4{*}\cos(x^4) - x^5{*}\cos(x^5)$ (F_5.5)
FP_6	$\sin(x^2) + x^2$ (F_6.1) $\Rightarrow \sin(\sin(x^2) + x^2)^2$ (F_6.2) $\Rightarrow \sin(\sin(\sin(x^2) + x^2)^2)^3$ (F_6.3) $\Rightarrow \sin(\sin(\sin(\sin(x^2) + x^2)^2)^3)^4$ (F_6.4) $\Rightarrow \sin(\sin(\sin(\sin(\sin(x^2) + x^2)^2)^3)^4)^5$ (F_6.5)

Table 2. Run and evolutionary parameters values

Parameters	Value
Population's size; No. of generations	500; 500 (100 gens for each sub-task)
Selection; tournament size	Tournament; 2
Crossover rate; mutation rate	0.9; 0.1
Min depth; max depth	6; 15
Non-terminals	+, −, *, / (protected one), sin, cos, exp, log
Terminals	$x_1, x_2, \ldots x_n$
Training set; testing set	60 points in $[-1, 1]$; 30 points in $[-3, 3]$
The fitness function	Average of absolute error on all fitness cases
No. of trials; transferred knowledge	30; $k = 25\%$

5.2 Generalization Ability Analysis

The generalization ability can be seen as the most desirable property of a learner. It can be measured by the learner's error on test sets. It means that the higher generalization ability,

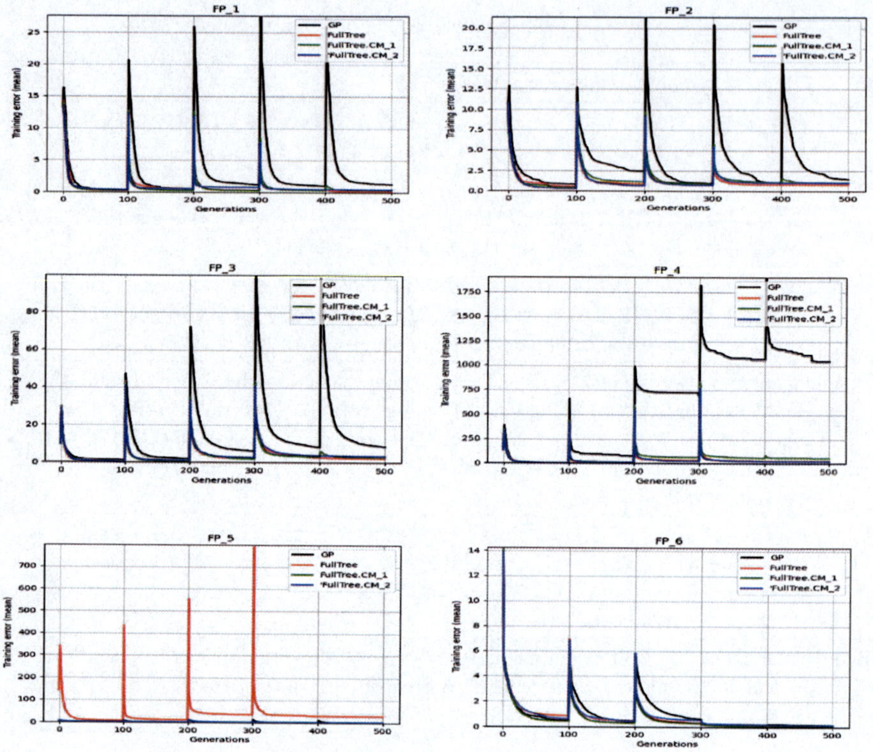

Fig. 2. The plot of mean best fitness on the training set.

the smaller testing error. Table 3 presents the generalization ability of experimental schemes.

With six problem families from *FP_1* to *FP_6*, we have a total of thirty corresponding sub-tasks. From the bold cases in Table 3, we see that: most of the tested problem families such as *FP_1, FP_2, FP_3, FP_4*, transfer GP methods (*FullTree, FullTree.CM_1, FullTree.CM_2*) learned solutions with smaller testing errors than that of *Standard GP*. *FullTree.CM_1* and *FullTree.CM_2* are seem better than *FullTree* (original). More detail: (1) *GP* is the best configuration with only on a 1/30 sub-tasks (*F6.5*); (2) *FullTree* is the best configuration on 9/30 sub-tasks (*F_1.1, F_1.2, F_1.5, F_2.4, F_2,5, F_4.2, F_4.4, F_4.5, F_6.2*); (3) *FullTree.CM_1* found the best solution on 10/30 sub-tasks and finally (4) *FullTree.CM_2* is the best generalization configuration on 10/30 sub-tasks. So, using the mutation operation combined with transfer knowledge makes *FullTree.CM_1* and *FullTree.CM_2* can learn the higher generalizability solutions than that of *GP* and *FullTree* on 20/30 subtasks, or seventy percent of all tested sub-tasks.

5.3 Computation Complexity Analysis

Computational complexity is measured by the effort (time) spent in evaluating fitness. It is similar to the size of the individual. To measure this metric, for each experimental

Table 3. Mean of testing error. Lower is better. Bold case is the best.

FP	Sub-task	GP	FullTree	FullTree.CM_1	FullTree.CM_2
FP_1	F_1.1	16,57638	**14,2296099**	41,60268152	14,64121143
	F_1.2	21,1753	**9,40459763**	35,96968611	12,8289908
	F_1.3	114,0047	63,49578573	45,5763237	**31,76261256**
	F_1.4	230,5506	129,9546638	108,0394575	**75,66534264**
	F_1.5	284,9919	**74,4011554**	177,091086	76,75693972
FP_2	F_2.1	0,312792	0,219023371	0,170454586	**0,110896944**
	F_2.2	0,570986	0,222708294	0,226553859	**0,128176867**
	F_2.3	0,369605	0,301559033	0,375614475	**0,274250143**
	F_2.4	0,406241	**0,31890151**	0,46530026	0,360507792
	F_2.5	0,666594	**0,31080106**	0,434844663	0,356290194
FP_3	F_3.1	0,420643	0,235074228	0,209674407	**0,195062017**
	F_3.2	0,677496	0,288736741	**0,213770372**	0,481742199
	F_3.3	3,002416	1,001376526	**0,848949604**	1,060605531
	F_3.4	2,98402	1,245109215	**1,20139477**	1,211062883
	F_3.5	8,18152	1,146575747	**1,023027449**	1,17193628
FP_4	F_4.1	1,867528	1,609621314	1,88533751	**1,556251576**
	F_4.2	219,2912	**15,0388629**	32,46949729	23,98778374
	F_4.3	261,4641	12,96027665	19,99757335	**12,15386641**
	F_4.4	319,7765	**5,895629**	7,925142964	8,916608034
	F_4.5	269,5858	**5,72072262**	9,428922774	8,666299293
FP_5	F_5.1	10,31598	10,5065756	8,936711644	**8,091849365**
	F_5.2	13,55056	10,41379883	**9,78382886**	13,75627771
	F_5.3	58,63186	29,21633203	**25,79455152**	60,60718983
	F_5.4	57,05046	73,11944277	**56,16118379**	58,30260549
	F_5.5	58,32654	68,19063266	**56,97937518**	59,32816556
FP_6	F_6.1	17,80764	17,86777095	**7,294710151**	14,14427297
	F_6.2	5,430692	**5,09039209**	242,9179766	5,470745228
	F_6.3	15,44835	48,56931554	**10,90547806**	23,55989122
	F_6.4	225,5403	124,6082591	19,17734494	**0,304277168**
	F_6.5	**1,16038**	102,8102804	15,03408377	4,48762614

configuration, we calculate the total effort for each run, then take the average of the 30 runs. Table 4 presents these results. From bold cases, we see that: *FullTree.CM_1*, *FullTree.CM_2* take less effort than *GP* and *FullTree* on 4/6 problem families (*FP_2*, *FP_4*, *FP_5*, *FP_6*). This also means that *FullTree.CM_1* and *FullTree.CM_2* ran faster than others, or they learned the solution with smaller size than that of *GP* and *FullTree*. Obtaining the s olutions with smaller size will help GP in reducing the issue of bload code. And bload code is a one of challenges of GP.

Table 4. Mean of computation complexity. Lower is better. Bold case is the best.

	FP_1	FP_2	FP_3	FP_4	FP_5	FP_6
GP	**117639**	117849	**117886**	117851	117849	117874
FullTree	118757	118977	118956	118995	118982	119008
FullTree.CM_1	118242	**117581**	118099	**116987**	**117595**	**117687**
FullTree.CM_2	118767	118723	118740	118754	118751	118746

6 Conclusions and Future Works

In this research, we propose two new schemes *FullTree.CM_1* and *FullTree.CM_2* They are extended from *FullTree* scheme [4] by reusing transfer knowledge in the mutating process. By running experiments, we can conclude that: Applying transfer learning can help our methods achieving better solutions than *GP* and *FullTree* on most of tested problem families in terms of training error, testing error, and computational complexity or size of solutions. This is very promising and motivating for further research to improve GP based on transfer mutation. And that, naturally, mutation operator should be self-correcting individuals, self-changing to be more and more perfect.

References

1. Pan, S.J., Yang, Q.: A survey on transfer learning. IEEE Trans. Knowl. Data Eng. **22**(10), 1345–1359 (2010)
2. Taylor, M.E., Stone, P.: Cross-domain transfer for reinforcement learning. In: 24th International Proceedings on Machine learning, pp. 879–886 (2007)
3. Ramon, J., Driessens, K., Croonenborghs, T.: Transfer learning in reinforcement learning problems through partial policy recycling. In: 18th International Proceedings on Machine Learning, pp. 699–707. Springer, Heidelberg (2007)
4. Dinh, T.T.H., Chu, T.H., Nguyen, Q.U.: Transfer learning in genetic programming. In: IEEE Congress on Evolutionary Computation (CEC), pp. 1145–1151 (2015)
5. Uy, N.Q., et al.: Semantic-based subtree crossover applied to dynamic problems. In: 3th International Conference on Knowledge and Systems Engineering, pp. 78–84 (2011)
6. Hoang, T.H., et al.: Building on success in genetic programming: adaptive variation and developmental evaluation. In: 2nd International Proceedings on Advances in Computation and Intelligence, pp. 137–146. Springer, Heidelberg (2007)
7. O'neill, D., et al.: Common subtrees in related problems: a novel transfer learning approach for genetic programming. In: 2017 IEEE Congress on Evolutionary Computation (CEC), pp. 1287–1294 (2017)
8. Helmuth, T., et al.: Genetic source sensitivity and transfer learning in genetic programming. In: 32 Proceedings on Artificial Life, pp. 303–311. MIT Press, Cambridge (2020)
9. Koza, J.R.: Genetic programming II: automatic discovery of reusable programs. MIT press (1994)

Improvement of Spectral Clustering Method in Social Network Community Detection

Nguyen Hien Trinh[1](✉) and Cap Thanh Tung[2]

[1] Thai Nguyen University of Information and Communication Technology, Thai Nguyen, Vietnam
nhtrinh@ictu.edu.vn

[2] Thai Nguyen University of Education, Thai Nguyen, Vietnam
tungct@tnue.edu.vn

Abstract. Nowadays, community discovery on a social network is an important research direction in the field of computer science. Social networks are usually represented in the form of graph data structures. Therefore, the discovery of community on social networks is mainly associated with the clustering problem on the graph. To solve the problem, there are many algorithms that are interested in research. In this paper, we will present an aggregate method, that is, clustering graphs based on the concept of spectrum to reduce the number of dimensions of the data set to be considered, thus reducing complexity. In addition, the modularity of the algorithm is focused on improving. Our algorithm is highly effective for large social networks. Computation quickly and resulting in community structure detection on social networks. Tests on a set of popular, standard social networks and certain real network have shown the high speed and high effiency in finding communities.

Keywords: Social network · Community structure · Clustering graphs; community detection · Spectrum · Spectral clustering

1 Introduction

Since the concept of a social network, there have been many community detection algorithms proposed by the method of clustering the G graph into smaller subgraphs with specific features. Subsequent algorithms like Tyler [1]; Gregory [2]; Brandes [3], … is proposed to improve the speed of Girvan-Newman's algorithm. However, they still have great complexity, $O(mn)$ or $O(mn + n^2 \log n)$ where n is the number of vertices, m is the number of edges of the graph. To find a way to reduce the computational complexity when performing clustering, we study, propose and improve the processing algorithm for detecting community structure on social networks with the foundation of spectral clustering technique [4–7]. This method approaches in the direction of reducing the number of dimensions of data (matrix form, maybe even a multidimensional matrix) to only vector form (real numbers series), combined with the idea of optimizing the function. Min-cut thanks to the use of Laplace matrix, so it is very effective for processing

P. T. Nghia et al. (Eds.): ICTA 2023, LNNS 847, pp. 269–279, 2023.
https://doi.org/10.1007/978-3-031-49529-8_29

community structure detection on social networks. Thus, by reducing the number of dimensions of the vector space (original data) thanks to spectral clustering techniques, it was possible to reduce the computational load when dealing with community structure detection for social networks. The method of applying Spectrum technique is not only effective in detecting community structure, but also shows that the approach accurately depicts the complex structure of real-life social network systems. The execution time of our algorithm is faster than that of UlrikeVon Luxburg [4] and SpcSA [5].The quality of the resulting communities is also better. The experimental results confirm that this method can ensure the selection of a reasonable number of communities for the network.

In this paper, we introduce a method to improve spectral clustering in social network community detection. The remaining of the paper: Sect. 2, presents the algorithms. Section 3, Experimental results. Finally, Sect. 4, Conclusion and contributions of the research, along with suggesting future research directions.

2 Recommended Algorithms

2.1 Distance Between Objects

Objects with many different attributes can be described by points in a multidimensional space, each dimension representing an attribute. The distance between objects allows us to evaluate the closeness between them. In theory and experiment, determining the equivalence between two objects X_i, X_j is evaluated according to the Gauss distribution:

$$W(i,j) = \exp(-\frac{\|X_i - X_j\|^2}{2\sigma^2}) \tag{1}$$

where σ is the standard deviation. In experiment, we choose this value to adjust the size of the cluster. In addition, the distance between two objects is also determined by Euclidean, Cosine, Hamming distance...Then, from a distance, we can determine the connection through many different methods.

2.2 Reduce the Graph

Given graph $G = (V, E)$, to solve the community detection problem if G is a large graph, to speed up calculation and improve community quality, we will perform some the preprocessing step is to temporarily leave out the graph G some special vertices and edges. A hanging vertex is a vertex of degree 1. Then the hanging vertex can only belong to the community of its neighbors, so we can temporarily remove it from the graph before conducting community search steps. We see that if two vertices $i, j \in V$ share many neighbors, the possibility that they are in the same community is very high. Therefore, we propose an algorithm that calculates the number of common neighbors of two adjacent nodes i, j when $(i,j) \in E$, treating it as an parameter to the weight represents the degree of connection between two nodes i and j.

Algorithm WTG (Weight Triangle)

```
Input : G=(E, V); i,j ∈ V
Output: a set WTG stores the number of common neighbors of each pair i,j=1.. n
  {  for i ∈ V do
          for j ∈ V do   WTG_i,j = 0;
     for i ∈ V do
          for j ∈ N_i do
             for k ∈ N_j do
                        if (i>k)  WTG_i,j  = WTG_i,j  + 1 ;
    return a set WTG;}
```

The complexity of the algorithm is $O(n^2)$, where n is the number of vertices of graph G. After running WTG with the graph G, we set a threshold value W_{th} and try to remove edges with calculation weights less than this threshold value.

Algorithm RE (Remove Edge)

```
Input: G = (V, E) is represented by matrix A; W_th; WTG_i,j; i,j ∈ V
Output: G' ⊆ G, represented by the matrix A'; Vector S_1
{ S_1 = ϕ;
  for i ∈ V do
     for j ∈ V do  {
                        a'_i,j  =  a_i,j + WTG_i,j;
                        if a'_i,j < W_th then
                           if  vertex i or j is removed then   S_1 ← i, j
                     }
return A', S_1; } //Vector S_1 keeps track of the removed vertices.
```

We perform graph G reduction by removing hanging vertices, then removing some edges (not important) by the above RE algorithm.

Algorithm MCG (Make Compact Graph)

Input $G=(E, V)$
Output $G'' \subseteq G$, vectors S_1, S_2 stores information about the removed vertices
{ $S_2 = \phi$;
  for i $\in$ V do
    if $D_i = 1$ {
      $S_2 = S_2 \cup \{i\}$; $V = V - \{i\}$; // Remove the hanging vertice
    }
  G', $S_1 \leftarrow$RE(G);
  return G', S_1, S_2;}

After performing clustering, the new graph $G_{new} = G_1 \cup G_2 \cup \ldots \cup G_k$ we need to perform re-accumulation of vertices and temporarily removed edges at the edges previous period. These vertices will belong to the community that has many neighbors in it.

Algorithm RCL (ReClustering)

Input : Set of $G_1, G_2, \ldots G_k$ with $G_{new} = G_1 \cup G_2 \cup \ldots \cup G_k$; S_1; S_2
Output $G = G_1' \cup G_2' \cup \ldots \cup G_k'$;
{ $S = S_1 \cup S_2$
  for v $\in$ S do {
    for i=1 to k do
      $C_i = |N_{v \in G_i}|$; // Number of neighbors of v in G_i
    $C_{max} = \max\{C_i\}$; // Select max(Ci)
    $G_{max} = G_{max} \cup \{v\}$ // Admit vertex v to the cluster with the most neighbors
    $G'_i = G_i$ }
  return $G = G' \cup G' \cup \ldots \cup G'$; }

2.3 Proposing a Community Detection Algorithm Based on Spectral Clustering

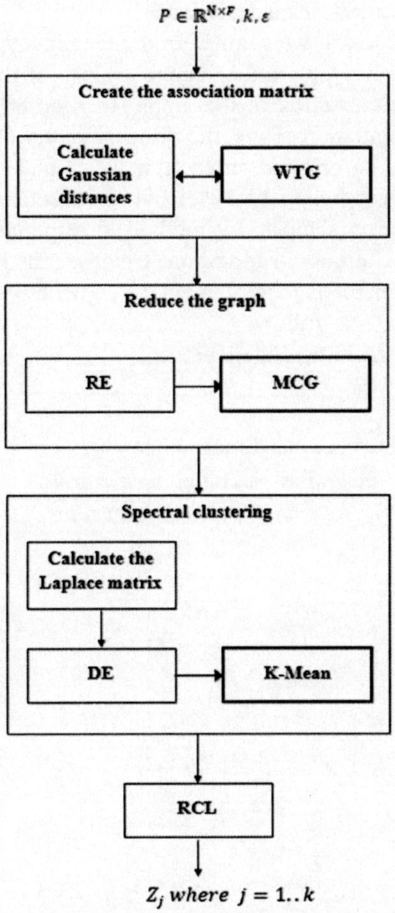

Fig. 1. Functional diagram of SCN

The difference between the algorithms in the spectral clustering group is that they calculate the Laplace matrix in different ways, and the feature vectors are calculated differently. The main trick is to change the representation of the abstract data points, how to construct the association matrix of the similarity graph differently (Fig. 1).

The main content of the proposed SCN algorithm is based on the principle of helping the eigenvectors of the Laplace matrix, transforming the original set of data objects into a set of points in coordinate space. Are the elements of the eigenvector, then the points are clustered using standard k-means techniques.

From the equation$Lu = \lambda u$, we can determine the eigenvalues and eigenvectors of L. The values of an eigenvector are the spectral values of the peak of the graph and are used to calculate the clustering of the graph. In addition, the eigenvector will be normalized for easy computation. Thus, the dimensionality of the spectrum is smaller than the dimensionality of the original vertex set in the graph because from the adjacency matrix A [*nxn*], we have switched to the n-element eigenvector treatment. The number of clusters to be split in the spectral clustering method corresponds to the k value chosen for the k-means clustering technique. In addition, we can use the Power Iteration algorithm [8]. We apply numerical methods to implement the algorithm to determine eigenvectors and eigenvalues of matrix A as follows:

Algorithm DE (Determining Eigenvectors)

Input : Matrix A has dimension $n \times n$; error value ε

Output: Eigenvalue λ and eigenvector x

{ $x = u^{(0)}$;// *x is the initialization vector*

t=0;

do {

$$x^{(t+1)} = \frac{Mx^{(t)}}{\left\| Mx^{(t)} \right\|};$$

t=t+1;

}

while $\left(\left\| x^{(t)} - x^{(t+1)} \right\| \le \varepsilon \right)$;

$\lambda = x^T Mx$;

return λ, x; }

Algorithm SCN (Spectral Clustering New)

. *Input*: Data set $P \in R^{nxF}$, n: number of data points, *F* number of spatial dimensions, *k* number of communities, error ε

Output:Communities $Z_1, Z_2, ..., Z_k$ where $Z_j = \{P_i | P_i \in C_j, i=1..n, j=1..k\}$

{

1. *for* $P_i \in P\ (i \in 1..n)$

   if P_i *is connected to* P_j *then*

   $$a(i,j) = \exp(-\frac{\|P_i - P_j\|}{2})\ ; i, j \in 1..n$$

2. $A_{WTG} \leftarrow$ WTG(A);
3. A, S_1, S_2 ←MCG(A);
4. *for* $P_i \in P\ (i \in 1..n)$

   $\deg(v_i) = \deg(P_i)$

5. *for* $i, j = 1..n$

   $(D_{ij}) = diag(v_1, v_2, ..., v_n)$ // *Calculate degree matrix*

6. *for* $i, j = 1..n$

   $(L_{ij}) = (D_{ij}) - (a_{ij})$ // *Calculate the Laplace Matrix*

7. $\lambda, u \leftarrow DE(L, \varepsilon)$

   // *Calculate the eigenvalue* λ, *set of eigenvectors u of L such that* $Lu = \lambda u$

8. $\{u_1, u_2, ..., u_k\} \leftarrow \{u_1, u_2, ..., u_n\}$

   // *Determine k eigen-vector* $\{u_1, u_2, ..., u_k\}$ *of L*

9. $w = (w_1, w_2, ..., w_n) \leftarrow \{u_1, u_2, ..., u_k\}$ // *select w from the set of* //*eigenvectors* $u_1, u_2, ..., u_k$

10. $y = \frac{w}{\|w\|}$ //*normalize w to get y*, $\|w\| = \sqrt{w_1^2 + w_2^2 + ... + w_n^2}$

11. $C_j = k\_means(y)$, *j=1..k*

    // *Clustering points* $y_i, i = 1..n$ *into k clusters* $C_1, C_2, ..., C_k$ *according to* // *k-Means*

12. $Z_j = \{y_i | y_i \in C_j\}$; // $Z_j = \{P_i | P_i \in C_j, i=1..n, j=1..k\}$

13. $Z'_j \leftarrow$ RCL(Z_j, S_1, S_2);

14. return Z'_j;

}

3 Experimental Results

Experimental environment is for configuration: Intel(R) Core i7, CPU @ 2.50 GHz, RAM 8.0 GB, Win 10, 64 bit. Programming language Matlab 2019, Python. Experimental algorithms were run separately on the data set, and in the computer were experimented with only a single program. The results were performed on the datasets published at [9] to evaluate the effectiveness of the SCN algorithm to detect the communities compared with the algorithms UlrikeVon Luxburg [4] and SpcSA [5] of Tang F, comparing the accuracy through the quality of the community (Modularity), the standard mutual information (NMI) of the community detected by the proposed algorithm. In the experimental data, the symbol *n*-number of vertices, *m*-number of edges of the social network under consideration.

Table 1. Experimental results on the execution time of SCN, SpcSA and UVonLB on the real network, in second (s)

No	Social Networks	n	m	SCN	SpcSA	UVonLB
1	AdjNoun	112	425	2.06	2.54	3.75
2	Caltech36	769	16700	19.75	23.31	27.16
3	Simmons81	1500	33000	59.22	62.08	78.01
4	Pages tvshow	3900	17200	124.85	127.62	134.98
5	Lehigh96	5100	198300	145.56	152.34	163.23

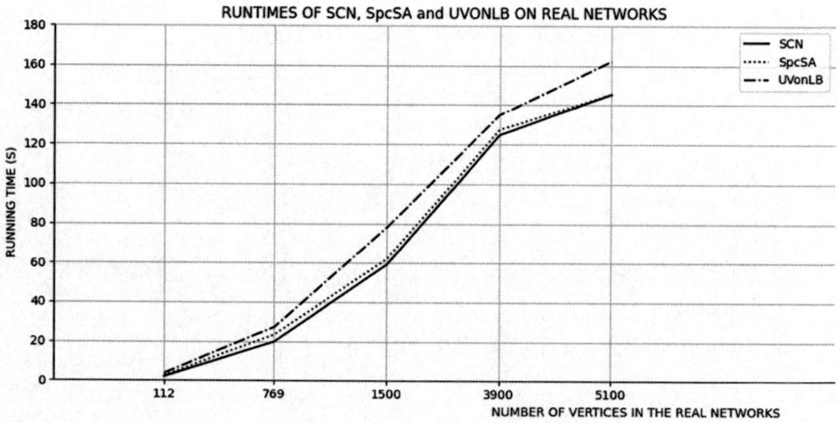

Fig. 2. Runtime comparison (seconds)

The results in Fig. 2 show that the execution time of the SCN algorithm is reduced by more than *7.5%*, SpcSA is reduced by more than *5.5%* compared to the execution time of UVonLB. The execution time of the SCN algorithm is reduced by more than *1.9%* compared to SpcSA. Thus, SCN has better execution time than SpcSA and UVonLB and SCN is quite effective for performing disjoint community structure detection.

Table 2. Experimental results on community quality (Modularity) with SCN, SpcSA and UVonLB

No	Social Networks	n	m	SCN	SpcSA	UVonLB
1	AdjNoun	112	425	0.57	0.53	0.48
2	Caltech36	769	16700	0.55	0.51	0.45
3	Simmons81	1500	33000	0.63	0.61	0.57
4	Pages tvshow	3900	17200	0.67	0.63	0.55
5	Lehigh96	5100	198300	0.71	0.69	0.63

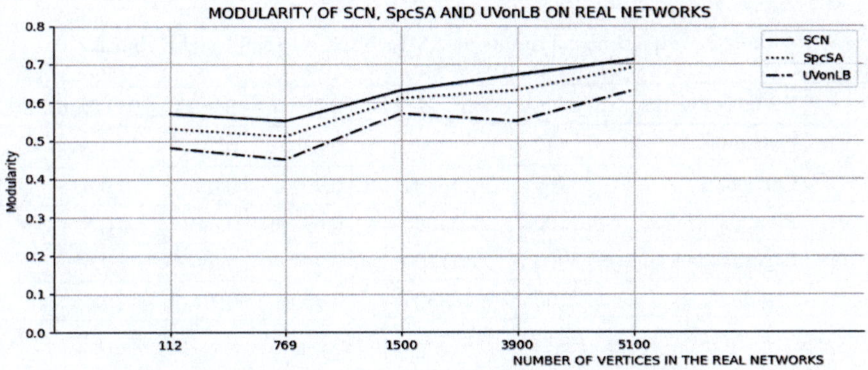

Fig. 3. Community quality comparison (modularity)

Modularity Q results in Table 2 and Fig. 3 show: modularity of SCN increased by more than *6%* of SpcSA by more than *4%* vs UVonLB. The modularity of SCN is more than *2%* compared to SpcSA. That confirms the quality of the community detected when using the SCN algorithm has improved, and there has been a closer connection in each result community.

In addition, with different community detection methods, the NMI value will also be used to evaluate the effectiveness of the methods. If the value of NMI is closer to 1, the method is better. Evaluation results of the NMI in Fig. 4 show that: If using SCN, it increases by more than *7%*, SpcSA increases by more than *5%* compared to UVonLB. SCN's NMI increased by more than *2%* compared to SpcSA. Thus, the experimental results show that the community detected by the SCN algorithm is more likely to coincide with the real community than the other algorithms.

Experimental results on large-sized real data sets are given in Tables 1, 2 and 3 and Figs. 2, 3 and 4, shows that the proposed SCN algorithm has performed quite well in identifying communities, the execution time of the proposed algorithm is faster than the SpcSA algorithm and UVonLB. The quality of the obtained communities is also higher, the normalized mutual information (NMI) is close to 1 and larger than the algorithms of UlrikeVon Luxburg and Tang F. This confirms the effectiveness of the proposed algorithm.

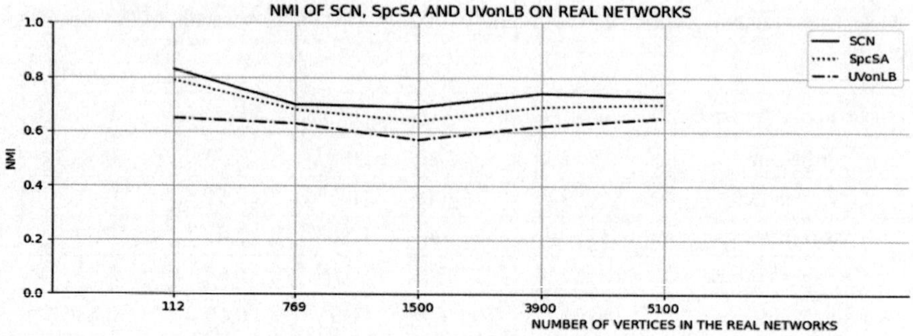

Fig. 4. Comparison of NMI

Table 3. Exp erimental results on NMI with SCN, SpcSA and UVonLB

No	Social Networks	n	m	SCN	SpcSA	UVonLB
1	AdjNoun	112	425	0.83	0.79	0.65
2	Caltech36	769	16700	0.70	0.68	0.63
3	Simmons81	1500	33000	0.69	0.64	0.57
4	Pages tvshow	3900	17200	0.74	0.69	0.62
5	Lehigh96	5100	198300	0.73	0.70	0.65

4 Conclusion

We have introduced a synthetic method, which approaches in the direction of reducing the number of dimensions of data (Matrix form, possibly multidimensional) to only vector form (real number series), in addition to integrating with The idea of optimizing Min-cut function by using Laplace matrix, so it is very effective for processing community structure detection on social networks. The foundation of the proposed method is the Spectrum technique. This method can ensure that a reasonable number of communities is selected for the network. We have proven the recommender algorithm to be effective in community detection, and also show that the proposed method can be used to unravel the complex structure of social network systems (which do not exist). Need to be described in multidimensional space). Algorithm execution time is faster than that of UlrikeVon Luxburg and SpcSA of Tang F. The quality of the resulting communities is also better. In the next time, we will continue to expand and perfect the algorithms (For directed graphs, real matrices, multidimensional...) and improve graph clustering techniques for fast detection. High-quality communities for the analysis and exploitation of information on social networks.

References

1. Tyler, J.R., Wilkinson, D.M.: Automated discovery of community structure within organization. Phys. Rev. E **15**, 723–739 (2003)

2. Gregory, S.: An algorithm to find overlapping community structure innetworks. Knowledge Discovery in Databases: PKDD. pp. 91 − 102 (2007).
3. Brandes, U.: A faster algorithm for betweenness centrality. J. Math. Sociol. **25**(2), 163–177 (2001)
4. Von Luxburg, U.: A tutorial on spectral clustering. Stat. Comput. **17**(4), arXiv:0711.0189 [cs.Ds] (2007)
5. Tang, F., Wang, Y., Su, J., Wang, C.: Spectral clustering-based network community detection with node attributes. Stat. Interface **12**, 123–133 (2019)
6. Ren, S., Zhang, S., Wu, T.: An improved spectral clustering community detection algorithm based on probability matrix. Discret. Dyn. Nat. Soc. **2020**, 6. Article ID 4540302. https://doi.org/10.1155/2020/4540302 (2020)
7. Ding, S., Jia, H., Du, M., Xue, Y.: A semi-supervised approximate spectral clustering algorithm based on HMRF model. Inf. Sci. **429**, 215–228 (2018)
8. Richard, L., Burden, J.: Douglas Faires: Numerical Analysis, 9th edn. Brooks/Cole (2011)
9. Leskovec, J., Krevl, A.: SNAP datasets tanford large network dataset collection. https://snap.stanford.edu (2014)

Probable Characteristics of Multi-channel Queuing Systems with "Impatient" and "Patient" Claims

Tran Quang Quy, Ngo Huu Huy(✉), and Nguyen Vu Hai

Thai Nguyen University of Information and Communication Technology, Thai Nguyen 24000, Vietnam

{tqquy,nhhuy}@ictu.edu.vn

Abstract. A mathematical model of an open multi-channel queuing sys-tem with a limited average residence time for claims in a queue having both "impatient" and "patient claims" is presented. With the help of the Kummer confluent hypergeometric function, a detailed mathematical formalization of the model was carried out and the basic probability char-acteristics of queuing systems of this type were computed. The main difference from the previously studied models of systems of this kind is the first introduced assumption that the so-called "impatient" customers can leave the queue only when the number of customers in the queue exceeds some predetermined fixed value. In this work, we used the general-purpose simulation system GPSS World based on the Monte Carlo method (random test method) and Mathcad as tools for simulation and visualization. The results show that the probability of leaving from the queue (impatient) of the service queue among the "patient" claims is less than the probability of service, it is the similarity of being in the queue among the claims that can exit it without waiting for the service to start.

Keywords: Queuing system · Demand stream · Queue · Service facility

1 Introduction

1.1 The Queuing Systems with "Patient" and "Impatient" Claims

In the theory of queuing, queuing systems (QS) with some kind of constraints on the control parameters of the system have always been of particular interest. The point in this case is that systems with various constraints are very often in demand when a variety of applied problems in different, sometimes very far from each other subject areas, are solved. In particular, logistics traditionally connected with the theory of mass service, as well as such innovative areas of modern applications as, for example, teletraffic theory, telecommunication theory, and many others can be referred to such subject areas.

Among problems with constraints, a class of problems with constraints imposed on the average residence time of a claim received by the system, both waiting for the service

Tran Quang Quy, Ngo Huu Huy: These authors contributed equally to this work.

P. T. Nghia et al. (Eds.): ICTA 2023, LNNS 847, pp. 280–290, 2023.
https://doi.org/10.1007/978-3-031-49529-8_30

to start in the queue and in queuing system as a whole (in the queue and under the service as well) is specified. In other words, some claims in the QS of theses types are the so called "impatient" ones which, after waiting for some time, may leave either the queue or the system, including those being at the service stage [1–3]. However, models of this type are still least studied classes among all types of QS.

From the point of view of mathematics, the key goal of studying systems of this kind is as follows. In the course of computation, even for the most convenient for research Markov-type models, sums of an infinite or finite number of summands appear, however, they are not reduced to the sums of corresponding geometric progressions. Thus, to solve problems of this kind, one has to resort to approximate numerical schemes in which each numerical characteristic of the problem is computed separately from others by summing several first terms of the corresponding finite or infinite series. In this case, it is impossible, of course, to obtain a closed analytical solution of the problem, though it is feasible, rather roughly, to evaluate main characteristics of a queuing system of this type. For a number of applied problems, this procedure is quite sufficient, however, as a rule, it is not possible to study processes occurring in systems of this kind in more detail. In modern conditions, the absence of a closed analytical solution within which all main numerical characteristics of the system would be computed with the required accuracy and at the same time associated with each other, is a significant gap in queuing theory, understood as an applied field of research.

In view of the foregoing, a more or less successful attempt of a new approach was made to study the queuing systems with constraints on the average residence time in the queue or in the system as a whole. The meaning of these works is as follows. Namely, [3] first proposed using a first-order Mittag-Leffler function (generalization of the exponential function) applied in the theory of the function of complex variable and integral transformations [5, 6] for summing series other than geometric progressions. Some of the results obtained in these papers were also presented in [7]. It was noted that this model formalization can also be carried out using Kummer confluent hypergeometric function to simplify a number of intermediate calculations and the type of final formulas. This idea was later implemented in the second edition of the monograph [8].

Full mathematical formalization of the appropriately assigned problem, including the calculation of the first and second moments of corresponding numerical characteristics of the system, was carried out for the first time in [3, 4, 8]. However, these works propose the solution of the problem with the constraint on the average time only in case when all the demands entering the queuing system can be called "impatient" demands (claims). Meanwhile, from the point of view of possible applications, the expanded formulation of the problem seems to be more interesting in a certain way.

Each QS consists of a certain number of parameters that determine the provision of service (these are machines, transport carts, robots, communication lines, cashiers, salespeople, etc.). Every QS exists to service some flows of applications (claims) arriving at some random moments in time.

From the point of view of practice, it would be very interesting to consider such a variant of problem statement in which the so-called "impatient" claims can leave a queue only when its length exceeds a predetermined fixed numerical value. Let us denote this value by the letter E in our further computations. After E claims are left before the

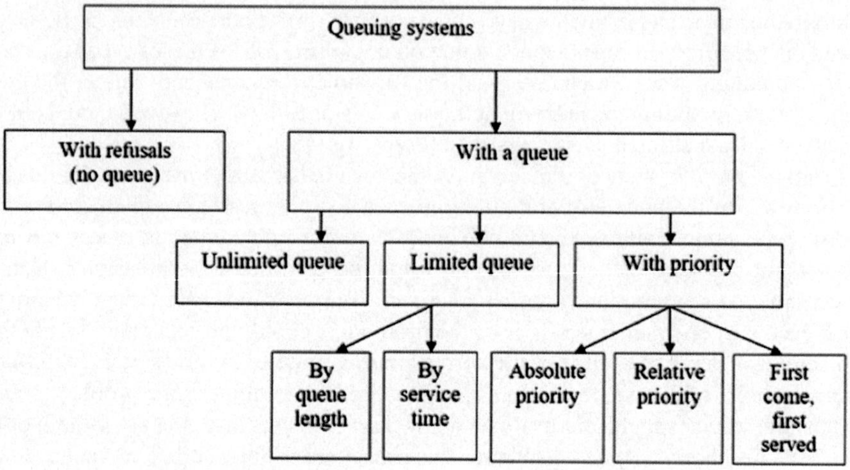

Fig. 1. Classification of QS according to the input claims

demand in the service queue, the demands cease to leave the queue, and in any case wait for the service to start, i.e. they leave the category of "impatient" demands and become "patient" ones.

1.2 The Characteristic Parameters of a Multi-channel Queuing System

Let us suppose that we have a multichannel QS with a homogeneous infinite simplest flow of claims and a queue of unlimited length. Let the intensity of the claims stream be equal to λ, and the intensity of service, i.e. the average number of claims served by one device per unit of time, be μ. The flow of service will also be considered the simplest (with the intensity μ). Let us assume now that the time spent by a single claim in a queue is limited to some random time t with an average value of $\bar{t}$. Thus, each claim in the queue is impacted by a kind of exits flow with the intensity $\nu = 1/\bar{t}$. It is clear that if this flow is of the simplest nature, then the process proceeding in the QS will be a Markov process. In accordance with the foregoing, we now assume that queued claims cease to leave the queue when E demands waiting for service are left before them. Thus, this QS has claims of two different types at the same time-both "impatient" and "patient" ones. Figure 1 presents the graph of such a queuing system. Let us carry out the mathematical formalization of the problem.

As we can see from the graph, we have a classical scheme of a death and reproduction process. Applying general expressions (e.g. [9]) for the probabilities of the limit (stationary) states in this scheme, we obtain (Fig. 2).

$$p_1 = \frac{\lambda}{\mu} p_0; \; p_2 = \frac{\lambda^2}{2!\mu^2} p_0; \ldots; p_m = \frac{\lambda^m}{m!\mu^m} p_0; \; p_{m+E} = \frac{\lambda^{m+E}}{m! m^E \mu^{m+E}} p_0 \quad (1)$$

or in the notation $\rho = \lambda/\mu$ (the reduced intensity of the claims stream, i.e. the average number of claims entering the system in the average service time of one claim)

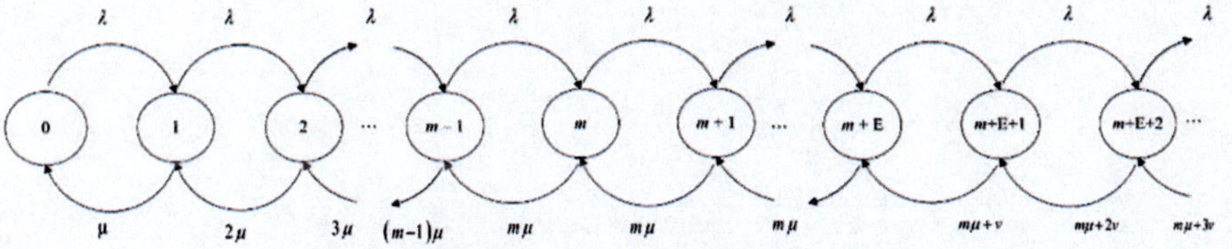

Fig. 2. The queuing system with E demands

and $\beta = \nu/\mu$ (the average number of claims leaving the queue unserved in the average service time of one claim)

$$p_1 = \rho\, p_0;\, p_2 = \frac{\rho^2}{2} p_0; \ldots; p_m = \frac{\rho^m}{m!} p_0; \ldots; p_{m+E} = \frac{\rho^{m+E}}{m!\, m^E} p_0; \tag{2}$$

As a result, we have the following formulas for p_k p_k

$$\begin{aligned} p_k &= \frac{\rho^k}{k!} p_0 \ at\ k \le m \\ p_k &= \frac{\rho^k}{m!\, m^{k-m}} p_0 \ at\ m \le k \le m+E \\ p_k &= \frac{\rho^k}{m!\, m^E\, (m+\beta)(m+2\beta)\cdots[m+(k-m-E)\beta]} p_0 \ at\ k > m+E \end{aligned} \tag{3}$$

According to the algorithm proposed in the above works, we divide the numerator and denominator of the third of these ratios by β^{k-m-E}. Then we obtain:

$$\begin{aligned} p_k &= \frac{\rho^{m+E}}{m!} \frac{\alpha^{k-m-E}}{m^E\, (m/\beta+1)(m/\beta+2)\cdots(m/\beta+k-m-E)} p_0 \\ &= \frac{\rho^{m+E}}{m!} \frac{\alpha^{k-m-E}}{m^E\, (m/\beta+1)_{k-m-E}} p_0 \ at\ k > m+E \end{aligned} \tag{4}$$

$$(a)_k = a(a+1)(a+2)\ \ldots\ (a+k-1)\ where\ (a)_0 = 1 \tag{5}$$

2 Average Number of Claims with Impatient Queues

2.1 General Formula for Calculation

The Eq. (5) is as the Pochhammer symbol [10] and the value $\alpha = \rho/\beta = \lambda/\nu$ obviously shows what average number of claims enters the system during the average time of one "impatient" claim being in the queue. Then the total sum of probabilities of stationary states of the system is:

$$p_0 + \rho p_0 + \frac{\rho^2}{2!} p_0 + \cdots + \frac{\rho^m}{m!} p_0 + \ldots + \frac{\rho^{m+E}}{m!\, m^E} p_0 +$$
$$\frac{\rho^{m+E}}{m!} \frac{\alpha}{m^E (m/\beta + 1)} p_0 + \frac{\rho^{m+E}}{m!} \frac{\alpha^2}{m^E (m/\beta + 1)(m/\beta + 2)} p_0 +$$
$$+ \frac{\rho^{m+E}}{m!} \frac{\alpha^3}{m^E (m/\beta + 1)(m/\beta + 2)(m/\beta + 3)} p_0 = \tag{6}$$
$$= \left\{ e_{m-1}(\rho) + \frac{\rho^m}{(m-1)!(m-\rho)} \left[1 - \left(\frac{\rho}{m}\right)^E \right] + \frac{\rho^{m+E}}{m!\, m^E} \sum_{k=0}^{\infty} \frac{\alpha^k}{(m/\beta + 1)_k} \right\} p_0$$

In this relation, $e_m(\rho) = 1 + \frac{\rho}{1!} + \frac{\rho^2}{2!} + \ldots + \frac{\rho^m}{m!}$ is an incomplete exponential function (incomplete exponent). Moreover, $e_0(\rho) = 1$. It is clear that $e_m(\rho) \to e^{\rho}$ with $m \to \infty$. The normalization condition for $\sum_{i=0}^{\infty} p_i = 1$ obviously gives in this case

$$\left\{ e_{m-1}(\rho) + \frac{\rho^m}{(m-1)!(m-\rho)} \left[1 - \left(\frac{\rho}{m}\right)^E \right] + \frac{\rho^{m+E}}{m!\, m^E} \sum_{k=0}^{\infty} \frac{\alpha^k}{(m/\beta + 1)_k} \right\} p_0 =$$
$$= \left\{ e_{m-1}(\rho) + \frac{\rho^m}{(m-1)!(m-\rho)} \left[1 - \left(\frac{\rho}{m}\right)^E \right] + + \frac{\rho^{m+E}}{m!\, m^E} M(1; m/\beta + 1; \alpha) \right\}$$
$$p_0 = 1 \tag{7}$$

2.2 The Kummer Confluent Hypergeometric Function

Where $M(a; b; z)$ is the Kummer confluent hypergeometric function defined by the relation [11, 12]

$$M(a; b; z) = \sum_{k=0}^{\infty} \frac{(a)_k}{(b)_k} \frac{z^k}{k!} \tag{8}$$

Since $(1)_k = k\,!$. Then for the probability p_0 of complete system downtime, we have

$$p_0 = \left\{ e_{m-1}(\rho) + \frac{\rho^m}{(m-1)!(m-\rho)} \left[1 - \left(\frac{\rho}{m}\right)^E \right] + \frac{\rho^{m+E}}{m!\, m^E} M(1; m/\beta + 1; \alpha) \right\}^{-1} \tag{9}$$

The latter expression, in turn, can be further simplified as follows. Noticing that by definition

$$(a)_k = \frac{\Gamma(a+k)}{\Gamma(a)}, \tag{10}$$

where Γ is a gamma function, let us rewrite the expression for the Kummer confluent hypergeometric function as

$$M(a; b; z) = \frac{\Gamma(b)}{\Gamma(a)} \sum_{k=0}^{\infty} \frac{\Gamma(a+k)}{\Gamma(b+k)} \frac{z^k}{k!} \tag{11}$$

and the quantity of interest in the way

$$M(1; b+1; z) = \Gamma(b+1) \sum_{k=0}^{\infty} \frac{z^k}{\Gamma(b+1+k)} \tag{12}$$

since $\Gamma(k+1) = k!$. Further, to simplify the latter expression, we use the following chain of formulas:

$$\sum_{k=0}^{\infty} \frac{z^k}{\Gamma(b+1+k)} = \sum_{k=1}^{\infty} \frac{z^{k-1}}{\Gamma(b+k)} = \frac{1}{z} \sum_{k=1}^{\infty} \frac{z^k}{\Gamma(b+k)}$$

$$= \frac{1}{z}\left[\sum_{k=0}^{\infty} \frac{z^k}{\Gamma(b+k)} - \frac{1}{\Gamma(b)}\right] \tag{13}$$

Thus, we obtain:

$$M(1; b+1; z) = \frac{\Gamma(b+1)}{z}\left[\sum_{k=0}^{\infty} \frac{z^k}{\Gamma(b+k)} - \frac{1}{\Gamma(b)}\right] \tag{14}$$

when, taking into account the known recurrence relation for the gamma function $\Gamma(b+1) = b\,\Gamma(b)$,

$$M(1; b+1; z) = \frac{b}{z}\left[\Gamma(b) \sum_{k=0}^{\infty} \frac{\alpha^k}{\Gamma(b+k)} - 1\right] = \frac{b}{z}\left[\sum_{k=0}^{\infty} \frac{z^k}{(b)_k} - 1\right] \tag{15}$$

$$M(1; b+1; z) = \frac{b}{z}\left[M(1; b; z) - 1\right] \tag{16}$$

The Eq. (16) is the recurrence formula for $M(1; b; z)$. In this case, it is obvious that we have $b = m/\beta$, $z = \alpha = \rho/\beta$, and $b/z = m/\rho$, then:

$$M(1; m/\beta + 1; \alpha) = \frac{m}{\rho}\left[M(1; m/\beta; \alpha) - 1\right] \tag{17}$$

As a result, the probability of complete system downtime will take the following form:

$$p_0 = \left\{e_{m-1}(\rho) + \frac{\rho^m}{(m-1)!(m-\rho)}\left[1 - \left(\frac{\rho}{m}\right)^E\right] + \frac{\rho^{m+E-1}}{(m-1)!m^E}\left[M(1; m/\beta; \alpha) - 1\right]\right\}^{-1} \tag{18}$$

3 Probability of System State

To monitor the correctness of the results obtained due to these computations, let us verify the obtained relation in two limiting cases for which the corresponding solutions are known. It is clear that in the first limiting case, when $E \to 0$, the relation (18) turns into a previously published dependence [8].

$$p_0(E = 0) = \left\{ e_{m-1}(\rho) + \frac{\rho^{m-1}}{(m-1)!} \left[M\left(1; m/\beta; \alpha\right) - 1 \right] \right\}^{-1} \tag{19}$$

It is interesting to note that the same formula will apparently remain valid for the general model $E \neq 0$, but in the degenerate case when $\rho = m$.

Let us regard the second limiting case when $\beta \to 0$. This obviously requires the consideration of how the Kummer confluent hypergeometric function $M\left(1; m/\beta; \alpha\right)$ behaves when the values of the parameter β are close to zero. The general formula for the expansion of the function $M\left(1; m/\beta; \alpha\right)$ in its K. Maclaurin series for small values of the parameter β is obtained in [7] and has the following form:

$$M\left(1; m/\beta; \alpha\right) \approx 1 + \frac{\rho}{m - \rho} \left[1 - \frac{\rho}{(m-\rho)^2} \beta + \frac{\rho\,(m + 2\,\rho)}{(m-\rho)^4} \beta^2 \right] \tag{20}$$

Thus, it is easy to verify that:

$$\lim M\left(1; m/\beta; \alpha\right) |_{\beta \to 0} = \frac{m}{m - \rho} \tag{21}$$

and then in the second limiting case when $\beta \to 0$, the relation (6) turns into a known dependence:

$$p_0(\beta = 0) = \left[e_{m-1}(\rho) + \frac{\rho^m}{(m-1)!\,(m-\rho)} \right]^{-1} = \left[e_m(\rho) + \frac{\rho^{m+1}}{m!(m-\rho)} \right]^{-1} \tag{22}$$

of the multi-channel QS models with expectation (according to Kendall's classification-M/M/m model).

The probability of waiting, i.e. the probability that incoming demands will find all the channels occupied (regardless of whether it will wait for service or not) is:

$$\begin{aligned} p_{\text{wait}} &= \sum_{k=m}^{\infty} p_k = \frac{\rho^m\, p_0}{m!} \sum_{k=m}^{m+E-1} \frac{\rho^{k-m}}{m^{k-m}} \\ &+ \frac{\rho^{m+E}\, p_0}{m!\, m^E} \sum_{k=m+E}^{\infty} \frac{\alpha^{k-m-E}}{(m/\beta + 1)_{k-m-E}} = \\ &= \frac{\rho^m\, p_0}{(m-1)!(m-\rho)} \left[1 - \left(\frac{\rho}{m}\right)^E \right] + \frac{\rho^{m+E}\, p_0}{m!\, m^E} M\left(1; m/\beta + 1; \alpha\right) \end{aligned} \tag{23}$$

that, taking into account the recurrence relation (16), suggests:

$$p_{\text{wait}} = \frac{\rho^m\, p_0}{(m-1)\,!(m-\rho)}\left[1-\left(\frac{\rho}{m}\right)^E\right] + \frac{\rho^{m+E-1}\, p_0}{(m-1)\,!\,m^E}\left[M\left(1; m/\beta\,;\,\alpha\right)-1\right] \tag{24}$$

In the first limiting case, when $E = 0$, the latter relation apparently becomes dependent [8]:

$$p_{\text{wait}}(E=0) = \frac{\rho^{m-1}\, p_0}{(m-1)\,!}\left[M\left(1; m/\beta;\,\alpha\right)-1\right] \tag{25}$$

Using the expansion (20), we can show that for small values, close to zero, of the parameter β:

$$p_{\text{wait}}(\beta) \approx p_{\text{wait}}(\beta=0)\left\{1-\left(\frac{\rho}{m}\right)^E\left[\frac{\rho}{(m-\rho)^2}\,\beta-\frac{\rho\,(m+2\,\rho)}{(m-\rho)^4}\,\beta^2\right]\right\} \tag{26}$$

where from it comes:

$$p_{\text{wait}}(\beta=0) - p_{\text{wait}}(\beta) \approx \frac{\rho^{E+1}\, p_{\text{wait}}(\beta=0)}{(m-\rho)^2\, m^E}\left[\beta-\frac{m+2\,\rho}{(m-\rho)^2}\,\beta^2\right] \tag{27}$$

In conclusion, it should be noted that formula (24) for the probability of waiting for the start of servicing a claim newly entering the system with the help of the probability of waiting $p_{\text{wait}}(0)$ can be written in the following most compact form:

$$p_{\text{wait}}(\beta) = p_{\text{wait}}(0)\left\{1-\left(\frac{\rho}{m}\right)^E+\left(\frac{\rho}{m}\right)^E\frac{m-\rho}{\rho}\left[M\left(1; m/\beta\,;\,\alpha\right)-1\right]\right\} \tag{28}$$

Or

$$p_{\text{wait}}(\beta) = p_{\text{wait}}(0)\left\{1-\left(\frac{\rho}{m}\right)^{E-1}+\left(\frac{\rho}{m}\right)^E\frac{m-\rho}{\rho}\,M\left(1; m/\beta\,;\,\alpha\right)\right\} \tag{29}$$

Here $p_{\text{wait}}(0)$ is an expression that coincides in form with the formula (10), however, unlike it, it contains within itself a dependence on the parameters β and E, which is contained in the probability of the complete absence of claims in the system $p_0 = p_0(\beta,\, E)$ according to the formula (18).

Summing up, let us note that in the limiting case, when $E = 0$, in compliance with the formulas (27, 28), we can propose a more elegant notation of the relation (25):

$$p_{\text{wait}}(E=0) = p_{\text{wait}}(0)\,\frac{m-\rho}{\rho}\left[M\left(1; m/\beta\,;\,\alpha\right)-1\right] \tag{30}$$

4 Simulation

In this work, we used the general-purpose simulation system GPSS World based on the Monte Carlo method (random test method) and Mathcad as tools for simulation and visualization. This tools are widely used for modeling both discrete and continuous systems like A.A. Markov. Such modeling systems have specialized tools that implement additional capabilities for organizing model experiments on a PC. What is especially valuable, they provide an opportunity to take into account the time factor in the models, that is, to build dynamic simulation models, which is extremely important when studying systems with queues. In the models written in GPSS World, it is possible to take into account a large number of different factors and at the same time abandon many restrictions and assumptions. Such software tools usually contain built-in high-level programming languages focused on a specific subject area, which makes it possible to create model programs for studying complex systems at the lowest cost, which fully applies to the GPSS World simulation system. Finally, the GPSS World system operates in the Windows operating environment and is maximally focused on the use of modern technologies that provide high interactivity and visual presentation of information. All of the above advantages led to the choice in this case in favor of GPSS World.

Here, a request entering the system is rejected, that is, it leaves the system without service, if two conditions are met at the time the transaction enters the system: the number of occupied places in the queue exceeds the specified number E and the average waiting time in the queue exceeds the specified value T.

Suppose that, we have multi channels for the service, then with $m = 3$ and $m = 4$ correlatively obtain (Figs. 3, 4 and 5).

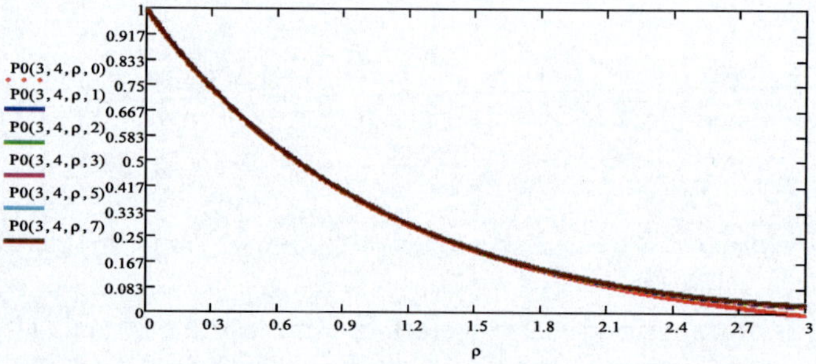

Fig. 3. Probabilities of system states for the case m = 3, E = 4 when only one channel for the service

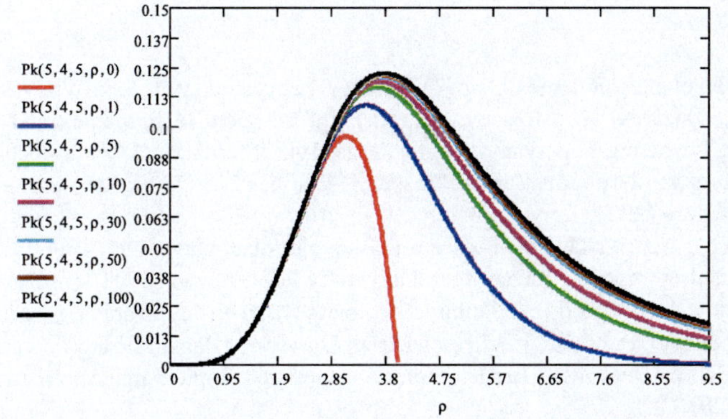

Fig. 4. Probabilities of system states for the case k = 3, m = 4, E = 5

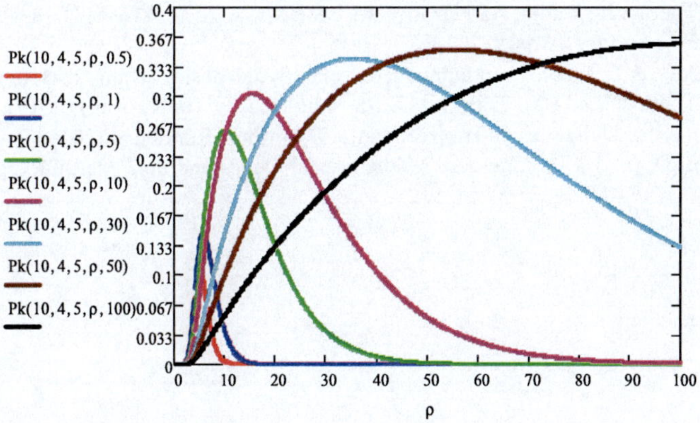

Fig. 5. Probabilities of system states for the case k = 10, m = 4, E = 5

5 Conclusion

Following the graphs, in the intermediate region of states of the system, when $m \leq k \leq m + E$, by absolute probability value, the probabilities of its states are minimal by the ratio to two other regions of states. Or, in other words, the probability of leaving from the queue (impatient) of the service queue among the "patient" claims is less than the probability of service, it is the similarity of being in the queue among the claims that can exit it without waiting for the service to start.

A mathematical model is proposed for an open multichannel queuing system with a limited average time of a claim being in the queue for service. The main difference from the previously studied models of systems of this kind is the first introduced assumption that the so-called "impatient" customers can leave the queue only when the number of customers in the queue exceeds some predetermined fixed value.

References

1. Shortle, J.F., et al.: Fundamentals of Queueing Theory, vol. 399. John Wiley & Sons (2018)
2. Dudin, A., Nazarov, A., Moiseev, A., (ed.): Information Technologies and Mathematical Modelling. Queueing Theory and Applications: 19th International Conference, ITMM 2020, Named After AF Terpugov, Tomsk, Russia, December 2–5, 2020, Revised Selected Papers. Springer Nature (2021)
3. Kirpichnikov, A.P.: Prikladnaya teoriya massovogo obsluzhivaniya (Applied Queuing Theory). Publishing Office of Kazan State University Publ, Kazan (2008). (in Russian)
4. Kirpichnikov, A.P.: Metody prikladnoy teorii massovogo obsluzhivaniya (Methods of Applied Queuing Theory). Publishing Office of Kazan University Publ., Kazan (in Russian) (2011)
5. Gorenflo, R., et al.: Mittag-Leffler Functions, Related Topics and Applications. Springer, Berlin (2020)
6. Fang, M., Yang, D., Liu, D.: Value distribution of meromorphic functions concerning rational functions and differences. Adv. Difference Equ. **2020**(1), 692 (2020)
7. Kirpichnikov, A., Titovtsev, A., Yakimov, I.: Mittag-Leffler Function in Applied Problems of Queuing Theory. In: Dudin, A., Nazarov, A., Moiseev, A. (eds.) ITMM 2018. CCIS, vol. 912, pp. 225–235. Springer, Sham (2018)
8. Kirpichnikov, A.P.: Metody prikladnoy teorii massovogo obsluzhivaniya (Methods of Applied Queuing Theory), 2nd edn. Editorial URSS Publ Moscow (in Russian) (2017)
9. Hua, X.H. (ed.): Dynamics of Transcendental Functions. Routledge (2019)
10. Zwillinger, D. (ed.): CRC Standard Mathematical Tables and Formulas. CRC press (2018)

Processing Overlapping Communities Using the Gradient Descent Optimization Method

Nguyen Hien Trinh[1(✉)] and Cap Thanh Tung[2]

[1] Thai Nguyen University of Information and Communication Technology, Thai Nguyen, Vietnam
nhtrinh@ictu.edu.vn

[2] Thai Nguyen University of Education, Thai Nguyen, Vietnam
tungct@tnue.edu.vn

Abstract. Community structure detection is an important study in social network analysis, especially with complex networks. In a community structure, usually, a node belongs to only one community. In reality, however, networks are built in different relationships and nodes can be shared by multiple communities. An overlapping node is a vertex that belongs to more than one community. The concept of overlapping community has emerged and has great practical significance for applications on social networks. To detect, analyze or process overlapping communities, many soft algorithms (probability-based computation, giving acceptable results…) are proposed to perform. In this study, we introduce the tool "Maximum Likelihood Estimation-MLE, a commonly used tool in machine learning, then apply Gradient descent to calculate the extrema for the proposed objective function. The experimental results prove the flexibility and effectiveness of the proposed method, which is good support for further studies on social network analysis and graph data mining.

Keywords: Social network graph · Overlapping communities · Maximum likelihood estimation · MLE · Gradient descent

1 Introduction

As we all know, overlapping communities on social networks is a practical problem that cannot be ignored. The problem of detecting overlapping communities in complex network graphs is a matter of intense research and development interest [1–6]. Among them, Gregory proposes the CONGO algorithm [2], to detect overlapping communities using the concept of community structure division according to local intermediate degrees. Steve Gregory [3] proposes the COPRA algorithm to detect overlapping communities by extending the label and the label propagation step to include information about more than one community. Saradha and Arul have focused their research on the overlapping community detection technique that is optimized by applying the improved community coefficient-based label propagation algorithm (IVICCOPRA) [6]. In addition, it is possible to add algorithms based on clustering theory (CPM [5]), algorithms based on the

P. T. Nghia et al. (Eds.): ICTA 2023, LNNS 847, pp. 291–300, 2023.
https://doi.org/10.1007/978-3-031-49529-8_31

local expansion (LFM [4]), algorithms based on the agent (COPRA [3]), algorithm based on link partition (LINK [7])…The processing algorithms for them are highly complex and time-consuming. In addition, it often requires more global data about the network, so it also consumes space when analyzing large-scale social networks. To detect, analyze or process overlapping communities, many soft algorithms (probability-based computation, giving acceptable results…) are proposed to perform. Starting from the idea: with overlapping communities, it is necessary to find a method of adjustment to make the problem easier and more suitable, we propose the following: The continued evaluation for graph data mining will be based on probability theory and make assumptions: Two objects (two nodes/vertices in the graph) that are members of more than one community are more likely to become close friends, so these two objects can be placed in the same community. In the process of analyzing and solving problems, we will use the MLE method [8, 9]−(Maximum-likelihood estimation). MLE is a method of estimating the parameters of an assumed probability distribution, given some observed data. This is achieved by maximizing a likelihood function so that, under the assumed statistical model, the observed data is most probable. We apply the MLE to generate the objective function, then use Gradient descent to perform parameter optimization for the entire social network graph, as a basis to convert the problem to simpler.

In this paper, we introduce the method of procesing overlapping communities by the MLE tool combined with the Gradient Descent optimization method. The remaining of the paper: Sect. 2, Discussing related research works. Section 3 presents the algorithms. Section 4, Experimental results. Finally, Sect. 5, Conclusion and contributions of the research, along with suggesting future research directions.

2 Related Works

MLE predicts the parameters of a statistical model based on available "observations", by finding a set of parameters that maximizes the probability that the model with that parameter set produces available observations. Assuming the model is described by the parameter set θ, the "observations" (or data points) are $x_1, x_2, \ldots, x_N$. Then we need to find:

$$\theta = \text{argmax}\{\, p(x_1, x_2, \ldots, x_N|\theta) \,\} \tag{1}$$

where: $p(x_1,x_2,\ldots,x_N|\theta)$ is the probability that the events $x_1, x_2, \ldots, x_N$ occur simultaneously, is called the likelihood. (That's why this method is called Maximum Likelihood). However, solving the problem (1) directly is often difficult. We can simplify the problem by assuming the data points occur independently of each other. Then, (1) becomes:

$$\theta = \text{argmax}\left\{ \prod\prod p(x_n|\theta) \right\} \tag{2}$$

Gradient Descent [10, 11] is an extremely effective method in optimization in general and in machine learning in particular. This method allows us to find the extreme of a function (minimum or maximum). The formula for Gradient descent is as follows:

$$\theta \leftarrow \theta - \eta \nabla_{\theta} f(\theta) \tag{3}$$

where θ is the set of variables to be updated, η is the update rate (reasonable jump); $\nabla_{\theta} f(\theta)$ is the gradient of the function f over the set θ (the derivative of the function f at). A stopping condition is a value that does not change (convergence) after a finite number of times.

3 Recommended Algorithms

3.1 Problem Analysis

For example, let's analyze a simple case: suppose there is a social network graph with 2 communities. In a company, employees are classified into two groups of training: one group is specialized in information technology (IT) and the other group is in economics. We see that nodes 4,5,6,7 are likely to be closer to each other than other nodes because of many common characteristics. Thus, if we use a join edge to represent the probability of becoming close friends (up to a threshold value), then the nodes in the common part of the two groups (the intersection of the two groups) will have a greater value of the connection. We will use the concept of probability to talk about this. That is, the model that we build will consider the possibility of a connection between the vertices of the graph and that possibility is characterized by the probability value. Through the above example, considering the general case, we have a completely random graph, including vertices and nodes. The possible edges, will carry the probability value p (of course the possibility of no edge will be 1-p). Thus, p is a small but non-zero number. According to the MLE method, we will calculate based on the parameter p to give the model the maximum aggregate value. If we give each edge a probability of existence p, then the probability that an edge does not exist is 1-p. With 11 vertices, the maximum possible number of edges will be:

$$C_{11}^{2} = \frac{11!}{2!(11-2)!} = \frac{11!}{2!9!} = 55$$

Then the probability of the whole system is:

$$f = p^{19}(1-p)^{55-19} = p^{19}(1-p)^{36}$$

The expression f reaches its maximum when its derivative is zero, i.e.:

$$19p^{18}(1-p)^{36} - 36p^{19}(1-p)^{35} = 0 \ hay \ p^{18}(1-p)^{35}(19(1-p) - 36p) = 0$$

When p = 1 or p = 0, then f = 0, for maximum f, then:

$$19(1-p) - 36p = 0$$

So: $p = \frac{19}{55} \approx 0.345455$, the probability of existence of edge of each pair of vertices in the given graph is 0.35. In general, if u, v are two concurrent members in a non-empty set M of communities, then the probability that there is an edge (u,v) between them is:

$$p_{u,v} = 1 - \prod_{C \in M} (1 - p_C) \tag{4}$$

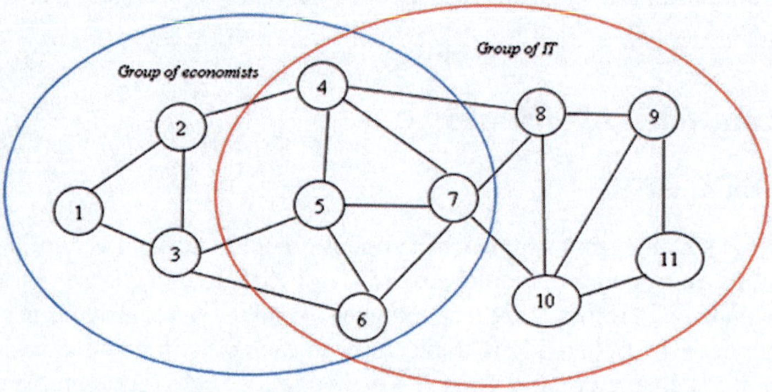

Fig. 1. Social network graph has 11 vertices, 2 communities

And especially when processing: If u and v are not together in any group, then $p_{u,v}$ will be some extremely small, non-zero number ε. Let M_{uv} be the set of communities (groups) that have the same node u and node v in the community, then the possibility of the existence of edge set E for the generated graph will be:

$$\prod_{(u,v) \in M} p_{u,v} \prod_{(u,v) \notin M} (1 - p_{u,v}) \tag{5}$$

3.2 Algorithm CEF (Comput Exponent for Facter)

Algorithm to calculate the exponent for the factors of the function that needs to calculate the maximum CEF.

Input: + Adjacent matrix A represents graph G. Communities C, D, OC.
+ The matrix B represents the attribute of the vertex

Output: The objective function *f* to be maximized (the exponents for the factors *mx, my, mcxy, mnx, mny, mepsi, mcxy_1* of the polynomial to be maximized)

```
n=length(A);
mx=0; my=0; mcxy=0;mnx=0;mny=0;mepsi=0;mcxy_1=0;
for i=1:n
for j=i:n
    a <- B(i,:)
    b<- B(j,:)
  if (A(i,j)= =1)
    if (node i and j ∉ CO)
          if (node i and j ∈ C)     mx = mx+1;
          if (node i and j ∈ D)     my=my+1;
          if (node i and j ∈ OC)    mcxy = mcxy+1;
  if (A(i,j)= =0)
    if (node i and j ∉ OC)
          if (noide i and j ∈ C)     mnx = mnx+1;
          if (node i and j ∈ D)      mny=mny+1;
          if (node i and j ∉ OC)     mcxy_1=m cxy_1+1;
          if (node i and j ∉ C, D)   mepsi=mepsi+1;
return mx,my,mcxy,mnx,mny,mepsi,mcxy_1;
```

The complexity of the CEF algorithm is $O(n^2)$.

3.3 Calculate the Derivative by the Difference Method

Differential method to approximate the derivative, with very small ε (eg 10^{-6}):

$$f'(x) \approx \frac{f(x+\varepsilon) - f(x-\varepsilon)}{2\varepsilon} \tag{6}$$

Algorithm **Ndf**

Input: Initialization vector w, function cost(w)

Output: Derivative value at w

```
eps = 1e-6
g=np.zeros_like(w)
for i in range(len(w)):
  w_p=w.copy()
  w_n=w.copy()
  w_p[i] +=eps
 w_n[i] - =eps
 g[i]=(cost(w_p)-cost(w_n))/(2*eps)
 return g
```

The complexity of the algorithm is O(k), where k is the size of the vector w.

3.4 Probability for the Expression to Maximize

Gradient Descent method:

1. Predict a starting point $\theta = \theta_0$
2. Update θ_n according to the formula $\theta \leftarrow \theta - \eta \nabla_\theta f(\theta)$ until an acceptable result is obtained.

Where:
+ θ is the set of variables to be updated, η is the update rate (reasonable jump);
+ $\nabla_\theta f(\theta)$ is the gradient of the function f over the set θ (the derivative of the function f at)

Algorithm MyGD // usage: (x2,y2,it2)= myGD(eta,x0,y0,precision)
// Can take precision = 0.0001; eta = 0.0015; k=400;

Input: The polynomial $f(p_c,p_d,...)$ needs to be maximized
Output: Probability values p_c, p_d, … of overlapping communities.

```
select [x0, y0];
x1 = [x0];
y1 = [y0];
for it in range(k)
{
   g1,g2 = Ndf(x1[-1],y1[-1],f); //index [-1] is to get the last value of
                                   the array
   x_new=x1[-1] - eta*g1
   y_new=y1[-1] - eta*g2
   if (abs(g1) < precision) and(abs(g2) < precision) :
   break
   x1.append(x_new)
   y1.append(y_new)
}
pc=x1[-1]
pd=y1[-1]
return (pc, pd,it)
```

The complexity of the algorithm is O(k), where k is the number of iterations.

Consider the Social Network Shown in Fig. 1, Which Has 11 Vertices and 23 Edges, Assuming that Using the Vertex Properties, Two Communities C and D Have Been Identified as Illustrated. Apply the General Formula (5), and the Algorithm CEF, We Have a Function that Needs to Calculate the Maximum:

$$f = x^6 y^8 (x + y - xy)^5 (1 - y)^{14} (1 - x)^9 (1 - x - y + xy) \quad (7)$$

The convergence process is shown on contour lines using Gradient Descent with (x_0 = 0.5, y_0 = 0.7), $\varepsilon = 1e - 13$, $\eta = 0.001$: : point (0.316, 0.573). With (x_0 = 0.15, y_0 = 0.7), we have the same result: Communities C = {1,2,3,4,5,6,7}; D = {4,5,6,7,8,9,10,11}; Overlapping community = {4,5,6,7}; Thus, the function f reaches its maximum point when $p_c = 0.32, p_d = 0.57$. Therefore, the most likely overlapping community is the *32%* chance that a member in set *C* has an edge and a *57%* chance that a member in a set *D* has an edge connecting them (Fig. 2).

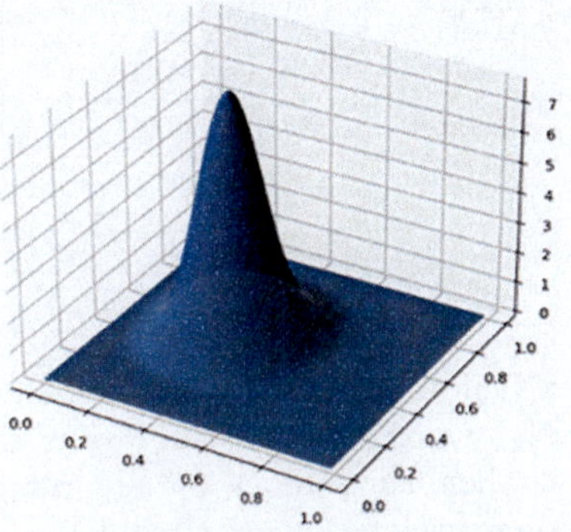

Fig. 2. 3D representation of the function f (Eq. 7)

3.5 Approach How to Process Overlapping Communities

Thus, we must build the expression to calculate the probability of the edges giving the maximum value for the whole system. In the optimization methods, we can apply the Gradient Descent method to solve the problem of finding the extrema (local or global) and get an approximate p (acceptable) value for the edges. This function has many variables, and we calculate the gradient by the difference method. The approximate results will be used to compare with the exact results calculated previously. After determining the edge probability value of the vertices in the overlapping communities of the whole network, we can recalculate the weights of the edges in the original network and use them as input data for the Community detection algorithm based on association matrix. We already have an algorithm like that (SC_NT [12]). So we present our new algorithm (OCSG - Overlapping Community detection using Spectra and Gradients) as follows:

(i) Use the gradient method to handle overlapping social networks.
(ii) Use the SC_NT algorithm to detect communities

4 Experimental Results

Experimental environment is for configuration: Intel(R) Core i7, CPU @ 2.50 GHz, RAM 8.0 GB, Win 10, 64 bit. Programming language Matlab 2019, Python. Experimental algorithms were run separately on the data set, and in the computer were experimented with only a single program. The results were performed on the datasets published at [13] to evaluate the effectiveness of the OCSG algorithm to detect the communities compared with the algorithms LCDNN [14] and PMOEA [15], comparing the accuracy through the quality of the community (Modularity), the standard mutual information (NMI) of the community detected by the proposed algorithm. In the experimental data, the symbol *n*-number of vertices, *m*-number of edges of the social network under consideration.

Table 1. Experimental results on community quality (Modularity) with OCSG, LCDNN and PMOEA

No	Social networks	n	m	OCSG	LCDNN	PMOEA
1	Karate club	34	78	0.55	0.49	0.51
2	Book amazon	105	441	0.58	0.51	0.54
3	Email-Eu-core	1005	25571	0.61	0.55	0.59
4	Hamsterster	2400	16600	0.63	0.58	0.61
5	Government	7100	89400	0.67	0.63	0.65

Modularity Q results in Table 1 and Fig. 3 show: Modularity of OCSG increased by more than 5.6% of LCDNN, by more than 2.8% of PMOEA. That confirms the quality of the community detected when using the OCSG algorithm has improved, and there has been a closer connection in each result community.

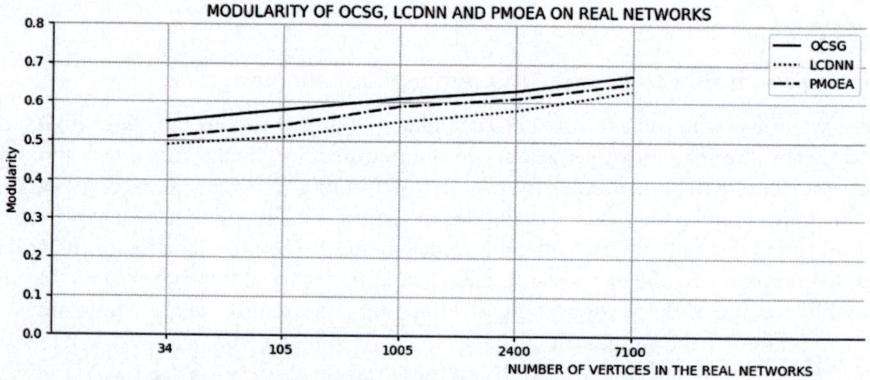

Fig. 3. Community quality comparison (Modularity)

In addition, with different community detection methods, the NMI value will also be used to evaluate the effectiveness of the methods. If the value of NMI is closer to 1, the method is better. Evaluation results of the NMI in Fig. 4 show that: If using OCSG, it increases by more than 5.57% compared to LCDNN. OCSG's NMI increased by more than 2.2% compared to PMOEA. Thus, the experimental results show that the community detected by the OCSG algorithm is more likely to coincide with the real community than the other algorithms (Fig. 4 and Table 2).

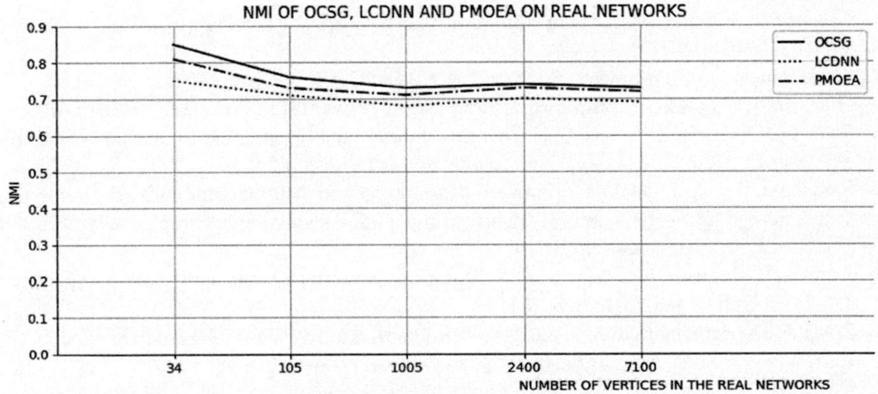

Fig. 4. Comparison of NMI

Table 2. Experimental results on NMI with with OCSG, LCDNN and PMOEA

No	Social networks	n	m	OCSG	LCDNN	PMOEA
1	Karate club	34	78	0.85	0.75	0.81
2	Book amazon	105	441	0.76	0.71	0.73
3	Email-Eu-core	1005	25571	0.73	0.68	0.71
4	Hamsterster	2400	16600	0.74	0.70	0.73
5	Government	7100	89400	0.73	0.69	0.72

5 Conclusion

In this paper, we present research results and propose ideas for analyzing, assessing, detecting, and dealing with the fast overlapping community problem with the ability to control overlapping attributes and evaluate the degree of overlapping nodes. We apply the MLE to generate the objective function, then use Gradient descent to perform parameter optimization for the entire social network graph, as a basis to convert the problem to simpler. We tested several existing social networks with well-known community structures to evaluate research feasibility and effectiveness. Our next research direction will be to continue to further improve the calculation speed, increase efficiency and reduce the complexity of the processing …

References

1. Devi, J.C., Poovammal, E.: An analysis of overlapping community detection algorithms in social networks. In: Twelfth International Multi-Conference on Information Processing IMCIP (2016)
2. Gregory, S.: A fast algorithm to find overlapping communities in networks Lect. In: ECMLPKDD'08, Volume Part I September 2008, pp. 408–423 (2008)

3. Gregory, S.: Finding overlapping communities in networks by label propagation. New J. Phys. **12**(10), 103018 (2010)
4. Lancichinetti, A., Fortunato, S., Kertesz, J.: Detecting the overlapping and hierarchical community structure in complex networks. New J. Phys. **11**(3). Article ID 033015 (2009)
5. Palla, G., Deresnyi, I., Farkas, I.: Uncovering the overlapping community structure of complex networks in nature and society. Nature **435**, 814–818 (2005)
6. Saradha, C.S., Arul, D.P.: An optimized overlapping and disjoint community detection techniques using improved community overlap propagation algorithm in complex networks. Adv. Sci. Res. JCR **7**(4), 782–790 (2020)
7. Ahn, Y.Y., Bagrow, J.P., Lehmann, S.: Link communities reveal multiscale complexity in networks. Nature **466**, 761–764 (2015)
8. Rossi, R.J.: Mathematical Statistics: An Introduction to Likelihood Based Inference, p. 227. John Wiley & Sons, New York. ISBN 978–1–118–77104–4 (2018)
9. David, H., Nielsen, B.: Econometric modeling: a likelihood approach. Princeton University Press, Princeton. ISBN 978–0–691–13128–3 (2007)
10. Glasmachers, T., Igel, C.: Gradient-based adaptation of general Gaussian kernels. Neural Comput. **17**(10), 2099–2105 (2005)
11. Ruder, S.: An overview of gradient descent optimisation algorithms. arXiv preprint: 1609.04747 [cs.LG] (2016)
12. Trinh, N.H., Quang, V.V.: The application of range clustering method in community detecting problem. TNU J. Sci. Technol. **255**(06), 303–310. eISSN: 2615–9562, ISSN: 1859–2171, 2734–9098 (2020)
13. Leskovec, J., Krevl, A.: SNAP datasets tanford large network dataset collection, [Online]. Available: https://snap.stanford.edu. Accessed 15 May 2023 (2014)
14. Luo, W., Lu, N., Ni, L., Zhu, W., Ding, W.: Local community detection by the nearest nodes with greater centrality. Inf. Sci. **517**, 377–392 (2020)
15. Su, Y., Zhou, K., Zhang, X., Cheng, R., Zheng, C.: A parallel multi-objective evolutionary algorithm for community detection in large-scale complex networks. Inf. Sci. **576**, 374–392 (2021)

Scheduling Production of High Economic Values Crops in Plant Factory

Nguyen Dinh Dai[1], Pham Chau Long[1], Trung-Nghia Phung[2], Minh-Trien Pham[1], Khanh Le Nguyen[1](✉), and Duc-Dong Do[1](✉)

[1] VNU University of Engineering and Technology, 144 Xuan Thuy, 100000 Hanoi, Vietnam
{trienpm,nl.khanh,dongdoduc}@vnu.edu.vn

[2] Thai Nguyen University of Information and Communication Technology, 250000 Thai Nguyen, Vietnam
ptnghia@ictu.edu.vn, huyvq@hanu.edu.vn

Abstract. Plant factories were designed to control environmental factors such as temperature, relative humidity, light, and carbon dioxide concentration to optimize growth conditions and enhance crop quality. Crop selection and scheduling are obligated to get a consistent harvest in the plant factory. This study uses a Mixed Integer Linear Programming (MILP) approach to model the crop scheduling problem in Plant Factory. The objective function is designed to determine the maximum revenue while taking into account the practical operating circumstances, including crop price per unit and time, crop family and lighting demand, the number of harvests, and number of growing racks. While the optimization potentially solved scheduling issues, the challenges in modifying many racks must be addressed to ensure practical feasibility. Therefore, this study improved the Heuristic Plant Factory Scheduler (HPFS) algorithm by incorporating Dynamic Programming (DP). The experimental results demonstrated that HPFS with DP accelerated computational performance and produced acceptable scheduling quality, mainly when the crop allocation space is enormous.

Keywords: Plant factory · Crop scheduling · Mixed integer linear programming · Heuristic scheduler · Dynamic programming

1 Introduction

Innovative agricultural technologies are facing to bridge the gap between global food security and insufficient production due to climate change. By controlling the cultivation conditions in plant factories, crops are grown faster, two to four times than conditional farming practice [1]. Plant factories could be constructed in various locations, especially urban areas, that minimized transportation expenses and preserved the freshness of products. Moreover, these crops are unaffected by the fluctuation of outside climate or soil fertility decline, rendering them available for year-round production. Crops were cultivated on the multi-shelve system that facilitated the yield per square meter and was appropriate for small spaces. Plants could receive the energy for photosynthesis from sunlight, artificial lights (LED lights), or both. Besides, the produce from the plant factories

P. T. Nghia et al. (Eds.): ICTA 2023, LNNS 847, pp. 301–309, 2023.
https://doi.org/10.1007/978-3-031-49529-8_32

is free of pesticides and has a lower bacterial load [2, 3]. Although plant factories offered many advantages, the selection and scheduling of crops still required being optimized to maximize the productivity and profitability of the space used [4, 5]. Crop selection and scheduling should optimize resource allocation by considering the abovementioned aspects to increase profitability [6]. Many attempts have been made to address the issue of crop scheduling in a controlled environment. Numerous precise and approximate algorithms were developed to maximize resource allocation and revenue that related to several objectives such as crop choice, planning time, quantity, product perishability, cultivation, and storage methods to meet the market demand [7]. A MILP problem in which a farmer need to decide on a time to plant, time to different harvest, and how to plan the inventory was solved by an improved heuristic approach instead of the enumeration method, ensuring computationally feasible results [8]. Yang [9] persevered in developing a heuristic algorithm to optimize profits for a factory that incorporated numerous crops and a solar-paneled roof. The authors formulated the plant factory scheduling problem as a MILP, accounting for various practical operating conditions, including the market value and time of each crop, the adjacency constraint, cleaning and maintenance needs, and the impact of solar panels on the roof on sunlight availability. Dynamic Programming was inserted to improve the computational performance when challenged with the movement of many trays. When a problem can be optimally solved by breaking it down into sub-problems and finding the optimal solutions to those sub-problems, it is said to possess an optimal substructure. The concept behind dynamic programming is to save the results of subproblems in memory so that we do not need to re-calculate them when required later on. This straightforward optimization reduces time complexity from exponential to polynomial.

In this paper, we focused on developing an optimal scheduling model to maximize profits based on practical vegetables' exported dataset in Vietnam for two years. The organization of this paper is as follows: Section 2 proposes the problem statement and MILP model as a baseline algorithm for finding the optimal rotation scheduling in a multi-crop plant factory. A new heuristic algorithm for the model is introduced in Sect. 3. Section 4 compares the performances of the baseline algorithm and the heuristic model. The conclusion and future research are presented in the final section.

2 Preliminaries

2.1 Problem Definition

This study aims to develop a model that can provide an ideal crop schedule to increase the income of a multi-crop plant factory. In other words, the problem at hand involves finding an optimal crop rotation schedule that maximizes total revenue while accounting for various practical factors, including (1) the market price of the crops, (2) the adjacency issue between crops, and (3) the cleaning and maintenance requirements. The problem statement includes determining a crop selection, available cultivation space, and a predetermined cultivation time. The proposed solution addresses the challenges of maximizing revenue while accounting for the practical constraints in crop cultivation, assuming that a factory's operational costs would remain constant. Figure 1 shows the distribution of different type of crops over the racks in the plant factory.

In this study, crop scheduling also considered the proximity of the crops and the upkeep and cleaning of the grow tray system. Table 1 provides an inclusive summary of the study's variables, parameters, and symbols employed to facilitate comprehension.

Rack #1	Week 1	Week 2	Week 3	Week 4	Week 5	Week 6	Week 7	Week 8	Week 9	Week 10
	mustard	mustard	Spinach	Spinach	Spinach	Spinach	Spinach	Spinach	tomato	tomato

...

Rack #n	Week 1	Week 2	Week 3	Week 4	Week 5	Week 6	Week 7	Week 8	Week 9	Week 10
	cabbage	cabbage	cabbage	cabbage	cabbage	cabbage	tomato	tomato	tomato	tomato

Fig. 1. Distribution of crops in racks

Table 1. Definition of a MILP model for the crop scheduling problem

Indices	Definition
$i \in \{1, 2, \ldots, I\}$	i is the index of crop; I is the total number of crops
$r \in \{1, 2, \ldots, R\}$	r is the index of cultivation rack; R is the total number of cultivation racks
$t \in \{1, 2, \ldots, T\}$	indicates time period (week); T is the total time horizon (week)
Parameters	
θ_i	Cultivation time of crop i in weeks
μ_i	Survival rate of crop i
W_r	Cultivation area of rack r in square meter
P_{it}	Price per kilogram of crop i at week t
U_{it}	Price per unit cultivation space and time of crop i at harves
Q_i	Harvested quantity of crop i in kilogram per cultivation area
Variables	
x_{irt}	$x_{irt} = 1$ if crop i is planted in rack r at week t; otherwise, $x_{irt} = 0$

2.2 MILP Model Formulation

The following is a description of a MILP model for the multiple crop scheduling problem in plant factories:

$$\text{Maximize: } \sum_{i=1}^{I}\sum_{r=1}^{R}\sum_{t=1}^{T} x_{irt} W_r U_{i(t+\theta_i-1)} \mu_i \theta_i \tag{1}$$

$$\text{where: } U_it = \frac{P_{it}Q_i}{\theta_i} \tag{2}$$

The study aims to optimize total revenue by identifying the crop that generates maximum revenue using an objective function (1). The crop value at harvest is calculated using expression (2), which considers the area and time of planting to determine the optimal crop for profit maximization. Detail about the constrains has been listed and explain in [9].

3 Proposed Method

While the model described previously can compute the optimal crop rotation schedule, it presents two limitations: (i) The magnitude of the model expands quickly as more racks, periods, and crops are considered. Large problem sizes will require more computation time, and it also limits the capability of the optimization solver. (ii) The lack of flexibility to reschedule after running the schedule for a certain time. Crop rescheduling may be required after applying the schedule due to unforeseen reasons. To address the issue mentioned above, Yang [9], introduced a heuristic algorithm called HPFS. This algorithm is built on tray-by-tray scheduling and allows rescheduling for specific trays only when needed without significant algorithmic changes.

3.1 Optimize Scheduling with Dynamic Planning

A new scheduling strategy is proposed after realizing that the current heuristic approach [9], which utilizes local search to arrange the growing schedule for each tray, may not yield the most optimal results. The proposed strategy involves HPFS using the DP algorithm to achieve maximal results at each scheduling time for every tray. The proposed approach is modeled through a specific objective function.

$DP_{t,k}$—the revenue of the optimal growth strategy when scheduling tray r ends at week t under condition k, with $t = 1, \ldots, W$, $k = 0$ maintainance was not inserted and $k = 1$ maintainance was inserted.

Denote that $revenue_{i,t,r} = W_r P_{it} Q_i \mu_i$. We will have the most optimal planting strategies for the r tray when considering the t week with the k condition as follows:

- There are no trees to be harvested at this week, so the optimal planting strategy up to week t satisfying the condition k is the optimal planting strategy until week $t - 1$, that is $DP_{t,k} = DP_{t-1,k}$.
- Insert the tray maintenance schedule this week (with $k = 1$), so the optimal planting strategy to week t satisfying the condition k is the optimal planting strategy to week $t - 1$, and before that, there is no maintenance schedule, that is $DP_{t,k} = DP_{t-1,0}$.
- Looking at each type of tree, if the i tree is harvested in the t week, it is planted in the t week $- \theta_i$. Therefore, the optimal strategy up to week t satisfying condition k is the optimal planting strategy up to a week. $t - \theta_i$, plus revenue from trees i at week t, that is: $DP_{t,k} = max(DP_{t-\theta_i,k} + revenue_{i,t,r})$.

Thus, at week t with condition k, the optimal growing strategy can be calculated as follows:

$$DP_{t,k} = max(DP_{t-1,0}, DP_{t-1,k}, DP_{t-\theta_i,k} + revenue_{i,t,r}) \tag{3}$$

with i satisfies: $i = 1, \ldots, I$; $\theta_i \le t$; Crop i is not in the same family with crops plan in adjacent racks of rack r from week $t - \theta i + 1$ to week t; Rack r satisfies a light condition of crop i.

By reducing the large problem (t large) to smaller problems (t small) with the above formula, we use the DP algorithm to solve from small problems to small problems as large as performing a loop with $t = 1, \ldots, W$ to compute the objective function $DP_{t,k}$ with initial value: $DP_{0,0} = 0$ and $DP_{0,1} = -\infty$.

Input: Crop properties, Price of crops, Rack configuration
Output: The optimal revenue - $DP_{W,1}$

```
initialize DP_{0,k} (k = 0,1)
for t = 1...W do
    DP_{t,0} ← DP_{t-1,0}
    DP_{t,1} ← max(DP_{t-1,1}, DP_{t-1,0})
    for i ← 1...I do
        if canInsert(i, t) = true then
            for k ← 0...1 do
                DP_{t,k} ← max(DP_{t,k}, DP_{t-θ_i,k} + revenue_{i,t,r})
            end for
        end if
    end for
end for
return DP_{W,1} The best schedule revenue is DP_{W,1}
```

Fig. 2. Algorithm 1 - DP for rack scheduling - DPRS

After the DP algorithm is completed, the optimal growth strategy for the r-tray is $DP_{W,1}$. The pseudocode is described in Algorithm 1 (Fig. 2).

3.2 Shuffled Racks Plant Scheduling

Given that the outcome of scheduling each tray depends on the moment when it is taken into account, the ultimate output will hinge on the sequence in which the trays are scheduled. In order to optimize the outcome, we propose to conduct multiple crop scheduling iterations, wherein the order of the trays is randomized at each iteration to determine the most optimal strategy. Further elaboration on the number of shuffles will be presented in the experiment section (Fig. 3).

```
Input: Crop properties, Price of crops, Rack configuration
Output: Algorithm best plant schedule revenue - bestRevenue
bestRevenue ← 0
Initialize amount of rack: ARack, number of shuffles: S
for Every time step do
    currentRevenue ← 0
    for r ← 1...ARack do
        currentRevenue ← currentRevenue + DPRS(r)
    end for
    bestRevenue ← max(bestRevenue, currentRevenue)
end for
```

Fig. 3. Algorithm 2 - Shuffled racks plant scheduling - SRPS

```
Input: Crop properties, Price of crops, Rack configuration
Output: The optimal revenue - DP_{W,1}
initialize DP_{0,k} (k = 0,1)
for t = 1...W do
    DP_{t,0} ← DP_{t-1,0}
    DP_{t,1} ← max( DP_{t-1,1}, DP_{t-1,0})
    for i ← 1...I do
        for m ← 1...M_i do
            if canInsert(i, t − (m + θ_i) + 1, t) = true then
                for k ← 0...1 do
                    DP_{t,k} ← max( DP_{t,k}, DP_{t-(m+θ_i)} + Σ_{j-0}^{m-1} revenue_{i,t-j,r})
                end for
            end if
        end for
    end for
end for
return DP_{W,1}   The best schedule revenue is DP_{W,1}
```

Fig. 4. Algorithm 3 - DP for crop rotation scheduling - DPCRS

Using the DP technique mentioned previously, in addition to solving the problem of crop optimization for single-harvested crops, we also expanded with multi-harvested crops. The crop schedule search will still be performed sequentially for each tray as in Fig. 4.

4 Experiments and Results

In this study, HPFS algorithms using a DP strategy were tested with a dataset on exported prices of 16 types of vegetable and fruit products in Vietnam for two years, with assumed lighting and tray conditions. The solar cells are arranged so that the tray is not exposed to sunlight surrounded by the tray exposed to direct sunlight. The properties of plants are summarized in Table 2.

HPFS, as well as HPFS combined with DP, are applied to create a crop schedule that maximizes revenue at the same time based on the following conditions: Plants of the same family are not planted in two trays next to each other; Plants that require much sunlight cannot be grown in trays that are not exposed to direct sunlight. However, the plants that can survive in low light are not limited; and the growing schedule should include one week for cleaning and maintaining the tray. Different trays can be serviced at different weeks.

Table 2. Dataset characteristic

Crop	Family	Planting time (weeks)	Requires sunlight	Yield (kg/m^2)
Lettuce Iceberg	Asteraceae	5	0	2
Watercress	Asteraceae	5	0	2
Lettuce Roll	Asteraceae	5	0	2
Lettuce lolo	Asteraceae	5	0	2
Lettuce Ezabel	Asteraceae	5	0	2
Lettuce Radicchio	Asteraceae	5	0	2
Lettuce Salanova	Asteraceae	5	1	2
Japanese Eggplant	Solanaceae	10	0	3.0
Melon	Cucurbitaceae	10	1	4.0
Green Bell Pepper	Solanaceae	13	0	1.0
Baby Bell Pepper	Solanaceae	13	0	1.0
Green pepper	Solanaceae	16	1	0.8
Peas	Fabaceae	7	1	1.2
Cove Beans	Fabaceae	7	1	1.2
Spinach	Fabaceae	4	1	0.4
Celery	Apiaceae	5	1	4.0

The experiment was performed on the same data set with a time setting of 40 weeks and random trays with the numbers 8, 12, 20, 50, 100, 200, 400. Since the order of the trays will affect the results obtained during the algorithm's running, we run the algorithm 10,000 times for each dataset. At each run, we shuffle the scheduling order of the trays together and choose the order in which running the algorithm gives the best results.

Table 3 displays a revenue comparison among the original HPFS algorithm using a greedy strategy, DP strategy, and DP strategy with the order of trays shuffled to schedule each crop. Evidently, the scheduling method utilizing DP yields approximately 15% superior revenue results compared to the original heuristic approach.

It has been observed that randomizing the tray arrangement to obtain the optimal scheduling order may yield superior outcomes compared to utilizing a solitary initial order. This result underscores the importance of selecting a proper tray order to optimize scheduling instead of relying solely on a predetermined order.

Table 3. Revenue from crop scheduling methods (*Price per unit cultivation space and time of crop at harvest*)

Number of racks	8	12	20	50	100	200	400
HPFS (heuristic) [9]	329.234	381.162	658.968	1597.04	3703.54	7071.11	13258.7
HPFS (DP)	366.76	415.18	738.84	1724.3	4062.05	7746.75	14429.1
HPFS (DP + shuffled)	366.76	448.98	806.44	1821.64	4252.54	7746.75	14429.1
MILP	447.20	550.63	987.01	2226.57	5215.65	9860.22	18549.75

5 Conclusion

In this study, an optimal algorithm has been developed for crop scheduling based on the DP algorithm in a cropping plant system equipped with a solar battery system on the factory roof to maximize efficiency and profit while satisfying plant operating conditions. Based on experimental results, the DP algorithm was found to provide more optimal results than the Heuristic algorithm, with no change in computation time. The proposed algorithm uses tray-by-tray planting scheduling, taking advantage of local search to allow for changes in the planting plan of each tray at any time without affecting other trays in the factory. This study represents the first application of DP to the plant scheduling problem. In future studies, alternative methods will be explored for optimizing the selection of scheduling order for trays and optimizing for a more significant number of trays per DP algorithm run instead of local optimization for each tray. Furthermore, it is worth noting that the DP algorithm holds promise for solving NP-hard problems.

Acknowledgement. This work has been supported by VNU University of Engineering and Technology under project number CN21.24.

References

1. Anpo, M.: Plant factory using artificial light: adapting to environmental disruption and clues to agricultural innovation. Elsevier Science (2018)
2. Van Delden, S.H.: Current status and future challenges in implementing and upscaling vertical farming systems. Nat. Food **2**(12), 944–956 (2021)

3. Goto, E.: Plant production in a closed plant factory with artificial lighting, ActaHorticulturae **956**, 37–49 (2012)
4. Cetegen, S.A.: Optimal design of controlled environment agricultural systems under market uncertainty. Comput. Chem. Eng. **149**, 107285 (2021)
5. Huang, K.-L.: Plant factory crop scheduling considering volume, yield changes and multi-period harvests using Lagrangian relaxation. Biosys. Eng. **200**, 328–337 (2020)
6. Santini, A.: The crop growth planning problem in vertical farming. Eur. J. Oper. Res. **294**(1), 377–390 (2021)
7. Schulman, B.: A production capacity investment decision-making tool for the indoor vertical farming industry. Smart Agric. Technol. **5**, 100244 (2023)
8. You, P.-S.: A computational approach for crop production of organic vegetables. Comput. Electron. Agric. **134**, 33–42 (2017)
9. Yang, C.-L.: Recursive heuristic scheduling method for multi-crop plant factory with solar panel roof. Comput. Electron. Agric. **165**, 104941 (2019)
10. dos Santos, L.M.R.: Crop rotation scheduling with adjacency constraints. Ann. Oper. Res. **190**, 165–180 (2011)

Sixth Order Numerical Method for Solving Second Order Nonlinear Boundary Value Problem

Dang Quang A[1](✉) and Truong Ha Hai[2]

[1] Center for Informatics and Computing, VAST, Hanoi, Vietnam
dangquanga@cic.vast.vn

[2] Thái Nguyên University of Information and Communication Technology, Thái Nguyên, Vietnam
thhai@ictu.edu.vn

Abstract. In this paper we propose a numerical method of sixth order of accuracy for solving the Dirichlet problem for second order nonlinear differential equation. It is based on the use of sixth order difference schemes for second order linear boundary value problems and approximation of the first derivative by difference formulas with sixth order of accuracy at each iteration of a constructed iterative method on continuous level. The total error of the numerical solution is obtained. Many examples confirmed the accuracy of sixth order of the method and its advantages over some other methods.

Keywords: Numerical method · Sixth order of accuracy · Nonlinear boundary value problem

1 Introduction

Consider the following second order nonlinear boundary value problem (BVP)

$$u''(t) = f\big(t, u(t), u'(t)\big), \; a < t < b, \tag{1}$$

$$u(a) = \mu_0, \; u(b) = \mu_1. \tag{2}$$

This is a mathematical model of many problems in mechanics, physics and other fields of science. Therefore, it has attracted attention from numerous researchers. The existence and uniqueness of solution of the problem was studied in [1]. The iterative scheme

$$u''_{m+1}(t) = f\big(t, u_m, u'_m\big), \; a < t < b,$$

$$u_{m+1}(a) = \mu_0, \; u_{m+1}(b) = \mu_1$$

P. T. Nghia et al. (Eds.): ICTA 2023, LNNS 847, pp. 310–316, 2023.
https://doi.org/10.1007/978-3-031-49529-8_33

also was considered there. Besides, some monotone iterative methods and quasilinear methods for solving the problem were studied. Recently, some interesting, fixed-point schemes for the problem (1)–(2) have been proposed (see [10, 11]). For numerical solution of the problem many methods were proposed. Here we mention some typical works. The first is the work of Tirmizi and Twizell [9], where they constructed higher difference methods for the equation

$$u''(t) = f(t, u), \tag{3}$$

with the same boundary conditions (2) under the assumption $\frac{\partial f}{\partial u} \geq 0$. For the same Eq. (3) associated with (2), in [2] Costabile et al. used collocation method based on cubic Lidstone splines. Recently, in 2020, Mohanty et al. [8] constructed a difference scheme of order 4, i.e., with the accuracy $O(h^4)$ for the problem (1)–(2) under the assumptions that $f(t, u, u')$ has continuous partial derivatives, $\frac{\partial f}{\partial u} > 0$ and $|\partial f/\partial u'| \leq K$, where K is a positive constant.

In this paper, we construct a new iterative method on continuous level for the problem (1)–(2) and then construct a corresponding discrete iterative scheme. Theoretically, we prove that the numerical solution has the accuracy $O(h^6)$. This conclusion is illustrated in some examples.

2 Construction of Sixth Order Numerical Method

For simplicity and without generality we consider the problem on the interval [0, 1] with homogeneous boundary conditions, i.e., $u(0) = u(1) = 0$.

Analogously as was done in [3–5] for fourth and third order BVPs we introduce some notations:

- $G(t, s)$ – the Green function of the operator $-u''(t)$ associated with homogeneous boundary conditions.
- $M_0 = max_{0 \leq t \leq 1} \int_0^1 |G(t, s)| ds = \frac{1}{8}$.
- $M_1 = max_{0 \leq t \leq 1} \int_0^1 \left| \frac{\partial G}{\partial t}(t, s) \right| ds = \frac{1}{2}$.
- $D_M = \{(t, u, v) | 0 \leq t \leq 1, |u| \leq M_0 M, |v| \leq M_1 M\}$.
- $B[0, M] = \{u(t) \in C[0, 1], \|u\| \leq M\}$.

Following the methodology in [3–5] it is not difficult to prove the following theorems.

Theorem 1. *Suppose that there exist numbers $M > 0, L_0, L_1 \geq 0$ such that*

(i) $|f(t, u, v)| \leq M$ *for any* $(t, u, v) \in D_M$.
(ii) $|f(t, u_2, v_2) - f(t, u_1, v_1)| \leq L_0 |u_2 - u_1| + L_1 |v_2 - v_1|$ *for any* $(t, u_i, v_i) \in D_M$, $i = 1, 2$.
(iii) $q = L_0 M_0 + L_1 M_1 < 1$.

Then the problem (1)–(2) with homogeneous boundary conditions has a unique solution satisfying $|u(t)| \leq M/8$. □

To solve the problem, we propose the following iterative method:

(i) Given starting approximation $\varphi_0(t) \in B[0, M]$, for example,

$$\varphi_0(t) = f(t, 0, 0).$$

(ii) Knowing $\varphi_k(t)(k = 0, 1, \dots)$ solve the linear problem.

$$u_k''(t) = \varphi_k(t), 0 < t < 1,$$

$$u(0) = 0, u(1) = 0,$$

and compute $u'_k(t)$.

(iii) Update

$$\varphi_{k+1}(t) = f(t, u_k(t), u'_k(t))$$

Theorem 2. *Under the assumptions of Theorem 1 the above iterative method converges and there holds the estimate*

$$||u_k - u|| \leq \frac{q^k}{8(1-q)} ||\varphi_1 - \varphi_0||, \tag{4}$$

where $||.||$ *is max-norm in the space of continuous functions,* u *is the exact solution.*

Due to the limitation of the framework of the paper we omitt the proof of the above theorems.

Now we devote attention on the construction of numerical schemes for realization of the above iterative method. We do this for the problem (1)–(2) (with nonhomogeneous boundary conditions) on the interval $[a, b]$.

Cover interval $[a, b]$ by the uniform grid $\overline{\omega_h} = \left\{t_i = ih, h = \frac{b-a}{n}, n = 0, 1, \dots, n\right\}$ and denote $\omega_h = \overline{\omega_h} \backslash \{0, 1\}$. Denote by $\Phi_k(t_i)$, $U_k(t_i)$ the grid functions approximating the functions $\varphi_k(t)$, $u_k(t)$ on the grid $\overline{\omega_h}$. Consider the following iterative method.

Discrete iterative method:

(i) Given starting approximation $\Phi_0(t_i) = f(t_i, 0, 0), i = 0, 1, \dots, n.$

(ii) Knowing $\Phi_k(t)(k = 0, 1, \dots)$ solve the linear difference problem

$$(\Lambda U_k)_i = T\Phi_k(t_i), i = 1, \dots, n-1,$$

$$U_k(t_0) = \mu_0, U_k(t_n) = \mu_1,$$

where

$$(\Lambda y)_i = \frac{y_{i-1} - 2y_i + y_{i+1}}{h^2}, y_i = y(t_i),$$

$$(Tp)(t_i) = p_i + \frac{h^2}{12} D_4^{(2)} p_i + \frac{h^4}{360} D_2^{(4)} p_i,$$

$D_4^{(2)} p_i$ and $D_2^{(4)} p_i$ are the difference derivative of fourth order accuracy for second derivative p'' and the difference derivative of second order accuracy for fourth derivative $p^{(4)}$, respectively. This difference scheme has sixth order of accuracy (see [6]).

(iii) Compute the first derivative of the obtained function U_k with sixth order of accuracy (see [7])

$$V_k = D_6^{(1)} U_k$$

(iv) Update

$$\Phi_{k+1}(t_i) = f(t_i, U_k, V_k), i = 0, 1, \ldots, n$$

In a similar way as in [6] we have proved the following result of the convergence of the above discrete iterative method.

Theorem 3. *Assume that the function $f(t, u, v)$ has continuous partial derivatives up to sixth order and all the conditions of Theorem 1 are satisfied. Then the above discrete iterative method for solving the problem (1)–(2) converges and for the grid solution U_k there holds the estimate*

$$||U_k - u|| \leq \frac{q^k}{8(1-q)} ||\varphi_1 - \varphi_0|| + O(h^6),$$

where u is the exact solution of the problem and $||.||$ is max-norm in the space of grid functions functions.

3 Numerical Examples

Consider some examples for confirming the theoretical results in the previous section.

In all examples the criterion for stopping the iterative process is $\|U_k - U_{k-1}\| \leq 10^{-12}$. The error of the obtained approximate solution is measured by $\|U_k - u\|$, where u is the exact solution.

Example 1. (Problem 7.4 in [8]) Consider the problem for Modified Emden's equation.

$$u''(t) = g(t) - 3uu' - u^3,$$

with the exact solution $u(t) = \sinh(t)$. The function $g(t)$ and boundary conditions are calculated from the equation and the given solution.

The errors of the approximated solution obtained by the proposed method are reported in Table 1 in comparison with the ones of [8].

Example 2. (Problem 7.5 in [8]) Consider the problem for Vanderpol's equation.

$$u'' = \mu\left(1 - u^2\right)u' + u + g(t), 0 < t < 1,$$

with the exact solution $u(t) = e^t$. The function $g(t)$ and boundary conditions are calculated from the equation and the given solution.

The errors of the approximated solution obtained by the proposed method are reported in Table 2 in comparison with the ones of [8] for the case $\mu = 1$.

Table 1. The errors of the numerical solution in Example 1

n	h^6	k	Prop. method	Method in [8]
20	1.5625e-08	22	3.5521e-11	4.6734e-07
40	2.4414e-10	22	9.7489e-13	3.2098e-08
80	3.8147e-12	22	3.9468e-13	1.3743e-09

Table 2. The errors of the numerical solution in Example 2

n	h^6	k	Prop. method	Method in [8]
20	1.5625e-08	29	8.0477e-11	8.5372e-07
40	2.4414e-10	29	1.6114e-12	5.9191e-08
80	3.8147e-12	29	2.4514e-13	3.6757e-09

Example 3. (Problem 7.12 in [8])

$$u'' = \frac{1}{2}(1 + t + u)^3, 0 < t < 1,$$

$$u(0) = u(1) = 0.$$

The exact solution of the problem is $u(t) = \frac{2}{2-t} - t - 1$.

The errors of the approximated solution obtained by the proposed method are given in Table 3 in comparison with the ones of [8].

Table 3. The errors of the numerical solution in Example 3

n	h^6	k	Prop. method	Order	Method in [8]
8	3.8147e-06	22	1.5858e-06		1.2302e-05
16	5.9605e-08	22	1.4551e-08	6.7679	7.8582e-07
32	9.3132e-10	22	1.9663e-10	6.2095	4.9464e-08
64	1.4552e-11	22	4.0829e-12	5.5898	2.8655e-09

In Table 3 Order is the order convergence calculated by the formula

$$Order = \log_2 \frac{\|U_k^{n/2} - u\|}{\|U_k^n - u\|}.$$

Here, the superscripts $n/2$ and n of U_k mean that U_k is computed on the grid with the corresponding number of grid points.

Example 4. (Problem 7.11 in [8])

$$u'' = -\exp(-2u), 1 < t < 2,$$

$$u(1) = 0, u(2) = \ln(2).$$

The analytical solution is given by $u(t) = \ln(t)$.

The errors of the solution obtained by the proposed method are reported in Table 4 in comparison with the ones of [8].

Table 4. The errors of the numerical solution in Example 4

n	h^6	k	Prop. method	Order	Method in [8]
8	3.8147e-06	12	1.2816e-07		6.2911e-06
16	5.9605e-08	12	9.9647e-10	7.0069	3.9921e-07
32	9.3132e-10	12	1.8579e-11	5.7451	2.4915e-08
64	1.4552e-11	12	3.4006e-13	5.7717	1.2490e-09

Example 5 (Example 3 in [2]) Consider the problem

$$u'' = -\left(1 + 0.01u^2\right) + 0.01\cos^3(t), -1 < t < 1,$$

$$u(-1) = \cos(-1), u(1) = \cos(1),$$

with the exact solution $u = \cos(t)$.

The errors of the approximate solution obtained by the proposed method are reported in Table 5 in comparison with the ones of [2].

Table 5. The errors of the numerical solution in Example 5

n	h^6	k	Proposed method	Method in [2]
10	6.4000e-05	31	9.4589e-09	2.623e-003
30	8.7791e-08	31	3.2799e-11	2.935e-004
99	6.7978e-11	31	3.3662e-13	2.644e-005
200	1.0000e-12	31	3.5694e-13	6.610e-006

4 Conclusion

In the paper we have proposed a numerical method of sixth order accuracy for solving a second order nonlinear BVP. The estimate of total error of the numerical solution is obtained. Many numerical experiments for the method on the examples taken from the works of some other authors show the accuracy $O(h^6)$ and the convergence rate equal to 6 meanwhile the accuracies of the methods in their works are much lower.

The proposed method can be used for the second order nonlinear equation with general linear boundary conditions. In combining with the methodology of [6] it can be applicable to construct numerical methods of higher order of accuracy for high even order ODEs with Lidstone boundary conditions. This is the subject of our research in the future.

Acknowledgements. This research is funded by Vietnam National Foundation for Science and Technology Development (NAFOSTED) under grant number 102.01–2021.03.

References

1. Agarwal, R.P.: Boundary Value Problems for Higher Order Differential Equations. World Scientific, Singapore (1986)
2. Costabile, F.A., Gualtieri, M.I., Serafini, G.: Cubic Lidstone-Spline for numerical solution of BVPs. Math. Comput. Simul **141**, 56–64 (2017)
3. Dang, Q.A., Dang, Q.L., Ngo, T.K.Q.: A novel efficient method for nonlinear boundary value problems. Numer. Algor **76**, 427–439 (2017)
4. Dang, Q.A., Ngo, T.K.Q.: Existence results and iterative method for solving the cantilever beam equation with fully nonlinear term. Nonlinear Anal. Real World Appl. **36**, 56–68 (2017)
5. Dang, Q.A., Dang, Q.L.: Simple numerical methods of second and third-order convergence for solving a fully third-order nonlinear boundary value problem. Numer. Algorithms **87**, 1479–1499 (2021)
6. Dang, Q.A., Nguyen, T.T.H.: Numerical method of sixth order convergence for solving a fourth order nonlinear boundary value problem. Appl. Math. Lett. **146**, 108813 (2023)
7. Li, J.: General explicit difference formulas for numerical differentiation. J. Comput. Appl. Math. **183**(1), 29–52 (2005)
8. Mohanty, R.K., Manchanda, G., Khan, A., Khurana, G.: A new high accuracy method in exponential form based on off-step discretization for non-linear two-point boundary value problems. J. Differ. Equ. Appl. **26**(2), 171–202 (2020)
9. Tirmizi, I.A., Twizell, E.H.: Higher-order finite difference methods for nonlinear second-order two-point boundary-value problems. Appl. Math. Lett. **15**, 897–902 (2002)
10. Hacıoğlu, E., Gürsoy, F., Maldar, S., Atalan, Y., Milovanović, G.V.: Iterative approximation of fixed points and applications to two-point second-order boundary value problems and to machine learning. Appl. Numer. Math. **167**, 143–172 (2021)
11. Khuri, S.A., Louhichi, I.: A novel Ishikawa–Green's fixed point scheme for the solution of BVPs. Appl. Math. Lett. **82**, 50–57 (2018)

Solving the Dirichlet Problem for Fully Sixth Order Nonlinear Differential Equation

Nguyen Thanh Huong[1] and Vu Vinh Quang[2(✉)]

[1] Thai Nguyen University of Sciences, Thai Nguyen, Viet Nam
[2] Thai Nguyen University of Information and Communication Technology, Thai Nguyen, Viet Nam
vvquang@ictu.edu.vn

Abstract. The problem which we study in this paper is described by the fully sixth order equation $y^{(6)}(s) = F\Big(s, y(s), y\prime(s), y\prime\prime(s), y^{(3)}(s), y^{(4)}(s), y^{(5)}(s)\Big), 0 < s < 1$ associated with the Dirichlet boundary conditions. Firstly, we propose a novel method for investigating the existence and uniqueness of solution of the problem. It is based on the reduction of the original problem to an operator equation for the right-hand side function. Under some easily verified conditions on the function F in a specified bounded domain, we establish the existence, uniqueness and positivity of a solution. Next, we propose the simple iterative method for finding the approximate solution on both continuous and discrete levels. We prove that the discrete method is of second-order of accuracy due to the use of appropriate formula for numerical integration. The total error estimate for the discrete iterative method is also obtained. Finally, some numerical examples, where exact solution of the problem is known, are presented for illustrating the efficiency of the numerical method.

Keywords: Iterative method · Fully sixth order nonlinear equation · Existence and uniqueness of solution

1 Introduction

Sixth order boundary value problems (BVPs) arise in astrophysics. The dynamo action in some stars may be modelled by sixth order BVPs [7]. The narrow convecting layers bounded by stable layers which are believed to surround A-type stars can be modelled by sixth order BVPs [11]. The numerical study of BVPs are discussed by many authors. The existence and uniqueness of solution of sixth order BVPs can be found in the book written by Agarwal [1], although no numerical methods are contained therein. Akaram [2] used non-polynomial spline technique for the numerical solutions of the sixth order BVPs. Siddiqi [9] presented the solution of sixth order BVPs using the septic spline. In [8], Kasi Viswanadham and Showri Raju used a finite element method involving collocation method with quintic B-splines to solve sixth order BVPs. Sohaib et al. [10] obtained the numerical solution of sixth order BVPs using Legendre wavelet collocation method. In this paper we study the problem

$$y^{(6)}(s) = F\Big(s, y(s), y\prime(s), y\prime\prime(s), y^{(3)}(s), y^{(4)}(s), y^{(5)}(s)\Big), 0 < s < 1,$$

P. T. Nghia et al. (Eds.): ICTA 2023, LNNS 847, pp. 317–327, 2023.
https://doi.org/10.1007/978-3-031-49529-8_34

$$y(0) = y(1) = 0, y\prime(0) = y\prime(1) = 0, y\prime\prime(0) = y\prime\prime(1) = 0 \tag{1}$$

We use the method in the works [3–6] to solve the problem (1). Namely, the problem (1) is reduced to an operator equation for the nonlinear term

$$\varphi(s) = F\Big(s, y(s), y\prime(s), y\prime\prime(s), y^{(3)}(s), y^{(4)}(s), y^{(5)}(s)\Big)$$

With some easily conditions of the function $F(s, y, u, v, z, w, r)$ in a bounded domain we prove the existence and uniqueness of a solution. We study the iterative methods on both continuous and discrete levels for finding the solution. We also demonstrate the validity of the theoretical results and the efficiency of the numerical method in some examples.

2 The Existence and Uniqueness of a Solution

For functions $\varphi(s) \in C[0, 1]$ consider the nonlinear operator defined by

$$A(\varphi)(s) = F\Big(s, y(s), y\prime(s), y\prime\prime(s), y^{(3)}(s), y^{(4)}(s), y^{(5)}(s)\Big) \tag{2}$$

where $y(s)$ is a solution of the problem

$$y^{(6)}(s) = \varphi(s), 0 < s < 1,$$

$$y(0) = y(1) = 0, y\prime(0) = y\prime(1) = 0, y\prime\prime(0) = y\prime\prime(1) = 0 \tag{3}$$

The following result can be deduced.

Proposition 1. *If the function $\varphi(s)$ is a solution of the operator equation*

$$\varphi = A(\varphi) \tag{4}$$

then the function $y(s)$ determined from the problem (3) satisfies the problem (1). Conversely, if $y(s)$ is a solution of the problem (1) then the function

$$\varphi(s) = F\Big(s, y(s), y\prime(s), y\prime\prime(s), y^{(3)}(s), y^{(4)}(s), y^{(5)}(s)\Big)$$

is a fixed point of the operator A defined as in (2) and (3).

Therefore, we can get the solution of the problem (1) by using the solution of the operator Eq. (4). We now represent the solution of the problem (3) as follows

$$y(s) = y_\varphi(s) = \int_0^1 g(s, t)\varphi(t)dt \tag{5}$$

where the Green function

$$g(s,t)=\frac{1}{120}\begin{cases} t^3(s-1)^3\big[\big(6t^2-15t+10\big)s^2+\big(3t^2-5t\big)s+t^2\big], \\ \qquad 0\le t\le s\le 1, \\ s^3(t-1)^3\big[\big(6s^2-15s+10\big)t^2+\big(3s^2-5s\big)t+s^2\big], \\ \qquad 0\le s\le t\le 1. \end{cases}$$

From (5) we obtain

$$y^{(m)}(s)=\int_0^1 \frac{\partial^m g(s,t)}{\partial s^m}\varphi(t)dt,\ m=1,2,3,4,5.$$

We have

$$\int_0^1\left|\frac{\partial^m g(s,t)}{\partial s^m}\right|dt\le C_{6,m},\ m=0,1,2,3,4,5,$$

where $C_{6,0}=\frac{1}{46080}$, $C_{6,1}=\frac{1}{6000\sqrt{5}}$, $C_{6,2}=\frac{1}{2400}$, $C_{6,3}=\frac{1540}{3125}$, $C_{6,4}=\frac{16}{81}$, $C_{6,5}=\frac{1}{2}$.

Next, denote

$$u_\varphi(s)=y_\varphi{\prime}(s),\ v_\varphi(s)=y''_\varphi(s),\ z_\varphi(s)=y'''_\varphi(s),\ w_\varphi(s)=y^{(4)}_\varphi(s),\ r_\varphi(s)=y^{(5)}_\varphi(s).$$

For any number $M>0$, we use the denotation

$$\overline{D}_M=\big\{(s,y,u,v,z,w,r)|0\le s\le 1,\ |y|\le C_{6,0}M,\ |u|\le C_{6,1}M,\ |v|\le C_{6,2}M,\ |z|\le C_{6,3}M,\ |w|\le C_{6,4}M,\ |r|\le C_{6,5}M\big\}.$$

In the space of continuous functions $C[0,1]$, we take the norm $\|\varphi\|=\max_{0\le s\le 1}|\varphi(s|$.

Following the methodology in [3–6], it is not very hard to get the following theorems.

Theorem 1. *Assume that the function F is continuous and there exist numbers $M>0$, $K_0,K_1,K_2,K_3,K_4,K_5\ge 0$ such that*

$$|F(s,y,u,v,z,w,r)|\le M \tag{6}$$

for all $(s,y,u,v,z,w,r)\in\overline{D}_M$;

$$|F(s,y_2,u_2,v_2,z_2,w_2,r_2)-F(s,y_1,u_1,v_1,z_1,w,r_1)|\le K_0|y_2-y_1|+K_1|u_2-u_1|$$
$$+K_2|v_2-v_1|+K_3|z_2-z_1|+K_4|w_2-w_1|+K_5|r_2-r_1| \tag{7}$$

for all $(s,y_i,u_i,v_i,z_i,w,r_i)\in\overline{D}_M\,(i=1,2)$ and

$$p=\sum_{m=0}^{5}K_mC_{6,m}<1 \tag{8}$$

then the problem (1) has a unique solution y satisfying

$$\|y\|<C_{6,0}M,\ \|y\prime\|<C_{6,1},\ M\,\|y\prime\prime\|<C_{6,2}M$$

$$\|y'''\| < C_{6,3}M\,, \|y^{(4)}\| < C_{6,4}M\,, \|y^{(5)}\| < C_{6,5}M$$

We now use the following denotation

$$\overline{D_M^+} = \{(s, y, u, v, z, w, r)|0 \le s \le 1, 0 \le y \le C_{6,0}M\,, |u| \le C_{6,1}M\,, |v| \le C_{6,2}M\,,$$

$$|z| \le C_{6,3}M\,, |w| \le C_{6,4}M\,, |r| \le C_{6,5}M\}$$

and $\overline{S}_M = \{\varphi \in C[0, 1]|-M \le \varphi(s) \le 0\}$.

Theorem 2. *Assume that*

$$-M \le F(s, y, u, v, z, w, r) \le 0$$

for all $(s, y, u, v, z, w, r) \in \overline{D_M^+}$ *and the conditions (7), (8) of Theorem 1 are satisfied. Then the problem (1} has a unique nonnegative solution.*

3 Solution Method

Iterative method at continuous level

To solve the problem (1) we establish the iterative method as follows:

Method A:

(i) Given starting approximation

$$\varphi_0 = F(s, 0, 0, 0, 0, 0, 0) \tag{9}$$

(ii) Knowing $\varphi_k(s), k = 0, 1, 2, \dots$

$$y_k(s) = \int_0^1 g(s, t)\varphi_k(t)dt, \quad u_k(s) = \int_0^1 g_1(s, t)\varphi_k(t)dt \tag{10}$$

$$v_k(s) = \int_0^1 g_2(s, t)\varphi_k(t)dt, \quad z_k(s) = \int_0^1 g_3(s, t)\varphi_k(t)dt \tag{11}$$

$$w_k(s) = \int_0^1 g_4(s, t)\varphi_k(t)dt, \quad r_k(s) = \int_0^1 g_5(s, t)\varphi_k(t)dt \tag{12}$$

where$g_m(s, t) = \frac{\partial^m g(s,t)}{\partial s^m}$, $m = 1, 2, 3, 4, 5$.

(iii) Update

$$\varphi_{k+1}(s) = F(s, y_k(s), u_k(s), v_k(s), z_k(s), w_k(s), r_k(s)) \tag{13}$$

Theorem 3. *Under the assumptions of Theorem 1, Method A converges with the rate of geometric progression and*

$$\|y - y_k\| \le C_{6,0}p_k,\ \|y\prime - y\prime_k\| \le C_{6,1}p_k,\ \|y\prime\prime - y\prime\prime_k\| \le C_{6,2}p_k$$

$$\|y\prime\prime\prime - y\prime\prime\prime_k\| \le C_{6,3}p_k,\ \|y^{(4)} - y_k^{(4)}\| \le C_{6,4}p_k,\ \|y^{(5)} - y_k^{(5)}\| \le C_{6,5}p_k$$

where y is the exact solution of the problem (1) and$p_k = \frac{p^k}{1-p}\|\varphi_1 - \varphi_0\|$.

Remark 1. $g(s,t) \in C^2([0,1] \times [0,1])$, $g_m(s,t) \in C^2([0,1] \times [0,1])$, $m = 1, 2, 3$, meanwhile the function.

$$g_4(s,t) = \begin{cases} t^3\left(-6t^2s + 15ts + 3t^2 - 8t + 6 - 10s\right), 0 \le t \le s \le 1, \\ (1-t)^3\left(6t^2s - 3t^2 + 3ts - t + s\right), 0 \le s < t \le 1 \end{cases}$$

only belongs to $C([0,1] \times [0,1])$ and the function

$$g_5(s,t) = \begin{cases} t^3\left(-6t^2 + 15t - 10\right), 0 \le t < s \le 1, \\ (1-t)^3\left(6t^2 + 3t + 1\right), 0 \le s < t \le 1 \end{cases}$$

is interrupted at $s = t$. However, we can use the trapezoidal formula with the second order accuracy to approximate the integrals in (10)–(12) because the smoothness of the functions is still guaranteed.

The conclusion of the above remark is obtained from the following lemma.

Lemma 1. *Suppose that in* $\overline{D}_M$ *the function* $F(s,y,u,v,z,w,r)$ *has continuous derivatives of second order for all variables. Then* $y_k(s) \in C^8[0,1]$, $u_k(s) \in C^7[0,1]$, $v_k(s) \in C^6[0,1]$, $z_k(s) \in C^5[0,1]$, $w_k(s) \in C^4[0,1]$, $r_k(s) \in C^3[0,1]$, $(k = 1, 2, \dots)$.

Discrete iterative method

To obtain the discrete version of the iterative method (9)–(13), we use the trapezoidal formula with the second order accuracy to approximate the integrals in (10)–(12).

On the interval [0, 1], consider the uniform grid $s_i = ih$, $i = \overline{0,N}$, $h = 1/N$. Set ρ_i, $i = \overline{0,N}$ is the weight

$$\rho_i = \begin{cases} 1, i \ne 0, N, \\ \frac{1}{2}, i = 0, N. \end{cases}$$

We have the following lemma.

Lemma 2. *Assume that* $\varphi(s) \in C^2[0,1]$. *Then*

$$\int_0^1 g_m(s_i,t)\varphi(t)dt = \sum_{j=0}^{N} h\rho_j g_m(s_i,s_j)\varphi(s_j) + O\left(h^2\right), m = 0, 1, 2, 3, 4,$$

$$\int_0^1 g_5^*(s_i,t)\varphi(t)dt = \sum_{j=0}^{N} h\rho_j g_5^*(s_i,s_j)\varphi(s_j) + O\left(h^2\right), \tag{14}$$

where $i = \overline{0, N}$, $g_5^*(s, t)$ is defined from $g_5(s, t)$ as follows

$$g_5^*(s, t) = \begin{cases} g_5(s, t), t \neq s, \\ \frac{1}{2}\big[g_5(s, s^+) + g_5(s, s^-)\big], t = s, \end{cases}$$

where $g_5(s, s^\pm) = \lim\limits_{t \to s \pm 0} g_5(s, t)$.

Now, denote by $Y_k, U_k, V_k, Z_k, W_k, R_k, \Phi_k$ the grid functions. From Lemma 2 we obtain the discrete version of Method A as follow.

Method B:

(i) Given starting approximation

$$\Phi_0(s_i) = F(s_i, 0, 0, 0, 0, 0, 0), \ \ i = \overline{0, N}. \tag{15}$$

(ii) KnowingΦ_k, $k = 0, 1, 2, \ldots$ calculate

$$Y_k(s_i) = \sum_{j=0}^{N} h\rho_j g\big(s_i, s_j\big)\Phi_k\big(s_j\big), i = \overline{0, N} \tag{16}$$

$$U_k(s_i) = \sum_{j=0}^{N} h\rho_j g_1\big(s_i, s_j\big)\Phi_k\big(s_j\big), i = \overline{0, N} \tag{17}$$

$$V_k(s_i) = \sum_{j=0}^{N} h\rho_j g_2\big(s_i, s_j\big)\Phi_k\big(s_j\big), i = \overline{0, N} \tag{18}$$

$$Z_k(s_i) = \sum_{j=0}^{N} h\rho_j g_3\big(s_i, s_j\big)\Phi_k\big(s_j\big), i = \overline{0, N} \tag{19}$$

$$W_k(s_i) = \sum_{j=0}^{N} h\rho_j g_4\big(s_i, s_j\big)\Phi_k\big(s_j\big), i = \overline{0, N} \tag{20}$$

$$R_k(s_i) = \sum_{j=0}^{N} h\rho_j g_5^*\big(s_i, s_j\big)\Phi_k\big(s_j\big), i = \overline{0, N} \tag{21}$$

(iii) Update

$$\Phi_{k+1}(s) = F(s, Y_k(s_i), U_k(s_i), V_k(s_i), Z_k(s_i), W_k(s_i), R_k(s_i)), i = \overline{0, N} \tag{22}$$

For the grid function G on $\overline{\omega}_h$, we use the uniform norm.

$$\|G\|_{\overline{\omega}_h} = \max_{s \in \overline{\omega}_h} |G(s_i|$$

Theorem 4.
Suppose that in the domain. $\overline{D}_M$ the function $F(s, y, u, v, z, w, r)$ has continuous

derivatives of second order for all variables and satisfies the Lipschitz condition (7). Then

$$\|\Phi_k - \varphi_k\|_{\overline{\omega}_h} = O(h^2),\ \|Y_k - y_k\|_{\overline{\omega}_h} = O(h^2),\ \|U_k - u_k\|_{\overline{\omega}_h} = O(h^2)$$

$$\|V_k - v_k\|_{\overline{\omega}_h} = O(h^2),\ \ \|Z_k - z_k\|_{\overline{\omega}_h} = O(h^2),\ \ \|W_k - w_k\|_{\overline{\omega}_h} = O(h^2)$$

$$\|R_k - r_k\|_{\overline{\omega}_h} = O\left(h^2\right).$$

Proof. Consider the case $k = 0$. We have.

$$\varphi_0(s_i) = \Phi_0(s_i) = F(s_i, 0, 0, 0, 0, 0, 0),\ \ \overline{0, N}$$

So

$$\|\Phi_0 - \varphi_0\|_{\overline{\omega}_h} = 0 = O(h^2) \tag{23}$$

From the definition of y_k, Y_k in (10), (16) and Lemma 2 we obtain

$$(Y_0 - y_0)(s_i) = \sum_{j=0}^{N} h\rho_j g\left(s_i, s_j\right)(\Phi_0 - \varphi_0)\left(s_j\right) + O(h^2),\ i = \overline{0, N} \tag{24}$$

Combining (23) and (24) we have $\|Y_0 - y_0\|_{\overline{\omega}_h} = O(h^2)$. By the same argument we also obtain.

$$\|U_0 - u_0\|_{\overline{\omega}_h} = O(h^2),\ \|V_0 - v_0\|_{\overline{\omega}_h} = O(h^2)$$

$$\|Z_0 - z_0\|_{\overline{\omega}_h} = O(h^2),\ \|W_0 - w_0\|_{\overline{\omega}_h} = O(h^2),\ \|R_0 - r_0\|_{\overline{\omega}_h} = O\left(h^2\right)$$

Now assume that the conclusion of the theorem is true for $k = \overline{0, n}$, we must prove the theorem for $k = n + 1$. Indeed, we have.

$$\begin{aligned} |(\Phi_{n+1} - \varphi_{n+1})(s_i)| = {} & |F(s_i, Y_n(s_i), U_n(s_i), V_n(s_i), Z_n(s_i), W_n(s_i), R_n(s_i)) - F(s_i, y_n(s_i), \\ & u_n(s_i), v(s_i), z(s_i), w(s_i), r(s_i))| \\ \le {} & K_0|(Y_n - y_n)(s_i)| + K_1|(U_n - u_n)(s_i)| + K_2|(V_n - v_n)(s_i)| \\ & + K_3|(Z_n - z_n)(s_i)| + K_4|(W_n - w_n)(s_i)| + K_5|(R_n - r_n)(s_i)| \end{aligned}$$

$i = \overline{0, N}$. Therefore according to the inductive hypothesis, we have.

$$\|\Phi_{n+1} - \varphi_{n+1}\|_{\overline{\omega}_h} = O(h^2)$$

By the same argument in case $k = 0$

$$\|Y_{n+1} - y_{n+1}\|_{\overline{\omega}_h} = O(h^2),\ \|U_{n+1} - u_{n+1}\|_{\overline{\omega}_h} = O(h^2),\ \|V_{n+1} - v_{n+1}\|_{\overline{\omega}_h} = O(h^2)$$

$$\|Z_{n+1} - z_{n+1}\|_{\overline{\omega}_h} = O(h^2),\ \|W_{n+1} - w_{n+1}\|_{\overline{\omega}_h} = O(h^2),\ \|R_{n+1} - r_{n+1}\|_{\overline{\omega}_h} = O(h^2)$$

Theorem 5.
Under the assumptions of Theorem 3 and Theorem 4, for the approximate solution of the problem (1) obtained by Method B we have

$$\|Y_k - y\|_{\overline{\omega}_h} \le C_{6,0}p_k + O(h^2),\ \|U_k - y\prime\|_{\overline{\omega}_h} \le C_{6,1}p_k + O(h^2)$$

$$\|V_k - y\prime\prime\|_{\overline{\omega}_h} \le C_{6,2}p_k + O(h^2),\ \|Z_k - y\prime\prime\prime\|_{\overline{\omega}_h} \le C_{6,3}p_k + O(h^2).\square$$

$$\|W_k - y^{(4)}\|_{\overline{\omega}_h} \le C_{6,4}p_k + O(h^2),\ \|R_k - y^{(5)}\|_{\overline{\omega}_h} \le C_{6,5}p_k + O(h^2)$$

where y is the exact solution of the problem (1).

4 Some Examples

Consider some examples for confirming the theoretical results in the previous section. In all the tables, k denotes the number of performed iterations, $E_0 = \|Y_k - y\|_{\overline{\omega}_h}$, $E_1 = \|U_k - y\prime\|_{\overline{\omega}_h}$, $E_2 = \|V_k - y\prime\prime\|_{\overline{\omega}_h}$, $E_3 = \|Z_k - y\prime\prime\prime\|_{\overline{\omega}_h}$, $E_4 = \|W - y^{(4)}\|_{\overline{\omega}_h}$, $E_5 = \|R_k - y^{(5)}\|_{\overline{\omega}_h}$. We will stop the iterative process if.

$$e(k) = \|Y_{k+1} - Y_k\|_{\overline{\omega}_h} \le 10^{-14}$$

Example 1. Consider the problem

$$y^{(6)}(s) = \begin{cases} \frac{y\prime(s)y^{(4)}(s)}{24} - \frac{y(s)y^{(5)}(s)}{8} - y\prime(s) - \frac{y\prime\prime(s)+y\prime\prime\prime(s)}{2} \\ +5s^4 + 10s^3 - 24s^2 + 11s + 239,\ 0 < s < 1, \\ y(0) = y(1) = 0,\ y\prime(0) = y\prime(1) = 0,\ y\prime\prime(0) = y\prime\prime(1) = 0. \end{cases}$$

The function $y(s) = \frac{s^6}{3} - s^5 + s^4 - \frac{s^3}{3}$, $0 \le s \le 1$ is the exact solution of the problem. We have

$$F(s, y, u, v, z, w, r) = \frac{uw}{24} - \frac{yr}{8} - u - \frac{v + z}{2} + 5s^4 + 10s^3 - 24s^2 + 11s + 239,$$

$$\overline{D}_M = \left\{ \begin{array}{c} (s, y, u, v, z, w, r) | 0 \le s \le 1,\ |y| \le \frac{M}{46080},\ |u| \le \frac{M}{6000\sqrt{5}}, \\ |v| \le \frac{M}{2400},\ |z| \le \frac{54M}{3125},\ |w| \le \frac{16M}{81},\ |r| \le \frac{M}{2} \end{array} \right\}$$

In $\overline{D}_M$,

$$|F(s, y, u, v, z, w, r)|$$

$\leq 241 + \frac{1}{24} \times \frac{M}{6000\sqrt{5}} \times \frac{16M}{81} + \frac{1}{8} \times \frac{M}{46080} \times \frac{M}{2} + \frac{M}{6000\sqrt{5}} + \frac{1}{2}(\frac{M}{2400} + \frac{54M}{3125})$.

It is not difficult to verify that with $M = 250$, the condition (6) is satisfied. In the domain $\overline{D}_{250}$, since.

$$|F_{y'}| = \frac{|r|}{8} \leq \frac{M}{16}, \ |F_{u'}| \leq \frac{|w|}{24} + 1 \leq \frac{2M}{243} + 1, \ |F_{v'}| = \frac{1}{2},$$

$$|F_{z'}| = \frac{1}{2}, \ |F_{w'}| = \frac{|u|}{24} \leq \frac{M}{144000\sqrt{5}}, \ |F_{r'}| = \frac{|y|}{8} \leq \frac{M}{368640}$$

we can take $K_0 = \frac{125}{8}$, $K_1 = \frac{743}{243}$, $K_2 = \frac{1}{2}$, $K_3 = \frac{1}{2}$, $K_4 = \frac{1}{576\sqrt{5}}$, $K_5 = \frac{25}{36854}$.

Then $p = \frac{K_0}{46080} + \frac{K_1}{6000\sqrt{5}} + \frac{K_2}{2400} + \frac{54K_3}{3125} + \frac{16K_4}{81} + \frac{K_5}{2} \approx 9.9 \times 10^{-3} < 1$.

Hence, the problem has a unique solution, Method A and Method B converge. From Table 1 we see that the errors of the approximate solution and its derivatives are of order 2.

Table 1. The convergence in example 1

N	E_0	E_1	E_2	E_3	E_4	E_5
10	1.0412e-6	8.0422e-6	2.8447e-5	9.5423e-5	3.7465e-4	9.1342e-4
100	1.0412e-10	9.1425e-10	4.6357e-9	9.6452e-9	5.1445e-8	1.6341e-7
200	6.6124e-12	9.0012e-12	5.4247e-11	5.0422e-10	6.0123e-9	3.0127e-8
500	2.2427e-13	6.4152e-13	1.0025e-11	4.4142e-10	3.5684e-9	1.2352e-8

Example 2. (Example 6 in [8]) Consider the problem

$$\begin{cases} y^{(6)}(s) = -e^{-s}(y(s))^2 + e^{-s} + e^{-3s}, 0 < s < 1, \\ y(0) = 1, y(1) = \frac{1}{e}, y'(0) = -1, y''(1) = -1/e, y''(0) = 1, y''(1) = 1/e. \end{cases} \tag{25}$$

The exact solution of the problem is $y(s) = e^{-s}, 0 \leq s \leq 1$.

Set $y(s) = \overline{y}(s) + Q(s)$, where

$$Q(s) = \left(\frac{19}{2e} - \frac{7}{2}\right)s^5 + \left(\frac{17}{2} - \frac{23}{e}\right)s^4 + \left(\frac{29}{2e} - \frac{11}{2}\right)s^3 + \frac{1}{2}s^2 - s + 1$$

is the polynomial of the fifth degree satisfying the boundary conditions in (25).

The problem for $y(s)$ is reduced to the problem for $\overline{y}(s)$:

$$\begin{cases} \overline{y}^{(6)}(s) = -e^{-s}(\overline{y}(s) + Q(s))^2 + e^{-s} + e^{-3s}, 0 < s < 1, \\ \overline{y}(0) = \overline{y}(1) = 0, \overline{y}'(0) = \overline{y}'(1) = 0, \overline{y}''(0) = \overline{y}''(1) = 0. \end{cases} \tag{26}$$

Analogously as in Example 1, we can take $M = 4$, $K_0 = 11521/5760$, $K_1 = K_2 = K_3 = K_4 = K_5 = 0$. Then $p \approx 4.3 \times 10^{-5} < 1$. Hence the problem (26) (and therefore the problem (25)) has unique solution, Method A and Method B converge. From Table 2 we see that the errors of the approximate solution and its derivatives are of order 4 although Theorem 6 ensures only order 2.

Table 2. The convergence in example 2

N	E_0	E_1	E_2	E_3	E_4	E_5
10	2.9524e-9	9.7423e-9	4.0165e-8	9.5374e-8	3.7465e-7	8.7345e-7
100	3.0440e-13	9.7427e-13	3.5854e-12	8.1458e-12	6.2495e-11	1.7342e-10
200	1.9174e-14	8.2052e-14	4.7287e-13	9.1426e-13	5.2762e-12	1.0107e-11
500	9.3212e-16	4.7154e-15	9.8023e-15	4.7156e-14	9.7632e-14	4.1268e-13

Remark 2. In [8], Kasi Viswanadham and Showri Raju used a finite element method involving collocation method with quintic B-splines to solve the problem (25). They did not obtain estimate for total error of solution and its derivatives. For the problem (25), they only calculated the approximate solution at the points $s_i = 0.0; 0.1; 0.2; ...; 0:9$ and then they compared the obtained results with the exact solution at these points.

5 Conclusion

In this paper we have considered the boundary value problem for fully sixth order nonlinear differential equation with the Dirichlet boundary conditions. By reducing the original problem to an operator equation for the nonlinear term, we established the existence and uniqueness of solution. Then, we have construct an iterative method for finding the solution at both continuous and discrete levels. We have obtained the total error estimate for the discrete iterative solution. Some examples illustrate the effectiveness of the proposed method.

References

1. Agarwal, R.P.: Boundary Value Problems for Higher-Order Differential Equations. World Scientific, Singapore (1986)
2. Akram, G., Siddiqi, S.S.: Solution of sixth order boundary value problems using non-polynomial spline technique. Appl. Math. Comput. **181**, 708–720 (2006). https://doi.org/10.1016/j.amc.2006.01.053
3. Dang, Q.A., Dang, Q.L., Ngo, T.K.Q.: A novel efficient method for nonlinear boundary value problems. Numer. Algor. **76**, 427–439 (2017). https://doi.org/10.1007/s11075-017-0264-6
4. Dang, Q.A., Ngo, T.K.Q.: Existence results and iterative method for solving the cantilever beam equation with fully nonlinear term. Nonlinear Anal. Real World Appl. **36**, 56–68 (2017). https://doi.org/10.1016/j.nonrwa.2017.01.001
5. Dang, Q.A., Ngo, T.K.Q.: New fixed point approach for a fully nonlinear fourth Order Boundary Value Problem. Bol. Soc. Paran. Math. **36**(4), 209−223 (2018). https://doi.org/10.5269/bspm.v36i4.33584
6. Dang, Q.A., Nguyen, T.H.: Solving the dirichlet problem for fully fourth order non-linear differential equation. Afr. Mat. **30**, 623–641 (2019). https://doi.org/10.1007/s13370-019-00671-6

7. Glatzmaier, G.A.: Numerical simulations of stellar convective dynamos III. At the base of the convection zone. Geophys. Fluid. Dyn. **31**, 137–150 (1985). https://doi.org/10.1080/03091928508219267
8. Kasi Viswanadham, K.N.S., Showri Raju, Y.: Quintic B-spline collocation method for sixth order boundary value problems. Glob. J. Res. Eng. **12**(1) (2012)
9. Siddqi, S.S., Akram, G.: Septic spline solutions of sixth-order boundary value problems. J. Comput. Appl. Math. **215**, 288–301 (2008). https://doi.org/10.1016/j.cam.2007.04.013
10. Sohaib, M., Haq, S., Mukhtar, S., Khan, I.: Numerical solution of sixth-order boundary value problems using Legendre wavelet collocation method. Results Phys. **8**, 1204–1208 (2018). https://doi.org/10.1016/j.rinp.2018.01.065
11. Toomre, J., Zahn, J.R., Latour, J., Spiegel, E.A.: Stellar convection theory II: single-mode study of the second convection zone in an a-type star. Astrophys. J. **207**, 545–563 (1976). https://doi.org/10.1086/154522

The Minmax Regret Longest Path of a Tree with Interval Edge Lengths

Tran Quang Quy[1], Nguyen Ha Cong Ly[2(✉)], Do Dinh Cuong[1], and Huy Minh Le[3]

[1] Thai Nguyen University of Information and Communication Technology, Quyet Thang Commune, Thai Nguyen City, Vietnam
tqquy@ictu.edu.vn, ddcuong@ictu.edu.vn

[2] Department of Mathematics, FPT University, Can Tho, Vietnam
lynhc@fe.edu.vn

[3] Faculty of Fundamental Sciences, Van Lang University, 69/68 Dang Thuy Tram Street, Ward 13, Binh Thanh District, Ho Chi Minh City, Vietnam
huy.lm@vlu.edu.vn

Abstract. This paper focuses on tackling the challenge of determining the longest path on a tree represented as $T = (V, E)$, where the edges are assigned interval lengths. We employ the minmax regret concept as a metric to quantify robustness. The longest path associated with this concept is referred to as the minmax regret longest path on the tree T. Initially, we establish certain structural properties pertaining to both the longest and second longest paths within the tree. Then we reset the problem as a formulation regarding the longest and second longest path w.r.t candidate scenarios. We finally devise a combinatorial algorithm capable of identifying the minmax regret longest path within the tree in $O(|\mathcal{L}|^2|V|)$ time with $\mathcal{L}$ being the set of leaf nodes.

Keywords: Minmax regret · Longest path · Tree · Algorithm

1 Introduction

Exploring the longest path on networks presents an NP-hard challenge, since it is possible to be transformed into the Hamiltonian path or the Travelling Salesman problem. Moreover, the longest path problem is applied in many combinatorial optimization problems such as vehicle routing, scheduling, ect.; see [6,18] and references therein. Hence, delving into the algorithmic study of the longest path problem holds significant value. Monien [12] discussed how to identify the longest path in an undirected/directed graphs. He proposed exponential algorithms, say $O(k!|V||E|)$, for all paths in the graph which has the length k. Gutin [7] devised an $O(|V|^3)$ algorithm for computing the longest path on directed complete multipartite graphs, where it requires $O(|V|^{2.5})$ time for both verifying the presence of a Hamiltonian path and finding it if one exists. Recently, Williams [19] devised a $O(2^k pol(|V|))$ time algorithm for examining whether there exists a path with

P. T. Nghia et al. (Eds.): ICTA 2023, LNNS 847, pp. 328–340, 2023.
https://doi.org/10.1007/978-3-031-49529-8_35

length at least k in a an undirected graph with $pol(|V|)$ being a polynomial of $|V|$. For the approximation scheme, Björklund et al. [3] asserted that comprehending the approximation hardness of the longest path problem on undirected graphs is notably challenging and demonstrated that approximating the longest path cannot achieve a ratio better than $O(n/\log^{2+\epsilon} n)$, where $\epsilon > 0$. Moreover, Karger et al. [9] explored the possibility of approximating the longest path in $2^{O(\log^{1-\epsilon}|V|)}$, where $0 < \epsilon < 1$, in undirected graphs.

In practice, the input data is not exactly known but it is uncertain. For example, we can just estimate the travel time of vehicles on a road, the fuel consumption of a car as they depends on many circumstances. If we do not know the exact distribution of input data but knows only the estimation, we model the concerning optimization problem as a robust optimization one; see Ben-Tal [2] and Kouvelis [11] for basic concepts, ideas, and algorithms. In particular, when assessing robustness based on the greatest deviation from an optimal solution across all conceivable scenarios, we come across what is termed the minmax regret optimization problem. Let us revisit some existing literatures concerning the robust shortest path problem (RSPP). The RSPP is NP-hard under discrete scenarios (see [21]) and under interval edge lengths (see [22]). Hence, it is necessary to further investigate efficient solution methods for tackling such a hard problem. Montemanni et al. [13] introduced a precise algorithm for the RSPP involving interval data, and further enhanced the methodology to expedite its execution. They demonstrated that these methods swiftly produce high-quality solutions on certain categories of networks. Furthermore, the identical group of researchers, Montemanni et al. [14], devised a branch and bound approach to address the minmax regret shortest path problem. They showed that the proposed approach is efficient by running the benchmarks. Pérez-Galarce et al. [17] then presented the exact branch and cut, and new heuristic approach. Computational experiments are undertaken to analyze various approaches and evaluate their performance. Furthermore, Pascoal and Resende [16] studied the minmax regret shortest path problem under a finite multi-scenario. They introduced three relevant methodologies. They showed that the labeling and the hybrid approaches are superior to the other existing methods. Recently, Ye and Wang [20] investigated the minmax regret path median problem on a tree with uncertain vertex weights. They devised an $O(n^2)$ algorithm that finds the path with the minimum worst-case loss compared to the median path under any scenario. Besides, researchers also showed their attention to the other robust combinatorial optimization problems and the minmax regret versions specifically; see [1,4,5,10,15] to name a few.

According to the previous analysis, the longest path problem and the robustness concept are both up-to-date topics that are interesting to the community. Although the shortest path problem under uncertainty is intensively studied as mentioned above, the counterpart uncertain longest path problem is not under study so far. In this article, we combine the longest path and the reality of uncertainty, measured by the minmax regret robustness, to construct the minmax regret longest path problem on trees.

The remains are structured as follows. In Sect. 2, we revisit the concept of the longest path within a tree $T = (V, E)$, exploring its inherent properties. We also outline the formulation of the minmax regret longest path problem concerning trees endowed with interval edge lengths. Moving on to Sect. 3, we present the development of a combinatorial algorithm designed to identify the minmax regret longest path within the tree, achieving a complexity of $O(|\mathcal{L}|^2|V|)$. Here, $\mathcal{L}$ denotes the set of leaves present in the tree.

2 Preliminaries and Problem Formulation

2.1 Longest Path on a Tree

Consider a tree $T = (V, E)$ with vertex set V and edge set E. The function $l : E \longrightarrow \mathbb{R}_+$ maps an edge $e \in E$ to a positive real length $l(e)$. Given u and v in V, the path connecting them is denoted by $P(u, v)$. Furthermore, the length $L(P(u, v))$ of $P(u, v)$ corresponds to the cumulative edge lengths along this path, i.e., $L(P(u, v)) = \sum_{e \in E(P(u,v))} l(e)$, where $E(P(u, v))$ is the edge set of $P(u, v)$. We use $\mathcal{P}$ to stand for the set of all paths in the tree. From here on, a path P in $\mathcal{P}$ means that P connects two vertices in the tree and we use the denotation $P(u, v)$ for the known ending vertices u and v of the path. The conventional diameter of T regards to the greatest length among all paths within the tree. Formally, we state the diameter $d(T)$ of T as

$$d(T) = \max_{P \in \mathcal{P}} L(P).$$

The computation of a tree's diameter is achievable in linear time; see Handler [8], who use the longest path to identify the 1-center (in location analysis) on a tree. Precisely, we first choose an arbitrary vertex u in the tree and find a leaf v_1 that has the maximum distance to the vertex u. Then we find another leaf v_2 with maximum distance to v_1. The length of $P(v_1, v_2)$ is indeed the diameter of the tree. A path with length being the diameter of T is indeed the longest path of the tree, mathematically saying, if $L(P) = d(T)$, then P is a longest path of T.

Next, we proceed to analyze specific attributes of the tree's longest path. In this context, each edge is regarded as an interval upon which tree points are situated. A point ρ situated on $e = (u, v)$ is uniquely represented by $\theta \in [0, 1]$, satisfying $d(\rho, u) = \theta l(e)$ and $d(\rho, v) = (1 - \theta)l(e)$, where $d(\rho, \star)$ signifies the distance between point ρ and vertex $\star \in \{u, v\}$. Moreover, we can extend the function $d(\rho_1, \rho_2)$ to calculate the distance between two points ρ_1 and ρ_2 by measuring the length of the continuous path connecting these two points.

Assuming that $P(u, v)$ is a longest path in T and m be the midpoint of $P(u, v)$, i.e., $d(m, u) = d(m, v) = \frac{L(P(u,v))}{2}$. We partition the tree T into two continuous subtrees, say T_1 and T_2, such that $u \in V(T_1)$, $v \in V(T_2)$, $T_1 \cap T_2 = \{m\}$ and $T_1 \cup T_2 = A(T)$. Let $\mathcal{L}$ denote the set of all leaves in the tree T from here on, we obtain the following characterization for the distances concerning the leaves of T in the two partitions.

Proposition 1. *For any leaf w in $\mathcal{L}$ and $w \notin \{u, v\}$, if $w \in V(T_1)$ and $P(u, m) \cap P(w, m) \cap P(u, w) = u'$, then $d(u', w) \leq d(u', u)$. Otherwise, if $w \in V(T_2)$ and $P(v, m) \cap P(w, m) \cap P(v, w) = v'$, then $d(v', w) \leq d(v', v)$.*

Proof. Let us take into account the case $w \in V(T_1)$, the other case $w \in V(T_2)$ can be proved similarly. Assuming by contradiction that $d(u', w) > d(u', u)$, then we get

$$d(v, w) = d(v, u') + d(u', w) > d(v, u) = d(v, u') + d(u', u).$$

The result contradicts the hypothesis that $P(u, v)$ is the longest path within the tree; see Fig. 1. □

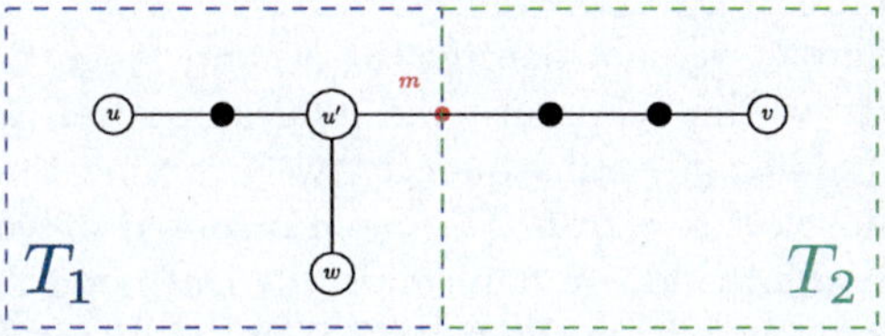

Fig. 1. A longest path $P(u, v)$ and a leaf $w \notin \{u, v\}$

In Proposition 1, if $w \in V(T_1)$ and $P(u, m) \cap P(w, m) \cap P(u, w) = u'$, we get $d(u', v) \geq \frac{L(P(u,v))}{2} \geq d(u', u)$. Otherwise, if $w \in V(T_2)$ and $P(v, m) \cap P(w, m) \cap P(v, m) = v'$, we also obtain $d(v', u) \geq \frac{L(P(u,v))}{2} \geq d(v', v)$. Therefore, we attain directly the following result.

Corollary 1. *For any leaf w in $\mathcal{L} \backslash \{u, v\}$ and $P(w, u) \cap P(w, v) \cap P(u, v) = w'$, then*

$$d(w', w) \leq \min\{d(w', u), d(w', v)\}.$$

We introduce the second longest path of the tree for later use.

Definition 1. A path P' is a second longest of T if $P' \neq P^*$ and $L(P') \geq L(P'')$ for all $P'' \in \mathcal{P} \backslash \{P^*\}$ for P^* being a longest path of the tree T.

Remark that, if the tree contains multiple longest paths, the length of the second longest path could potentially become the tree's diameter. Furthermore, we know that the second longest path is not unique. In what follows we specify the relation between the longest and the second longest paths of the tree.

Proposition 2. *Within tree T, there exists a second longest path that includes either vertex u or vertex v, given that $P(u, v)$ represents the longest path in T.*

Proof. Assume that $P(u', v')$ is the second longest path in T with $u', v' \notin \{u, v\}$. We examine the two cases.

In the first case $P(u', v') \cap P(u, v) \neq \emptyset$. Let $w = P(u, v) \cap P(u', v') \cap P(u, u')$. We know that $d(w, u') \leq \min\{d(w, u), d(w, v)\}$ according to Corollary 1. Therefore, the path $P(u, v')$ satisfies:

$$d(u, v') = d(u, w) + d(w, v') \geq d(u', w) + d(w, v') = d(u', v').$$

Hence, $P(u, v')$ is also the second longest path of the tree as $L(P(u, v')) \geq L(P(u', v'))$.

In the other case, $P(u', v') \cap P(u, v) = \emptyset$, we call $w = P(u', v) \cap P(v', u) \cap P(u, v)$. Then we get the similar argument.

As we can exchange the role of u and v without changing the result of the proof, there exists a second longest path that contains either u or v. □

Based on Proposition 2, we can identify the second longest path in the tree in linear time. Indeed, we first compute the longest path $P(u, v)$ of the tree by the algorithm of Handler [8]. Subsequently, we examine the tree T^u, which is induced by removing vertex v from V, and it's rooted at vertex u. Utilizing a breadth-first search algorithm, we determine the leaf node u_1 with the greatest distance from u. Similarly, within the tree T^v induced by removing vertex u from V and rooted at vertex v, we employ the same approach to identify the leaf node v_1 with the maximum distance from v. To ascertain the second longest path of the tree, we then compare the lengths of $P(u, u_1)$ and $P(v, v_1)$.

2.2 Minmax Regret Longest Path on a Tree

In practice, the length of edges in T are not exactly determined but they are estimated within intervals and no exact distribution is known. Then a robust approach is taken into account to deal with the longest path of the tree. We often investigate the choice of a decision maker w.r.t the worst-case loss. This generates the so-called minmax regret concept for the longest path in the tree. Precisely, we assume that $l(e) \in [\underline{l}(e), \overline{l}(e)]$ for e in E. We denote $\mathcal{S}$ by the set of all scenarios, i.e., for a scenario s in $\mathcal{S}$ we assign the values of edge lengths $l^s(e) \in [\underline{l}(e), \overline{l}(e)]$ for e in E. For a path $P(u, v)$, the length of this path w.r.t scenario s is denoted by

$$L^s(P(u, v)) = \sum_{e \in E(P(u,v))} l^s(e).$$

In special situation, if $l^s(e) = \overline{l}(e)$ and $l^s(e) = \underline{l}(e)$ for all $E \in E(P(u, v))$, we respectively use the abbreviation $\overline{L}(P(u, v)) = \sum_{e \in E(P(u,v))} \overline{l}(e)$ and $\underline{L}(P(u, v)) = \sum_{e \in E(P(u,v))} \underline{l}(e)$. Consequently, $d^s(u, v)$ is the distance between two vertices u and v w.r.t scenario s. Let us denote by $d^s(T) = \max_{P \in \mathcal{P}} L^s(P)$ the diameter of the tree w.r.t scenario s. Then the difference $R^s(P) = d^s(T) - L^s(P)$ is the maximum regret paying for choosing the path P w.r.t scenario s. Hence, $R(P) = \max_{s \in \mathcal{S}} R^s(P)$ is the maximum regret of choosing the path P

w.r.t all scenarios in $\mathcal{S}$. The minmax regret longest path on T is a path $P^* \in \mathcal{P}$ such that it minimizes the largest loss with respect to all scenarios. We formulate the minmax regret longest path problem on T as follows.

$$\min_{P \in \mathcal{P}} R(P) = \min_{P \in \mathcal{P}} \max_{s \in \mathcal{S}} R^s(P). \tag{1}$$

We first obtain the trivial but useful result.

Proposition 3. *The minmax regret longest path in T contains two leaves of the tree.*

Proof. Consider $P(a,b)$ the minmax regret longest path and s^* be the optimal scenario. Then the maximum regret value $R^{s^*}(P(a,b)) = d^{s^*}(T) - L^{s^*}(P(a,b))$. If a or b or both are not leaves of T, there exists a path $P(a',b')$ such that $P(a,b) \subset P(a',b')$ and $P(a,b) \neq P(a',b')$. This implies $R^{s^*}(P(a',b')) < R^{s^*}(P(a,b))$ as $L^{s^*}(P(a,b)) < L^{s^*}(P(a',b'))$. This contradicts the optimality of $P(a,b)$. □

Let $\mathcal{P}'$ denote the set of paths with their ending vertices being leaves of the tree T. By Proposition 3, we reformulate (1) as

$$\min_{P \in \mathcal{P}'} \max_{s \in \mathcal{S}} R^s(P). \tag{2}$$

Given a fixed path $P(u,v)$ in $\mathcal{P}'$, one considers the scenario s that leads to the maximum regret, i.e., $R(P(u,v)) = R^s(P(u,v))$.

Proposition 4. *For a fixed path $P(u,v)$ in $\mathcal{P}'$, the scenario s identified as*

$$l^s(e) = \begin{cases} \underline{l}(e), & \text{if } e \in E(P(u,v)) \\ \overline{l}(e), & \text{if } e \in E \backslash E(P(u,v)) \end{cases}$$

satisfies $R^s(P(u,v)) \geq R^{s'}(P(u,v))$ for all $s' \in \mathcal{S}$ or $R(P(u,v)) = R^s(P(u,v))$.

Proof. Let $d^s(T)$ be the diameter of T w.r.t scenario s and $P(a,b)$ be the corresponding longest path, i.e., $d^s(T) = L^s(P(a,b)) \geq L^s(P)$ for any path P in $\mathcal{P}$. We denote by $\mathcal{M} = P(a,b) \cap P(u,v)$, then

$$R^s(P(u,v)) = \sum\nolimits_{e \in E(P(a,b)) \backslash E(\mathcal{M})} \overline{l}(e) - \sum\nolimits_{e \in E(P(u,v)) \backslash E(\mathcal{M})} \underline{l}(e).$$

Any scenario s' in $\mathcal{S}$ yields

$$R^{s'}(P(u,v)) = \sum\nolimits_{e \in E(P(a',b')) \backslash E(\mathcal{M}')} l^{s'}(e) - \sum\nolimits_{e \in E(P(u,v)) \backslash E(\mathcal{M}')} l^{s'}(e),$$

where $P(a',b')$ is the longest path w.r.t scenario s' and $\mathcal{M}' = P(a',b') \cap P(u,v)$. We aim to prove $R^s(P(u,v)) \geq R^{s'}(P(u,v))$ by analyzing the following cases.

We consider $\mathcal{M} = \emptyset$. If $\mathcal{M}' = \emptyset$, then let $\hat{a} = P(a',a) \cap P(a',b) \cap P(a,b)$. We know that $d^s(\hat{a},a') \leq d^s(\hat{a},a)$ and $d^s(\hat{a},b') \leq d^s(\hat{a},b)$ by Corollary 1. It implies

$d^s(\hat{a},a)+d^s(\hat{a},b) \geq d^s(a',b')$ or $d^s(a,b)-d^s(a_1,b_1) \geq d^s(a',b')-d^s(a_1,b_1)$. This fact yields

$$\begin{aligned} R^s(P(u,v)) &\geq (d^s(a',b') - d^s(a_1,b_1)) - (d^s(u,v) - d^s(a_1,b_1)) \\ &\geq d^s(a',b') - d^s(u,v), \end{aligned}$$

As $d^s(a',b') \geq d^{s'}(a',b')$, $d^s(u,v) \leq d^{s'}(u,v)$, and $P(u,v) \cap P(a',b') = \emptyset$, we obtain

$$R^s(P(u,v)) \geq d^{s'}(a',b') - d^{s'}(u,v) = R^{s'}(P(u,v)).$$

If $\mathcal{M}' = P(a_1',b_1')$, we also call $\hat{a} = P(a',a) \cap P(a',b) \cap P(a,b)$. Similar to the previous argument, one attains

$$\begin{aligned} R^s(P(u,v)) &\geq d^s(a',b') - d^s(u,v) \\ &= \sum\nolimits_{e \in E(P(a',b')) \backslash E(\mathcal{M}')} l^s(e) - \sum\nolimits_{e \in E(P(u,v)) \backslash E(\mathcal{M}')} l^s(e). \end{aligned}$$

As $l^s(e) = \bar{l}(e) \geq l^{s'}(e)$ for $e \in E(P(a',b')) \backslash E(\mathcal{M}')$ and $l^s(e) = \underline{l}(e) \leq l^{s'}(e)$ for $e \in E(P(u,v)) \backslash E(\mathcal{M}')$, we get

$$\begin{aligned} R^s(P(u,v)) &\geq \sum\nolimits_{e \in E(P(a',b')) \backslash E(\mathcal{M}')} l^{s'}(e) - \sum\nolimits_{e \in E(P(u,v)) \backslash E(\mathcal{M}')} l^{s'}(e) \\ &= R^{s'}(P(u,v)). \end{aligned}$$

In the second case $\mathcal{M} = \emptyset$, one can apply the analogous argument to get the result. □

Let s be the scenario such that $R(P(u,v)) = R^s(P(u,v))$ and $P(a,b)$ be the longest path w.r.t scenario s, i.e., $R(P(u,v)) = R^s(P(u,v)) = L^s(P(a,b)) - L^s(P(u,v))$. Then, the length $l^s(e)$ can attain any value in $[\underline{l}(e); \bar{l}(e)]$ for all edges e in $E(P(a,b)) \cap E(P(u,v))$. Indeed, the values of edge lengths in $P(a,b) \cap P(u,v)$ does not contribute to the regret function $R^s(P(u,v))$. For the sake of simplicity, we focus on the case $l^s(e) = \bar{l}(e)$ for $e \in E(P(a,b)) \cap E(P(u,v))$. Hence, we know that $l^s(e) = \bar{l}(e)$ for e in $E(P(a,b))$. This observation yields the following following corollary directly.

Corollary 2. *There exists an optimal scenario s^* in $\mathcal{S}$ and $P(a,b)$ is the corresponding longest path such that $l^{s^*}(e) = \bar{l}(e)$ for $e \in P(a,b)$ and $l^{s^*}(e) = \underline{l}(e)$ for $e \in E \backslash P(a,b)$.*

Assuming that the optimal scenario s^* is known, we further investigate the minmax regret longest path of the tree.

Lemma 1. *If s^* is an optimal scenario as in Corollary 2, then the second longest path w.r.t scenario s^* is the minmax regret longest path of the tree.*

Proof. As s^* is the optimal scenario, we find the path P^* to minimize $R^{s^*}(P) = d^{s^*}(T) - L^{s^*}(P)$ for P in $\mathcal{P}'$, then P^* is indeed the minmax regret longest path of the tree. Obviously, the second longest path w.r.t scenario s^* minimizes $R^{s^*}(P)$ for P in $\mathcal{P}'$. This yields the result. □

3 A Combinatorial Algorithm

Given a path P in $\mathcal{P}'$, the scenario s_P satisfies

$$l^{s_P}(e) := \begin{cases} \bar{l}(e) & \text{for } e \in E(P), \\ \underline{l}(e) & \text{for } e \notin E(P). \end{cases} \tag{3}$$

Denote $\mathcal{S}'$ the set of all scenarios s_P for P in $\mathcal{P}'$. Furthermore, the diameter of the tree and the second longest path of the tree w.r.t scenario s_P is abbreviated as $d^{s_P}(T)$ and P^2. We also restrict to the constraint $L^{s_P}(P^2) \leq D^{s_P}$ as the second longest path has the length at most the diameter of the tree. Based on Corollary 2 and Lemma 1, we can reformulate (2) as

$$\min_{P^2 \in \mathcal{P}'} \max_{s_P \in \mathcal{S}'} R^{s_P}(P^2) = \min\{d^{s_P}(T) - L^{s_P}(P^2) : P^2 \in \mathcal{P}' \text{ and } L^{s_P}(P^2) \leq d^{s_P}(T)\}. \tag{4}$$

As $|\mathcal{P}'| = O(|\mathcal{L}|^2)$, there are exactly $O(|\mathcal{L}|^2)$ candidate scenarios in $\mathcal{S}'$ and second longest paths in $\mathcal{P}'$. Therefore, the principle of the algorithm is to calculate $R^{s_P}(P^2)$ for all scenarios s_P in $\mathcal{S}'$. The path P^2 in $\mathcal{P}'$ w.r.t the minimum objective value is the minmax regret longest path of the tree.

Let us first root the tree at a centroid r of T. Here, the centroid of the tree partitions the tree into smaller subtrees with their number of vertices being at most $|V|/2$ and it can be identified in linear time. Given a vertex v in V, let $Ch(v)$ be the set of its children. We label each vertex v in V by

$$L_1(v) = \overline{L}(P(r,v)) = \sum\nolimits_{e \in E(P(r,v))} \bar{l}(e)$$

and

$$L_2(v) = \underline{L}(P(r,v)) = \sum\nolimits_{e \in E(P(r,v))} \underline{l}(e).$$

We set $L_1(r) = L_2(r) = 0$. Assuming that the labels $L_1(v)$ and $L_2(v)$ are already known, we can compute $L_1(u) = L_1(v) + \bar{l}(v,u)$ and $L_2(u) = L_2(v) + \underline{l}(u,v)$, for u being a child of v. By the recursive approach, we can compute all labels $L_1(v)$ and $L_2(v)$ for $v \in V$ in linear time.

Let us consider the pair of leaves u and v, we consider the path $P = P(u,v)$ and the corresponding scenario s_P as described in (3). It means $l^{s_P}(e) = \bar{l}(e)$ for $e \in E(P(u,v))$ and $l^{s_P}(e) = \underline{l}(e)$ for $e \in E \backslash E(P(u,v))$. Then, if $V(P(r,u)) \cap V(P(r,v)) \cap V(P(u,v)) = w$ then

$$\overline{L}(P(u,v)) = L_1(u) + L_1(v) - 2L_1(w).$$

Note that the vertex w can be coincident with r. Moreover, we know that the second longest path w.r.t the scenario s_P contains either u or v. Hence, for each vertex p in V, we consider the paths $P(u,p)$ and $P(v,p)$ and denote by

$$\Delta^{s_P}(u,v,p) = \min\{L^{s_P}(P) - L^{s_P}(P(u,p)), L^{s_P}(P) - L^{s_P}(P(v,p))\}. \tag{5}$$

Then the maximum regret regarding the scenario s_P where P being the longest path of T can be written as

$$R^{s_P}(P^2) = \begin{cases} \min_{p\in\mathcal{L}} \Delta^{s_P}(u,v,p), & \text{if } \Delta^{s_P}(u,v,p) \geq 0 \, \forall p \in \mathcal{L} \\ +\infty, & \text{otherwise.} \end{cases} \tag{6}$$

We note that $R^{s_P}(P^2) = +\infty$ if the specified path P is not the longest path w.r.t scenario s_P. We now compute (5) according to the labels $L_1(v)$ and $L_2(v)$ for $v \in V$ as follows:

1. If $P(r,p) \cap P(u,v) = \emptyset$, we call $w = P(p,u) \cap P(p,v)$. Then

$$\begin{aligned} L^{s_P}(P) - L^{s_P}(P(u,p)) &= \overline{L}(P(w,v)) - \underline{L}(P(w,p)) \\ &= (L_1(v) - L_1(w)) - (L_2(w) + L_2(p)). \end{aligned}$$

   By the similar argument, we also get $L^{s_P}(P) - L^{s_P}(P(v,p)) = (L_1(u) - L_1(w)) - (L_2(w) + L_2(p))$. Assuming that $L_1(u) \leq L_1(v)$, we can write

$$\Delta^{s_P}(u,v,p) = (L_1(u) - L_1(w)) - (L_2(w) + L_2(p)).$$

2. If $P(r,p) \cap P(u,v) = P(w',w)$, where w' is in $P(w,u)$, we know that $P(r,u) \cap P(r,v) \cap P(u,v) = w$ and $P(u,p) \cap P(v,p) \cap P(u,v) = w'$. Then

$$\begin{aligned} L^{s_P}(P(u,v)) - L^{s_P}(P(u,p)) &= L^{s_P}(P(w',v)) - L^{s_P}(P(w',p)) \\ &= [(L_1(v) - L_1(w)) + (L_1(w') - L_1(w))] \\ &\quad -(L_2(p) - L_2(w')). \end{aligned}$$

   and

$$\begin{aligned} L^{s_P}(P(u,v)) - L^{s_P}(P(v,p)) &= L^{s_P}(P(w',u)) - L^{s_P}(P(w',p)) \\ &= (L_1(u) - L_1(w')) - (L_2(p) - L_2(w')). \end{aligned}$$

   Therefore, we can compute

$$\begin{aligned} \Delta^{s_P}(u,v,p) = \min\{&[(L_1(v) - L_1(w)) + (L_1(w') - L_1(w))] \\ &- (L_2(p) - L_2(w')), (L_1(u) - L_1(w')) - (L_2(p) - L_2(w'))\}. \end{aligned}$$

We can find $\Delta^{s_P}(u,v,p)$ for all leaves p in $\mathcal{L}$ and then find $R^{s_P}(P^2)$ based on equation (6). However, computing $\Delta^{s_P}(u,v,p)$ for all s_P in $\mathcal{S}'$ and p in $\mathcal{L}$ costs $O(|\mathcal{L}|^2|V|^2)$ time. Indeed, there are $O(|\mathcal{L}|^2)$ candidate scenarios s_P in $\mathcal{S}'$ and in each scenario, for a leaf p, we have to search along the path $P(r,p)$ for case analysis which leads to the total complexity $O(\sum_{p\in\mathcal{L}} |V(P(r,p))|) = O(|V|^2)$ to consider all p in $\mathcal{L}$. We next attempt to find the value $R^{s_P}(P^2)$ more efficiently.

Assuming that $V(P(u,v)) = \{u, v_1, v_2, \ldots, v_n, v\}$, we induce n subtrees T_i with $v_i \in V(T_i)$ for $i = 1, \ldots, n$ by deleting all edges in $P(u,v)$. In the subtree T_i, let

$$p_i = \arg\max\{d^{s_P}(v_i, v) : v \in V(T_i)\}.$$

Trivially, one obtains the second longest path w.r.t the scenario s_P

$$P^2 = \arg\max\{L^{s_P}(P') : P' = P(u, p_i) \text{ or } P' = P(v, p_i) \text{ for } i = 1, \dots, n\}.$$

Due to this fact, we can write

$$\begin{aligned} R^{s_P}(P^2) &= \min_{i=1,\dots,n} \{d^{s_P}(T) - L^{s_P}(P(u, p_i)), d^{s_P}(T) - L^{s_P}(P(v, p_i))\} \\ &= \min_{i=1,\dots,n} \{\overline{L}(P(v_i, v)) - \underline{L}(P(v_i, p_i)), \overline{L}(P(v_i, u)) - \underline{L}(P(v_i, p_i))\}. \end{aligned}$$

We can calculate $\overline{L}(v_i, \star) = L_1(\star) - L_1(v_i)$ for $\star \in \{u, v\}$. It remains to calculate $\underline{L}(v_i, p_i)$ based on the vertex labels in the tree.

Recall that after deleting all edges of the path $P(u, v)$, one obtains the subtrees $T_1, T_2, \dots, T_n$ which contains $v_1, v_2, \dots, v_n$ on $V(P(u, v))$. We call $\mathcal{L}(T_i)$ the set of leaves in T_i for $i = 1, \dots, n$. For the subtree T_i, $i \in \{1, \dots, n\}$, which does not contains the root r, we can compute

$$\underline{L}(P(v_i, p_i)) = \max\{L_1(p) - L_1(v_i) : p \in \mathcal{L}(T_i)\}.$$

For the subtree T_j, $j \in \{1, \dots, n\}$, which contains the root r, we can split the subtree into two smaller subtrees T_j^1 and T_j^2 at v_j, i.e., $V(T_j^1) \cap V(T_j^2) = \{v_j\}$ and $V(T_j^1) \cup V(T_j^2) = V(T_j)$. Note that there are exactly one subtree T_j that contains the root r. Furthermore, we assume that $r \in V(T_j^2)$. For the leave p in $\mathcal{L}(T_j^1)$, we know $\underline{L}(P(v_j, p)) = L_1(p) - L_1(v_j)$. For the subtree T_j^2, we reroot it at v_j and the apply the breadth first search algorithm to compute $\underline{L}(P(v_j, p))$ for p in $\mathcal{L}(T_j^2)$ in $O(|V(T_j^2)|)$ time. Hence, we can find

$$\underline{L}(P(v_j, p_j)) = \max\{\underline{L}(P(v_j, p)) : p \in \mathcal{L}(T_j)\}.$$

The total complexity for computing $\underline{L}(P(v_i, p_i))$ for $i = 1, \dots, n$ is $\sum_{i=1}^{n} |V(T_i)| = O(|V|)$.

In summary, we formulate a combinatorial algorithm aimed at identifying the minmax regret longest path within the tree T. The idea is to consider in each iteration the longest path P w.r.t scenario s_P in $\mathcal{S}'$ and compute the regret $R^{s_P}(P^2)$ as defined previously. Finally, the minimum values among $R^{s_P}(P^2)$ for $P \in \mathcal{P}'$ is the optimal objective and the corresponding path is the optimal solution. For a comprehensive understanding, please refer to Algorithm 1.

As there are $|\mathcal{L}|^2$ many paths in the set $\mathcal{P}'$, there are $|\mathcal{L}|^2$ iterations in Algorithm 1. Moreover, in each iteration we can compute $\overline{L}(v_i^P, u)$, $\overline{L}(v_i^P, v)$, and $\underline{L}(v_i^P, p_i)$ in $O(|V(T_i)|)$ time for $i = 1, \dots, n_P$. Hence, the time complexity within each iteration amounts to $O(\sum_{i=1}^{n_P} |V(T_i)|) = O(|V|)$ time. This results the overall complexity of $O(|\mathcal{L}|^2|V|)$.

Theorem 1. *The minmax regret longest path on problem trees is solvable in $O(|\mathcal{L}|^2|V|)$ time.*

We demonstrate Algorithm 1 through the following.

Algorithm 1 Finds the minmax regret longest path of a tree.

Input: An instance of the tree $T = (V, E)$ with interval edge lengths.
Root the tree T at a centroid r.
Compute labels $L_1(v)$, $L_2(v)$ for all v in V.
for each path P in $\mathcal{P}'$ **do**
  Set $flag = 0$.
  Induce subtrees T_1, T_2, $\ldots$, T_n which contains v_1^P, v_2^P, $\ldots$, $v_{n_P}^P$ on $V(P)$ by deleting edges of P.
  for $i = 1, \ldots, n_P$ **do**
    Compute $\overline{L}(P(v_i^P, u))$, $\overline{L}(P(v_i^P, v))$, and $\underline{L}(P(v_i^P, p_i))$.
    if $\overline{L}(P(v_i^P, u)) < \underline{L}(P(v_i^P, p_i))$ or $\overline{L}(P(v_i^P, v)) < \underline{L}(P(v_i^P, p_i))$ **then**
      Set $flag = 1$.
    end if
  end for
  if $flag = 0$ **then**
    Set $R^{s_P}(P^2) = \min_{i=1,\ldots,n_P}\{\overline{L}(P(v_i, v)) - \underline{L}(P(v_i, p_i)), \overline{L}(P(v_i, u)) - \underline{L}(P(v_i, p_i))\}$ and find the corresponding P^2.
  else
    Set $R^{s_P}(P^2) = +\infty$.
  end if
end for
Output: The minmax regret longest path $P^* = \arg\min_{P^2 \in \mathcal{P}'} R^{s_P}(P^2)$ of a tree.

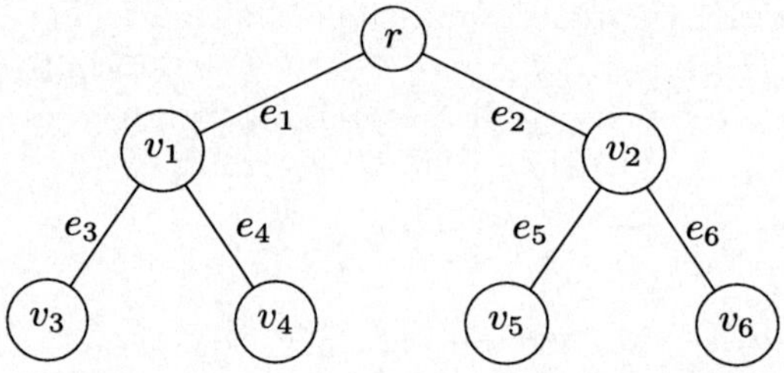

Fig. 2. An instance of the tree T rooted at r

Example 1. We examine the tree T as depicted in Fig. 2, the edge lengths are given as in Table 1.

We set $L_1(r) = L_2(r) = 0$ and label each vertex of the tree. Table 2 shows the result of computations.

Applying Algorithm 1, let us take into account the path $P = P(v_3, v_4)$, then $\overline{L}(P(v_1, v_3)) = 10$, $\overline{L}(P(v_1, v_4)) = 6$, and $\underline{L}(P(v_1, p_1)) = \underline{L}(P(v_1, p_1)) = 10$. Hence, $R^{s_P}(P^2) = +\infty$. Similarly, we get the iterations as in Table 3.

The minmax regret longest path in T is either $P(v_3, v_5)$ or $P(v_3, v_6)$.

Table 1. The inteval data of edge lengths

i	1	2	3	4	5	6
$l(e_i)$	[4, 8]	[3, 5]	[6, 10]	[2, 6]	[3, 8]	[4, 6]

Table 2. Labels of vertices in T

i	1	2	3	4	5	6
$L_1(v_i)$	8	5	18	14	13	11
$L_2(v_i)$	4	3	10	6	6	7

Table 3. The inteval data of edge lengths

Iter	1	2	3	4	5	6
P	$P(v_3, v_4)$	$P(v_3, v_5)$	$P(v_3, v_6)$	$P(v_4, v_5)$	$P(v_4, v_6)$	$P(v_5, v_6)$
P^2	–	$P(v_3, v_6)$	$P(v_3, v_5)$	$P(v_3, v_5)$	$P(v_3, v_6)$	–
$R^{s_P}(P^2)$	$+\infty$	4	3	0	0	$+\infty$

4 Conclusions

We have tackled the minmax regret longest path problem on a tree $T = (V, E)$ characterized by interval edge lengths. Leveraging the properties of the deterministic longest path on a tree, we delved into the structural aspects of the minmax regret longest path. Subsequently, we devised a combinatorial algorithm capable of solving this problem in $O(|\mathcal{L}|^2|V|)$ time, where $\mathcal{L}$ represents the set of leaves within the tree. Future research endeavors hold promise in extending the study of the minmax regret longest path problem to other graph types such as cycles, cacti, interval graphs, and more.

Acknowlegement. The authors thank the anonymous referees for valuable comments that helped to improved the paper. The fourth author (H.M. Le) would like to thank Van Lang University, Vietnam, for funding this work.

References

1. Bachtler, O., Krumke, S.O., Le, H.M.: Robust single machine makespan scheduling with release date uncertainty. Oper. Res. Lett. **48**, 816–819 (2020)
2. Ben-Tal, A., Nemirovski, A., El-Ghaoui, L.: Robust optimization. Princeton University Press, Princeton, New Jersey (2009)
3. Björklund, A., Husfeldt, T., Khanna, S.: Approximating longest directed paths and cycles. In: Proceedings of the International Colloquium on Automata, Languages and Programming (ICALP 2004), Lecture Notes in Computer Science, vol. 3142, pp. 222–233. Springer-Verlag, Berlin (2004)
4. Burkard, R.E., Dollani, H.: Robust location problems with pos/neg weights on a tree. Networks **38**, 102–113 (2001)

5. Chassein, A., Goerigk, M.: Variable-sized uncertainty and inverse problems in robust optimization. Eur. J. Oper. Res. **264**, 17–28 (2018)
6. Garey, M.R., Johnson, D.S.: Computers and Intractability: A Guide to the Theory of NP Completeness. Freeman, San Francisco, CA (1979)
7. Gutin, G.: Finding a longest path in a complete multipartite digraphs. SIAM J. Disc. Math. **6**(2), 270–273 (1993)
8. Handler, G.Y.: Minimax location of a facility in an undirected tree networks. Transp. Sci. **7**, 287–293 (1973)
9. Karger, D., Motwani, R., Ramkumar, G.D.S.: On approximating the longest path in a graph. Algorithmica **18**, 82–98 (1997)
10. Kasperski, A., Makuchowski, M., Zieliński, P.: A Tabu search algorithm for the minimax regret minimum spanning tree problem with interval data. J. Heuristics **18**(4), 593–625 (2012)
11. Kouvelis, P., Yu, G.: Robust Discrete Optimization and Its Applications. Kluwer Academic Publishers (1997)
12. Monien, B.: How to find long paths efficiently. Ann. Discrete Math. **25**, 239–254 (1985)
13. Montemanni, R., Gambardella, L.M.: An exact algorithm for the robust shortest path problem with interval data. Comput. Oper. Res. **31**, 1667–1680 (2004)
14. Montemanni, R., Gambardella, L.M., Donati, A.V.: A branch and bound algorithm for the robust shortest path problem with interval data. Oper. Res. Lett. **32**, 225–232 (2004)
15. Nguyen, K.T., Hung, N.T.: The minmax regret inverse maximum weight problem. Appl. Math. Comput. **407**, 126328 (2021)
16. Pascoal, M.M.B., Resende, M.: The minmax regret robust shortest path problem in a finite multi-scenario model. Appl. Math. Comput. **241**, 88–111 (2014)
17. Pérez-Galarce, F., Candia-Vejar, A., Astudillo, C.A., Bardeen, M.D.: Algorithms for the minmax regret path problem with interval data. Inf. Sci. **462**, 218–241 (2018)
18. Schrijver, A.: Combinatorial Optimization: Polyhedra and Efficiency. Springer-Verlag, Berlin Heidelberg (2003)
19. Williams, R.: Finding paths of length k in $O^*(2^k)$ time. Inf. Process. Lett. **109**, 315–318 (2009)
20. Ye, J.H., Wang, B.F.: On the minmax regret path median problem on trees. J. Comput. Syst. Sci. **81**(7), 1159–1170 (2015)
21. Yu, G., Yang, J.: On the robust shortest path problem. Comput. Oper. Res. **25**(6), 457–468 (1998)
22. Zieliński, P.: The computational complexity of the relative robust shortest path problem with interval data. Eur. J. Oper. Res. **158**, 570–576 (2004)

Author Index

P. T. Nghia et al. (Eds.): ICTA 2023, LNNS 847, pp. 341–342, 2023.
https://doi.org/10.1007/978-3-031-49529-8